THE SELF-MADE TAPESTRY

THE SELF-MADE TAPESTRY

Pattern formation in nature

Philip Ball

OXFORD • NEW YORK • TOKYO
OXFORD UNIVERSITY PRESS
1999

Oxford University Press, Great Clarendon Street, Oxford OX2 6DP

Oxford   New York

Athens   Auckland   Bangkok   Bogota   Bombay
Buenos Aires   Calcutta   Cape Town   Dar es Salaam
Delhi   Florence   Hong Kong   Istanbul   Karachi
Kuala Lumpur   Madras   Madrid   Melbourne
Mexico City   Nairobi   Paris   Singapore
Taipei   Tokyo   Toronto   Warsaw

and associated companies in
Berlin   Ibadan

Oxford is a trade mark of Oxford University Press

Published in the United States
by Oxford University Press Inc., New York

A catalogue record for this book is available from the British Library

Library of Congress Cataloging-in-Publication Data
Ball, Philip, 1962–
The self-made tapestry: pattern formation in nature / Philip
Ball.
Includes bibliographical references.
1. Pattern formation (Biology)   2. Symmetry.   I. Title.
QH491.B35   1999   571.3—dc21   98–16650
ISBN 0 19 850244 3 (Hbk)

Typeset by EXPO Holdings, Malaysia

Printed in Great Britain by
Bath Press Ltd, Bath

PREFACE

As I was close to completing this book, I found myself watching the sun go down from an empty beach in west Wales. The sky was livid with salmon-coloured bands of clouds. The shore was being washed with the steady pulse of the sea, and a stream threaded its braided course across the wrinkled brow of the sandy beach to the water's edge, where a white foam frothed on top of the turbulent eddies. Behind me rose rugged cliffs, each cradling countless miniature replicas of itself in a craggy hierarchy. Along the cliff path I had noticed earlier in the day the spiral arrangement of spikes on the gorse bushes, the five-petalled wild flowers. And I don't think it was until that moment that I truly appreciated how the patterns that I had spent the last several months describing were far from the arcane curiosities of laboratories or the virtual creations of a computer cyberspace, but indeed the blueprints for nature.

I had just read Brian Appleyard's response, in *Understanding the Present*, to the famous remark of physicist Richard Feynman on how understanding a flower scientifically can only increase our appreciation of it. Appleyard is unmoved: 'We are supposed to be *grateful*', he scoffs. Reactions to scientific inquiry will ever be diverse, I suppose; but I know that at that moment I was grateful that the patterns of west Wales's wild coast can be understood and appreciated, not just experienced. It made me feel at home there.

I hope that you too will acquire from this book the kind of excitement that I now feel when I observe the lace-work of the sky or the outrageous designs of a butterfly's wing. When a little mystery is dispelled, the wonder and beauty need not go with it.

As ever, my accuracy (not to say clarity) has been improved immeasurably by the generous advice of those who really know about this stuff. For comments on the text, I am deeply indebted to Robert Anderson, John Barrow, Michael Batty, Eshel Ben-Jacob, Elena Budrene, Scott Camazine, Pierre Hohenberg, Jim Kirchner, Rolf Landauer, Michael Marder, Hans Meinhardt, Jim Murray, Geoffrey Ozin, Pejman Rohani, Katepali Srinivasan, Tom Mullin, Udo Seifert, Gene Stanley, Tamás Vicsek, Art Winfree and George Zaslavsky. For providing reference material, I should like to thank Michele Emmer, Michael Gorman, Alan Mackay, Alan Newell, Peter Ortoleva, Juan Manuel Garcia-Ruiz, Robert Phelan, Luciano Pietronero, Lee Smolin, Harry Swinney, Steven VanHook and Dennis Weaire. And I am most grateful to all the others who generously loaned me photographs and illustrations. I would like to express particular thanks to Graeme Hogarth, Andrea Sella and the chemistry department of University College, London, for assistance with the chemical experiments in the appendices.

I have been greatly encouraged in this project by the enthusiasm of Cathy Kennedy at Oxford University Press, and also by that of many friends who I have regaled with the just-so stories of natural patterns. My partner Julia has listened patiently and helped me to see where my enthusiasm outruns my lucidity.

London   P. B.
October 1997

CONTENTS

The plates section falls between
pages 24 and 25

PATTERNS

The waves of the sea, the little ripples on the shore, the sweeping curve of the sandy bay between the headlands, the outline of the hills, the shape of the clouds, all these are so many riddles of form, so many problems of morphology, and all of them the physicist can more or less easily read and adequately solve.

D'Arcy Wentworth Thompson
On Growth and Form

There was always something a little different about meteorite ALH84001, found in 1984 on the icy Allan Hills of Antarctica. For one thing, it came from Mars—like only 11 other meteorites found around the world. But unlike these others, ALH84001 was *old*—and I mean four-and-a-half billion years old. The rock was formed when the Red Planet was newly born. But the most extraordinary aspect of this little lump of Mars did not emerge until August 1996, when scientists from NASA announced that it might contain signs of fossil life from our cosmic neighbour.

Maybe my years at *Nature* magazine have exposed me to too many amazing 'discoveries' that vanish like morning mist under close scrutiny; but I felt in my bones that this claim would not stand the test of time. If I'm wrong (and I rather hope I am), this is one of the most significant discoveries of the twentieth century. But although the jury is still out while scientists clamour for more pieces of the meteorite to carry out exacting tests, already there are signs that this evidence for ancient life on Mars is on shaky ground.

One of the lines of argument particularly caught my attention. Within the Martian rock the NASA team found microscopic wormlike features about a tenth of a micrometre in width, which they suggested might be the fossilized remains of bacteria (Fig. 1.1). What leapt to my mind was a book called *Earth's Earliest Biosphere*, in which Californian geologist William Schopf lists and depicts countless examples of curious, bacteria-like structures in ancient rocks from Earth's early history. Schopf explains that, while some of these are indeed microfossils of primitive bacteria dating back to around a billion years after the Earth was formed, many others are not fossils at all, but most probably structures formed in the rocks by purely geological processes.

Prospectors for early life on Earth are in constant danger of being fooled by these mineral structures, which in some cases look barely distinguishable from well-established microfossils (Fig. 1.2). There is a recognized class of objects called 'dubiofossils', which are microscopic rock structures whose origin one cannot unambiguously ascribe either to organic or inorganic causes. I should say that the NASA scientists were familiar with these pitfalls, and were also uncomfortably aware that their putative Martian fossils were much smaller than any known from Earth. But they felt that the several other suggestive chemical characteristics of meteorite ALH84001 added weight to the idea that the worm-like structures were indeed the mineralized casts of primitive organisms from Mars.

You might think that it should be an easy matter to distinguish a fossilized remnant of a living organism from some rock feature formed by physical forces alone. Surely we can, at even a brief glance, tell a crystal from a living creature, an insect from a rock?

Yet what is it that encourages us to make these distinctions, based on superficial features alone? I suspect that most of us at some level identify a kind of characteristic form that we associate with living things; but it is hard to put that into words. Living organisms come in all shapes and sizes—a tree, a rabbit, a spider—but there is something purposeful about these forms. They are complex (and I shall shortly have to be a little more

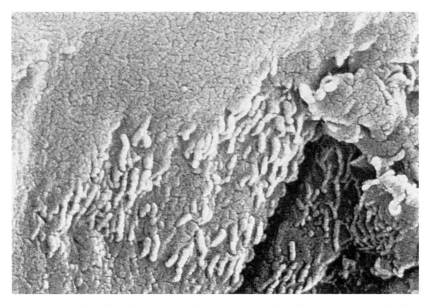

Fig. 1.1 These microscopic structures found in a Martian meteorite have been presented as evidence for ancient bacterial life on Mars. Are they the fossilized remnants of tiny worm-like organisms? (Photo: NASA.)

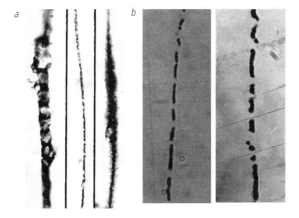

Fig. 1.2 How do you tell a fossil from a rock? The formations shown here have all been identified in ancient rocks; but whereas those in (a) are probably genuine fossilized bacteria, several billions of years old, it is possible that those in (b) were formed by purely geological processes. (Photos: from W. Schopf (ed.) (1991). *Earth's Earliest Biosphere*. Reprinted with permission of Princeton University Press.)

precise in using this word), but not random. They have a kind of regularity—evident, for instance, in the bilateral symmetry of our bodies or in the branching pattern of a tree—but it is not the geometric regularity of crystals. Somehow it seems natural, when we see forms like those in Fig. 1.1, to associate them with the subtle and delicate forces of life, not with the coldly geometrical exigencies of physics.

If there is one thing I hope to do in this book, it is to shake up these assumptions. I wish to show in particular that pattern and organized complexity of form need not arise from something as *complicated* as life, but can be created by simple physical laws. This idea of complexity from simplicity has become almost a new scientific paradigm in recent years, and most probably a cliché too. Yet I hope here to tie it down, to show that it is not a recondite solution to all of life's mysteries, nor a result of a newly acquired facility for tricky computer-modelling, nor even a particularly new discovery—but a theme that has featured in scientific enquiry for centuries. Some of the complex patterns that I shall consider in this book pose questions that are truly ancient: from where come the stripes of a tiger, the procession of 'mare's tail' clouds, the undulations of sand dunes, the vortex of a whirlpool, the shapes and decorative adornments of sea shells?

Imposters

Let me delve further into our preconceptions about form and pattern. If you saw through the microscope mineral formations like those in Fig. 1.3a, would you suspect that these are the shells or skeletons of some tiny creatures? That would be an understandable assump-

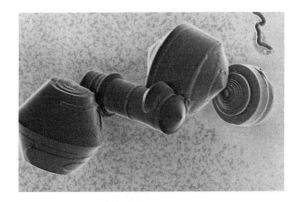

a

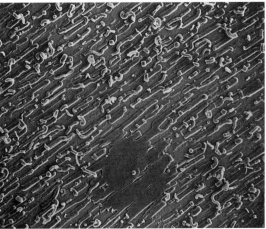

b

Fig. 1.3 (*a*) Are these complex, patterned mineral structures the shells or skeletons of tiny organisms? On the contrary, they are the product of a purely synthetic chemical process carried out in the laboratory. (*b*) A similar chemical process generates these surface patterns, which bear some (coincidental) resemblance to those in Fig. 1.1. (Photos: Geoffrey Ozin, University of Toronto.)

around the world. Ever since these curious, spongy structures were discovered in the nineteenth century, their origin has been disputed. The prevailing interpretation is that they represent the fossil remains of mat-like structures created by marine microorganisms such as cyanobacteria, which are amongst the oldest known forms of life on Earth. Fossil microbes *have* been found in some stromatolites, but the argument for their biological origin finds its most crucial evidence in the similarity in form between ancient stromatolites and modern analogues that are demonstrably still being

a

b

Fig. 1.4 (*a*) Modern-day stromatolites in Shark Bay, Western Australia. (*b*) The complex, laminated structure of a 2.7-billion-year-old stromatolite from Western Australia. The image shows an area of 3 × 4 cm. (Photos: Malcolm Walter, Macquarie University, Sydney.)

tion, yet they are the products of a purely chemical process involving the precipitation of silica from a soluble salt. Much the same chemical brew can produce the surface patterns in Fig. 1.3*b*, strikingly reminiscent in both shape and scale of the putative Martian fossils in Fig. 1.1! What on earth sculpts these mineral bodies into such odd and apparently 'organic' forms?

A particularly striking cautionary tale of this association between life and complex form—and one that reverberates through the story of the Martian meteorite—concerns the rock formations known as stromatolites that are found in ancient reef environments constructed from cyanobacterial and algal mats (Fig. 1.4a). If this association holds, stromatolites provide some of the oldest evidence for life on Earth, since they have been dated back to three-and-a-half billion years ago. Researchers have even proposed that searches for life on Mars itself should include the option of looking for stromatolite-like features around the dried-up lakes and springs of the Red Planet.

But in 1996 John Grotzinger and Daniel Rothman from the Massachusetts Institute of Technology showed that a comparison based on form alone cannot provide unambiguous evidence for the handiwork of biology. They demonstrated that the characteristic features of the irregular layers of a typical stromatolite (Fig. 1.4b), whose bumps and protrusions look for all the world like the product of biological growth, can be generated by simple physical processes of sedimentation and precipitation of minerals from the overlying water. This does not prove that stromatolites *are* purely geological structures (and it is virtually certain that at least some are not), but it shows that arguments based on form alone are not sufficient to rule out that possibility.

We can play this game the other way around. What are the objects shown in Fig. 1.5—living organisms or crystals? Their geometric regularity suggests the latter, but these are viruses, and all too dangerously alive. Complex form may not require an organic origin, but similarly geometric form does not exclude it. There are, in other words, forces guiding appearances that run deeper than those that govern life.

Look—no hands

Our prejudice says otherwise. The most striking examples of complex pattern and form that we encounter tend to be the products of human hands and minds—shaped with intelligence and purpose, constructed by *design*. The convolutions of a traditional patchwork fabric, the intertwining knots of a Celtic symbol, the horizon-spanning stepped terraces of Asian rice fields, the delicate traceries of microelectronic circuitry (Fig. 1.6)—all bear the mark of their human makers. The subconscious message that we take away from all this artifice is that patterning the world—shaping it into the forms of our needs and our dreams—is hard work. It requires a dedication of effort and a skill at manipulation. Each piece of the picture must be painstakingly put into place, whether by us or by nature. This, we have come to believe, is the way to create any complex form.

So when they found complexity in nature, it is scarcely surprising that many theologians throughout time have refused to see anything other than the signature of divine guidance. From the action of nature's most basic physical laws, on the other hand, such as Newton's inverse-square law of gravity, we have learnt to expect nothing but the geometric sterility of a planet's elliptical orbit around the Sun.

Would it not be extraordinary, however, if these laws could by themselves contrive to generate rich and beautiful patterns? If we could decorate a table cloth by using dyes that spontaneously segregate into a multicoloured design? Or to scatter a hillside with topsoil and watch it arrange itself into terraces ready to receive water and seed? But experience teaches us that this is not the way things go. On the contrary, dyes mix, don't they? Soil gets distributed randomly by the wind and rain, right?

The astonishing thing is that sometimes apparent reversals *do* happen. Fluids unmix of their own accord; landscapes become sculpted by the elements into regular patterns. Through such processes, nature's tapestry embroiders its own pattern. And by studying these strange and counter-intuitive processes, we discover that some of nature's patterns recur again and again in

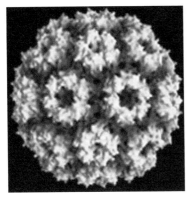

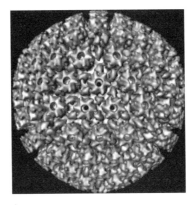

a   *b*

Fig. 1.5 These geometric, ordered forms are in fact living organisms—viruses. (*a*) The cowpea chlorotic mottle virus; (*b*) the herpesvirus. (Images: (*a*) Jean-Yves Sgro, University of Wisconsin; (*b*) Hong Zhou, University of Texas at Houston.)

a

c

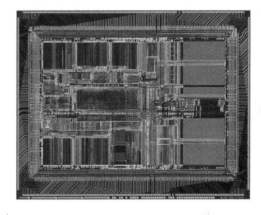

b

d

Fig. 1.6 Most of the complex patterns that we create are the products of painstaking labour: (*a*) a Kuna *mola* tapestry from Panama; (*b*) paddy fields in China; (*c*) Celtic design on a stone cross; (*d*) circuitry on a microprocessor chip. (Photos: (*b*) Getty Images; (*d*) Michael W. Davidson and the Florida State University.)

situations that appear to have nothing in common with one another. You can't avoid concluding, once you begin to examine this tapestry, that much of it is woven from a blueprint of archetypes, that there are themes to be discerned within the colourful fabric. Nature's artistry may be spontaneous, but it is not arbitrary.

Form and life

Biologists are used to the idea that form follows function. By this I mean that the shape and structure of a biological entity—a protein molecule, a limb, an organism, perhaps even a colony—is that which best equips the organism for survival. (In today's gene-centred view of biology, we should instead strictly say that it is the survival of the gene that is paramount, the organism

being merely a convenient vehicle for this.) This is the Darwinian paradigm: form is selected from a palette of possibilities, and by selected I mean favoured by *natural* selection. A form that gives the organism an evolutionary advantage tends to stick.

This is a simple idea, but phenomenally powerful. The objection that it would take an unreasonably long time to find the best form from the range of alternatives—a favourite argument for evolutionary sceptics—crumbles beneath the extraordinary and demonstrable efficiency of natural selection. We can watch the process take place in a matter of days for generations of bacteria bred in culture. In 1994, Swedish researchers performed computer experiments showing that even a biological device as sophisticated as an eye will evolve from a flat sandwich of photosensitive cells in a matter of around 400 000 generations—perhaps half-a-million years, a blink in geological terms—if one makes conservative assumptions about such factors as the rate of mutation between each generation. Even getting life started in the first place, from a brew of simple organic chemicals on the young Earth, seems to have been astonishingly easy: it may have taken less than 200 million years from the time that the planet first had a solid surface, and would presumably have involved competition and consequent selection amongst generations of replicating molecules and small molecular assemblies.

But as an explanation for natural form, natural selection is not entirely satisfying. Not because it is wrong, but because it says nothing about mechanism. In science, there are several different *kinds* of answer to many questions. It is like asking how a car gets from London to Edinburgh. One answer might be 'Because I got in, switched on the engine, and drove'. That is not so much an explanation as a narrative, and natural selection is a bit like that—a narrative of evolution. An engineer might offer a different scenario: the car got to Edinburgh because the chemical energy of the petrol was converted to kinetic energy of the vehicle (not to men-

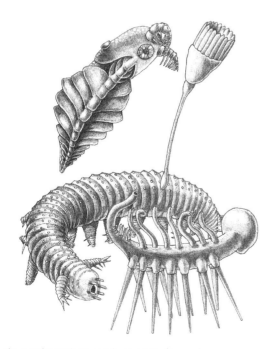

Fig. 1.7 The Cambrian period was a time of tremendous experimentation in nature's body plans. Here are just a few of the bizarre creatures reconstructed from remains found in the Burgess shale. Clockwise from *top left*: *Anomalocaris*, *Aysheaia*, *Hallucigenia* and *Dinomischus*. (Drawn by the author, after Marianne Collins.)

tion a fair amount of heat and acoustic energy). This too is a correct answer, but it will be a bit abstract and vague for some tastes. Why did the car's wheels go round? Because they were driven by a crankshaft from the engine… and before long you are into a mechanical account of the internal combustion engine.

Some biologists want to know about the internal combustion engine of biological form. They will accept that the form is one that conveys evolutionary success, that a fish shaped like a giraffe wouldn't exactly have the edge on its competitors. But this form has nonetheless to be put together from a single cell. What are the mechanical ins and outs of that process?

From a naive evolutionary perspective, anything seems possible. You assume that nature has at its disposal an infinite palette, and that it dabbles at random with the choices, occasionally hitting on a winning formula and then building mostly minor variations on that theme: for fish, the torpedo-body-and-fins theme, for land predators the four-legs-and-muscle idea. To judge from the astonishing diversity of form apparent in fossils from the Cambrian period (Fig. 1.7; see also Stephen Jay Gould's book *Wonderful Life*)—a diversity far exceeding anything we find in today's organisms—you might imagine that this is precisely what happens. But is the palette truly infinite? Once you start to ask the 'how?' of mechanism, you are up against the rules of chemistry, physics and mechanics, and the question becomes not just 'is the form successful?' but 'is it physically possible?'

Questions of this sort were what prompted the Scottish zoologist D'Arcy Wentworth Thompson in 1917 to write a beautiful book whose influence is still felt today. In *On Growth and Form*, Thompson gave an engineer's answer to the Darwinism that was rushing like a deluge through the biology of his time. Still in its first flush, Darwin's theory was propounded as the answer to every question that someone in Thompson's community might want to ask. The shape of a goat's horn, of a jellyfish's protoplasmic body, of a sea shell—all have the form they do because natural selection has sculpted them that way.

D'Arcy Thompson saw such ideas as an affront to one of science's guiding principles: economy of hypotheses, exemplified by the approach to problem solving expounded by the fourteenth-century philosopher William of Ockham and now known as Ockham's (or Occam's) razor. Put simply, this approach demands that we set aside complicated explanations for things when a simpler one will do. The principle is not much fun—there would be no UFOs, no paranormal phenomena, if we had all learnt to observe it—but it prevents the proliferation of unnecessary ideas.

What, suggested Thompson, could be more unnecessary than invoking millions of years of selective fine tuning to explain the shape of a horn or a shell when one could propose a very simple growth law, based on *proximate physical* causes, to account for it? The sabre-like sweep of an ibex horn does not have to be selected from a gallery of bizarre and ornate alternative horn shapes: we can merely assume that the horn grows at a progressively slower rate from one side of the circumference to the other, and hey presto—you have an arc.

There is no inconsistency here with the Darwinian scheme of things, within which it is quite possible for such a growth law to arise. But Thompson's point was that it need not have been *selected*—it was inevitable. Either the horn grew at the same rate all around the circumference, in which case it was straight, or there was this imbalance from one side to the other, giving a smooth curve. It just did not make sense to invoke other shapes: nature's palette contains just these two. Even the more elaborate spiral form of a ram's horn need be only the manifestation of a stronger degree of imbalance, causing the horn's tip to curve through several complete revolutions.

In D'Arcy Thompson's view, some biological forms, the shapes of amoeba say, can no more be regarded as 'selected for' than can the spherical form of a water droplet; rather, they are dictated by physical and chemical forces. To support this assertion he evinced many organisms whose shapes could be explained as a more or less inevitable corollary of the forces at work. What was the point, he asked, in accounting for the shape of a bone in evolutionary terms (which 'explained' nothing) when it could be rationalized through the same engineering principles that engineers use to design bridges? Skeletons are then seen not as arbitrary structures moulded this way and that by natural selection, but as constructions that must satisfy engineering requirements. The same is true of trees, and of all living forms whose stability is dominated by gravity. When small size reduces the influence of gravity, surface tension takes over and a new set of forms can result.

Despite, or perhaps because of, Thompson's erudition and facility with other disciplines (he was also a professor of Ancient Greek), *On Growth and Form* has a quixotic air. It sometimes veers in spirit towards the ideas of the Frenchman Jean Baptiste Lamarck, who argued before Darwin that evolution was a response to the environment, in which adaptation is not the result of random mutations but is guided along a preordained path by the environmental forces to which organisms are subject. Today this idea is biological heresy.

On Growth and Form came close to heresy too, and Thompson was conscious of it. 'Where it undoubtedly runs counter to conventional Darwinism', he said when submitting the manuscript, 'I do not rub this in, but leave the reader to draw the obvious morals for himself.' And so they did: the English biologist Sir Peter Medawar called the book 'Beyond comparison the finest work of literature in all the annals of science that have been recorded in the English tongue'. Without a doubt, it is beautifully written and deeply scholarly. But to what extent was Thompson right?

The black box of genetics

In its most basic form, D'Arcy Thompson's thesis was that biology cannot afford to neglect physics, in particular that branch of it that deals with the mechanics of matter. (He was far less concerned with chemistry, the other cornerstone of the physical sciences, but that seems to have been because he did not consider it sufficiently mathematical. Today there is much in the field of chemistry that would have served Thompson well.) His complaint was against the dogma of selective forces as the all-pervasive answer to questions in biology. For him this did not answer questions about causes; it merely relocated the question. A physicist, on the other hand, 'finds "causes" in what he has learned to recognize as fundamental properties... or unchanging laws, of matter and of energy'.

Today, Thompson would surely have to take up arms against the modern manifestation of the same Darwinian idea: genetics. It is not hard to become persuaded that in modern biology, all questions end with the gene. The pages of *Nature* and *Science* are filled with papers reporting the identification of a gene (or the protein derived from a gene) that is responsible for this or that biological phenomenon—for the development of a forearm, the predisposition to breast cancer, even for intelligence. The climate of the culture in molecular biology (although not, I think, the expressed belief of its individuals) is that, by understanding the roles of genes and the mutual interactions of the proteins derived from them, we will understand life.

This attitude finds expression, for instance, in the Human Genome Project, the international effort to map out every one of the 100 000 or so genes in the 23 chromosome pairs of the human cell. This project might be completed by the turn of the century, and to hear some speak about it, you would think that it will provide us with a complete instruction manual for the human body. But biologists know that it will not provide this at all. We can certainly expect to learn an awful lot about the way our cells work, and perhaps more importantly, we will obtain a tool that will greatly aid researchers studying genetically related diseases. That kind of information will be tremendously valuable for biomedical science.

Yet biological questions do not really end in the gene at all: they *start* there. It is easy to get the impression that once a gene for a particular congenital disease has been located, the problem is solved. But most genes are just blueprints for proteins, and the physiological pathology associated with the gene often results from some biochemical transformation that the protein does or does not facilitate. It might even result from some malfunction that shows up only several steps down the line from the behaviour of the gene product itself. Very often, if we are to make effective use of the information that genetics provides, we must figure out how the gene's protein product works, not just where the gene is. Biologists know this, of course, but I am constantly struck at how much of molecular biology advances at a 'black box' level, with little concern for the physical or chemical details of a biochemical process and an interest only in the *identity* of the genes and protein gene products that control it. The rest is, of course, truly the 'hard part' of biology (cynics might suggest that, now that chunks of human chromosomes can be patented and sold off, it is also the less profitable part). The crucial point, though, is that a gene itself might provide precious few clues about what this hard part entails.

Furthermore, organisms are *not* just genes and proteins made from them. There is goodness knows what else in the cell: sugars, soap-like molecules called lipids, non-protein hormones, oxygen, small inorganic molecules like nitric oxide used for cell communication, and minerals like the calcium hydroxyapatite of bone and tooth. None of these substances are encoded in DNA, and you would never guess, by looking at DNA alone, what role they play in the body. There are, furthermore, physical properties that biological structures possess, such as surface tension, electrical charge and viscosity. These are all relevant to the way that cells work, but gene-hunting cannot tell us much at all about what their role is.

In short, questions in biology of a 'How?' nature need more than genetics—and frequently more than a reductionist approach. If nature is at all economical (and we

have good reason to believe that this is usually so), we can expect that she will choose to create at least some complex forms not by laborious piece-by-piece construction but by utilizing some of the organizational and pattern-forming phenomena we see in the non-living world. If that is so, we can expect to see similarities in the forms and patterns of living and purely inorganic or physical systems, and we can expect too that the same ideas can be used to account for them both. It is in the undoubted truth of this idea that the spirit of *On Growth and Form* lies, and this is where the true prescience of D'Arcy Thompson's achievement resides. Although I shall focus only occasionally on pattern and form in biology, I feel that this spirit pervades all of what I shall say in this book.

Is biology just physics?

It is not often that biologists develop simple models based on physical laws in attempting to explain what they see. And with good reason: it is very hard to take account of all of the multifarious factors that are important in living organisms. Biological systems are usually too delicate to rely on crude, general physical principles, and so biologists are wary of trusting to broad physical phenomena for explanatory purposes. To them, it feels uncomfortably like driving a car with no hands on the wheel, hoping that friction and air resistance will somehow conspire to guide the vehicle down a tortuous road.

It can be tempting, once one starts to appreciate the stunning variety of complex pattern and form in the natural sciences, to let the pendulum swing too far to the other extreme. A popular accusation against modern genetics is that it is too reductionistic, that one cannot understand all of the rich complexity of biology by breaking it down to genetic influences. One hears this again and again from proponents of 'holistic' science, who have no shortage of arguments to support their point of view—for certainly, one can find emerging from large populations of interacting 'units' (be they living organisms or non-living entities) a kind of large-scale organization and structure that one would never be able to deduce from a close inspection of the individual units or their mode of interaction. Such ideas, which have now become fashionable under the banner of 'complexity,' are often lauded as an injection of richness and mystery into the sterility of a reductionist world view.

I applaud a perspective that broadens the horizons of 'black-box' biology, but there is no getting away from the fact that most of biology, particularly as a molecular science, is hideously *complicated*, which, in distinction to *complex*, means that the particulars matter: leave out one part of the chain, and the whole thing falls apart. In such a case, one gains rather than loses understanding as the magnification is increased. Until we get reductionistic about the immune response, let's say, we won't know much about it, and neither will we have much idea how to tackle pathological conditions such as AIDS. Reductionism can certainly be aesthetically unappealing, but it can also be fantastically useful. In addition, reductionism is not always the dogma it is cracked up to be. Richard Dawkins, whose books *The Selfish Gene* and *The Blind Watchmaker* are often invoked as the epitome of genetic reductionism, has stressed that his ideas by no means imply a kind of genetic determinism of biological form, characteristics and behaviour. Dawkins says only that it is the gene on which evolutionary forces ultimately act: that is, selection acts on the genotype (the organism's genetic composition), not the phenotype (the physical expression of that composition). There is nothing in what I shall say about biological form that is inconsistent with these notions.

But a few biologists, riding against the mainstream of current thought, hold a more extreme opposition to the genetic orthodoxy. Brian Goodwin from Britain's Open University has argued that a gene's eye view of modern biology cannot be complete, and that there are some fundamental aspects of an organism's form that persist *in spite of* natural selection, not because of it. Goodwin suggests that the pattern-forming principles seen in some non-living systems operate as strongly in living organisms, giving them features that evolution is powerless to erode away. I have to say that not many people believe Goodwin, although it seems to me that his arguments become weakened only when extended from specific instances to the status of a new developmental principle in biological growth. Insofar as I shall talk about biological form at all, the position I take is rather different. I don't think we know very much yet about whether natural selection has the power to modify or suppress certain pattern-forming principles that occur in nature. But I would suggest that, in the here and now, such principles undoubtedly exist—and do so in sharp distinction from the idea that genes are like a *deus ex machina* that holds all biological processes in thrall, building organisms in a laborious, brick-by-brick manner. To that extent, I don't believe I am saying anything that will disturb molecular biologists (although I think it a pity that they do not always regard these pattern-forming processes with a greater sense of wonder).

What is form?

This book is about the development of pattern and form, and so it is as well to have an indication of what I mean when I use these words. I cannot give either term a definition of mathematical rigour, however, nor can I always maintain a clear distinction between the two. There is always an element of subjectivity in perceiving patterns. On the whole I shall be concerned with patterns and forms in space, ones that we can see and perhaps touch. But of course there are all sorts of patterns—in a time sequence of events, in human behaviour and interactions, in stories and myths. The word is a very plastic one.

There are surely certain spatial images that most people would categorize as patterns—the repeating designs of wallpaper or carpets, for example. This prompts the idea that a pattern might be regarded as a regularly repeating array of identical units. I want to broaden that concept slightly, and include in my definition arrays of units that are similar but not necessarily identical, and which repeat but not necessarily regularly or with a well-defined symmetry. An example is the ripples of sand at the seaward edge of a beach or in a desert (Fig. 1.8). No two ripples are identical, and they are not positioned at exactly repeating intervals (which is to say, *periodically* in space)—but nonetheless, I don't think it is too hard to persuade ourselves that this might be reasonably called a pattern, as we can recognize within it elementary units (the ripples) that recur again throughout space. The ripples are usually all of more or less the same width; but I don't feel that even this need be essential to qualify as a pattern. A mountain range has features of all sizes, from little crevasses to huge sweeping valleys, but there is still something about the way it looks to us, from out of an aeroplane window, that allows us to see a pattern there.

Form is a more individual affair. I would define it loosely as the characteristic shape of a class of objects. Like the elements of a pattern as described above, objects with the same form do not have to be identical, or even similar in size; they simply have to share certain features that we can recognize as typical. Shells of sea creatures are like this. The shells of organisms of the same species all tend to have a certain form that can be recognized and identified even by a relatively untrained eye, despite the fact that no two shells are identical. The same is true of flowers, and of the shapes of mineral crystals. The true form of these objects is that which remains after we have averaged away all the slight and inevitable variations between individuals.

Fig. 1.8 Ripples in sand are self-organized patterns formed by wind-blown sand transport. (Photo: Nick Lancaster, Desert Research Institute, Nevada.)

Patterns, then are typically extended in space, while forms are bounded and finite. (But take this as a guideline, not a rule.)

Symmetry and order are related but not synonymous. Complex natural forms commonly have the appearance of a kind of order even when, mathematically speaking, they have very little symmetry. An oak tree, for instance, has as little symmetry as it is possible for an object to possess, but is it disorganized? It is often said of symmetry that our intuition is at odds with a mathematical description. Which is more symmetrical—a kaleidoscopic image like that in Fig. 1.9a or a six-pointed Star of David (Fig. 1.9b)? We might say that the kaleidoscope pattern has more organization, more repeating features; but mathematically the symmetries of both images are the same. And are either of these more symmetrical than a circle? No, they're not—the circle has the highest possible degree of symmetry for a two-dimensional (flat) object. It's just that we don't perceive the symmetry so readily when it becomes as great as it is in a circle—to us that just looks bland and featureless. I don't propose to say much more about symmetry *per se*, because there are many splendid books that deal with this endlessly fascinating topic, of which Hermann Weyl's *Symmetry* is a classic and *Fearful Symmetry* by

Fig. 1.9 The formal symmetry of these two patterns is the same, even though (*a*) looks much more complex than (*b*).

Ian Stewart and Martin Golubitsky is one of the most up-to-date and lucid.

The natural language of pattern and form is mathematics. This may dismay those of you who never quite made friends with this universal tool of science, and it may seem a little disappointing too—for patterns and forms can be things of tremendous beauty, whereas mathematics can often appear to be a cold, unromantic and, well, calculated practice. But mathematics has its own very profound beauty too, and this is something that you do not any longer have to take on trust. The now familiar images of fractal forms and patterns demonstrate that mathematics is perfectly able to produce and describe structures of immense complexity and subtlety.

The main point is that mathematics enables us to get to grips with the *essence* of pattern and form—to describe it at its most fundamental level, and thereby to see most clearly what features need to be reproduced by an explanation or a model. In short, the mathematical description of a form can be considered to pertain to that which is left after the particular irregularities or anomalies of any individual example of that form (for example, the small imperfections or bumps on a shell) are averaged out. To explain how the form of the shell arises, there is no point in trying to explain all the little bumps, since these will be different for each shell; we need instead to focus on the 'ideal' mathematical form. This concept of an ideal, perfect form behind the messy particulars of reality is one that is generally attributed to Plato.

Why does maths help us in this endeavour? For a start, it provides a very concise and precise description of a form. Try to describe a circle in words, without using any of the pre-existing associations of circular objects (such as 'the shape of a full moon'). It is 'round all over', but isn't that also true of an egg or a sea-smoothed pebble? And if you had to *construct* a circle without exploiting its mathematical features (which I am just coming too), you'd have an even harder time. The French mathematician Pierre Laplace was famously able to draw a perfect circle freehand, but this ability is not granted to most of us.

The mathematical description of a circle, meanwhile, can be expressed in words as 'a line in a flat plane that is everywhere an equal distance from a single point'. If that doesn't strike you as particularly concise, let me quickly indicate the symbolic depiction of this definition in mathematics:

$$x^2 + y^2 = R^2. \tag{1.1}$$

Not only does this help us to express exactly what a circle is; it also suggests immediately how we might construct one. You need only to keep your pen a constant distance from a point on the paper, for example by attaching its end to a piece of string anchored at the other end by a pin. The way to 'grow' a form often becomes obvious once the form is described mathematically.

This is a point that comes out with great force and clarity from D'Arcy Thompson's work. If you look at a mollusc shell (Fig. 1.10*a*) and try to imagine how the cluttered frenzy of the cell could put together such a gracefully spiralling object, or conversely how it might have arisen by chance through evolution and natural selection, you may be forced to conclude that the problem is profound. But once you recognize that the shell has a precise mathematical form—that of a so-called logarithmic spiral—then you begin to see that nothing more than a simple and plausible growth law is required.

The logarithmic spiral (Fig. 1.10*b*) was first characterized mathematically by René Descartes in 1638. It is

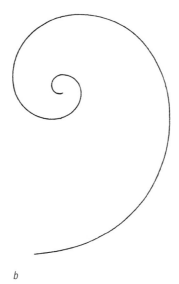

a *b*

Fig. 1.10 The shells of snails and other molluscs (*a*) trace out logarithmic spirals (*b*). (Photo: Scott Camazine, Pennsylvania State University.)

defined as the curve traced out by a point rotating around a point of origin with constant angular velocity (that is, it always takes the same time to rotate through the same angle), while its linear speed at any instant increases in direct proportion to its distance from the point of origin. That probably does not sound particularly elegant or concise, but the symbolic expression can be made extremely neat:

$$r = a\theta. \qquad (1.2)$$

This three-symbol formula might leave you none the wiser, but you can't deny its concision. And to a mathematician, it conveys precisely the same information as the curve in Fig. 1.10*b*.

The logarithmic spiral has the unique property that the curve is everywhere 'similar', differing in size but not in shape. In other words, as the curve rotates through a fixed angle, it grows uniformly in scale. This, and the description above, help us to see what are the fundamental generating mechanisms of such a form. Some things remain constant, for example the angular speed of the curve's tip, and the shape of the curve, while other things, for example the linear (tangential) speed of the tip, change in a well-defined way. We can then generate a form like this by proposing that the deposition of new fabric at the shell's rim follows a growth mechanism that produces these characteristics. A mechanism of this sort that generates a three-dimensional mollusc shell with the cross-section of a logarithmic spiral is as follows: the existing shell rim provides a template on which new shell material is laid down, so it stays the same shape, but the rim is expanded in *scale* at a constant rate. If, in addition, the growth happens initially to be slightly faster on one side of the embryonic rim than the other, this imbalance is maintained proportionately as the shell gets bigger, and it curves into a spiral. It does not take too much imagination to see that a mechanism like this is a rather 'natural' one to be expected from a creature making a shell that needs to keep pace with its own growth, and doesn't require any mysterious geometrical knowledge or an ability to figure out what on earth equation 1.2 means. The imbalance that leads to spiral growth could come from any source—*any* imbalance will produce a logarithmic spiral. If there is no imbalance, the shell instead has a cone shape, just as one can find in some species of mollusc.

You might be able to appreciate too that a growth mechanism this simple need not be restricted to shells, but could apply to any hard tissue whose shape is determined purely by the deposition rate at the growing edge.

Fig. 1.11 Many animal horns, like those of this male Dall's sheep, are logarithmic spirals.

Horns too are commonly logarithmic spirals, albeit often more gently curving (Fig. 1.11). So we can anticipate these forms too as the expected result of an obvious growth mechanism, rather than as a form selected at random from a huge range of others by natural selection.

Let's look at another way that the logarithmic spiral, the ideal form of sea shells, illustrates how mathematics helps us to get to the essence of form and to make its explanation a much simpler process. D'Arcy Thompson realized that even very complex shells have a form that can be generated by the logarithmic spiralling of a certain fixed two-dimensional shape (later called the generating curve). He said,

> The surface of any shell may be generated by the revolution about a fixed axis of a closed curve, which, remaining always geometrically similar to itself, increases its dimensions continually ... . The scale of the figure increases in geometric progression [exponentially in time] while the angle of rotation increases in arithmetical [at a constant rate].

This is illustrated in Fig. 1.12. Thompson noted that the form of the generating curve 'is seldom open to easy mathematical expressions'—but the way in which the shell shape is created by sweeping this curve in a spiral through space *is* mathematically well-defined, and can be imagined to be a consequence of a simple growth law. With this concept in mind, an explanation for the form of any particular kind of shell reduces to an explanation of the shape of the generating curve—the whole myriad of shell forms can be produced by the same kind of spiral evolution of these two-dimensional boundaries. Deborah Fowler and Przemyslaw Prusinkiewicz at the University of Regina in Canada have used computer-modelling to depict some of the shapes that result by taking a given

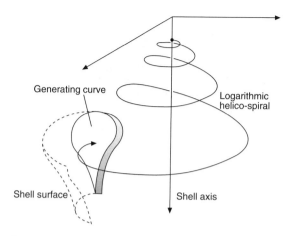

Fig. 1.12 A shell surface can be constructed by sweeping a two-dimensional 'generating curve' through a logarithmic spiral pulled out into a helix.

generating curve through the mathematical paces outlined by Thompson (Fig. 1.13).

The real advantage of using mathematics to describe form is that it makes the problem *algorithmic*. An algorithm is a sequence of logical steps that a computer program, say, must execute to carry out a certain task. Complex shapes like those seen here are often most easily described not in terms of 'what goes where' but by an algorithm that generates them. Once we know the mathematical algorithm, we can start to ask what kind of physical processes might provide a form-generating

rule to which the algorithm is a good approximation. In this way, the mystery of complex form and pattern becomes much more clearly defined.

Model making

I'll say much in this book about models. In everyday terms the word commonly implies 'a small-scale replica of the real thing' (although fashion models might take exception to this). To scientists, a model is something rather different. When science seeks explanations, it doesn't usually expect them to be exhaustive. On occasion that's simply because we don't know everything that is going on in the system we're looking at. In other cases it may be because we know that some influences are of little significance, so that their inclusion would just make the equations harder to solve without altering the solutions very much. Or again, it may be that certain influences are known to be important but we don't know *how* to include them, or how to solve the equations if we do. So then we look for good approximations that we *can* solve, knowing that the answers might not correspond quite so well with observations. Biology operates quite a lot of the time within the first of these scenarios—we don't know enough about the details. A lot of engineering is conducted in the spirit of the second situation—one might ignore the effect of air resistance or of friction in figuring out the way an object moves. (Galileo did the same in his famous—and apocryphal—experiment from the Tower of Pisa.) But engineers who worry about problems of

Fig. 1.13 Shells created on a computer by applying the 'algorithm' depicted in Fig. 1.12 to a variety of generating curves. (Images: Deborah Fowler and Przemyslaw Prusinkiewicz, University of Calgary, Canada; from Meinhardt (1995).)

Box 1.1: Exponents

Although mathematics provides the natural language for talking about patterns, I will not need to use in this book any more mathematics than can be expressed in words rather than in abstruse equations. You will need to know little more than the definition of squares and cubes of numbers. The square of 2, written 2^2, is 2×2, and the cube (2^3) is $2 \times 2 \times 2$. The superscripted number is often called the *power* or *exponent*: 2^3, for instance, might be called 'two to the power three'. We will at a later stage be confronted with rather more curious mathematical entities in which the powers are not whole numbers, such as $2^{2.26}$. It is not clear how to write that in multiplicative longhand, but you need know only that this quantity has a well-defined numerical value, that there are rather simple mathematical ways (involving logarithms) to calculate it, and that its value is greater than 2^2 (= 4) and less than 2^3 (= 8).

---

fluid flow (a topic discussed in Chapter 7) commonly find themselves in the third situation—making big approximations and accepting the consequences.

The point is that scientific descriptions of phenomena in all of these cases do not fully capture reality—they are models. This is not a shortcoming but a strength of science—much of the scientist's art lies in figuring out what to include and what to exclude in a model, and this ability allows science to make useful predictions without getting bogged down by intractable details.

Now, the thing that is not often stressed or appreciated about scientific model-building is that there are very many natural phenomena (one could make a strong case for this being true of them all, in fact) for which there is not a single, unique model that is 'right'. This is more than a matter of models differing by the choice of what to put in and leave out, as though all are assembled in a modular fashion. Rather, some phenomena can be tackled successfully from more than one entirely different perspective. This is true of several of the phenomena that I shall discuss in this book.

A particularly common distinction is that between *numerical* and *analytical* models. Computers are so sophisticated nowadays that complex physical processes, such as the freezing of a liquid, can be simulated computationally by simultaneously solving the equations of motion for thousands or even millions of simulated molecules. This is a numerical model, in which the behaviour of the entire system emerges from the piecemeal enumeration of the behaviour of each of its component parts. An analytical model of the same process, in contrast, might make no attempt to describe the motion of individual particles, but will involve mathematical expressions for the relationships between different bulk properties of the medium, such as temperature, density and energy. It may even be possible to solve such a model with pen and paper (which was all theorists had at their disposal up to half-a-century ago).

Although these two approaches are very different, they might both include (and exclude) in their recipes exactly the same physical forces and parameters. But other models might differ in their essential ingredients—we'll encounter later, for instance, two models for the growth of bacterial colonies, one that assumes that the bacteria repel each other and another that assumes only attractive interactions between cells. The problem is that is it not uncommon to find that two models differing like this will generate more or less the same apparent behaviour! The trick is then to find under what conditions the models *do* generate different results, and to try to conduct experiments under those conditions in order to decide which model to favour. This is a familiar challenge in the sciences of pattern formation.

Perhaps the strongest point that I want to make about models in the present context is that they can often generate the complex patterns seen in nature from remarkably few ingredients, which are themselves of striking simplicity. OK, you might say, it's not obvious that this should be so—but what does it mean? Well, on one level it means that growth and form need not be mysterious—we do not have to resign ourselves to thinking that the shape of a flower will be forever beyond our abilities to explain, or even that an explanation (at some level) will require years of dedicated research on plant genetics. On the other hand, it carries at least an implication that there exist universal patterns and forms which remain robust to the fine details of a particular system. For the simple rules of these models are typically of a general nature: 'Assume that the particles move about at random', for instance, not 'Assume that the ETS domain protein encoded by the P2 transcript of the *pointed* gene is a nuclear target of a signalling cascade involving Ras1 and Raf which acts downstream of R1/MAP kinase'.*

---

* That's a direct quote, by the way, more or less. I don't mean to imply that this kind of research is absurd, but just that nature can be, as far as we are concerned, absurdly complicated.

Thus a single model, and its adherent patterns and forms, may turn out to be applicable to a number of different real phenomena.

Breaking the monotony

It's not unusual to associate pattern with order: creating a recognizable pattern rather than a mess requires an orderly process of putting the pieces in place. It is, then, possibly a little alarming to discover that in nature the most highly symmetrical systems are also the most random.

As I mentioned earlier, we often don't perceive any pattern in the most symmetrical systems: they are uniform and, to our minds, featureless. In nature, however, nothing is truly featureless if you look close enough. All matter is made up of atoms, and at the scale of less than a millionth of a millimetre this graininess becomes apparent and the illusion of a uniform medium is lost. Physicists generally regard gases and (to a lesser degree) liquids as uniform, fully symmetrical systems. Yet on the atomic scale all one sees is random disorder, atoms and molecules whizzing about with no apparent symmetry at all. The uniformity and high symmetry become apparent only by considering the *average* features of these systems, which we can do either by focusing our attention on one region and averaging the molecular motions over time or by comparing a large number of different regions at any instant. In both cases, a gas then appears to have a completely uniform density of molecules, on average, at all points in space (it is *homogeneous*); and they travel in all directions with equal probability (the gas is *isotropic*).

When this randomness is absolute, the highest symmetries are observed. (By 'higher symmetry' I mean that the system has a greater number of transformations, such as rotations around an axis or reflections in a mirror plane, that leave it looking the same.) That is why a soap bubble is spherical: its perfect symmetry is a consequence of the fact that the pressure of gas inside the bubble is equal in all directions, because on average an identical number of gas molecules collide with the bubble walls at all points.

The problem of creating patterns and forms that we tend to recognize as such is therefore not one of how to generate the symmetry that they often possess, but of how to *reduce* the perfect symmetry that total randomness engenders, to give rise to the lower symmetry of the pattern. How do the water molecules moving at random in the atmosphere coalesce into a six-petalled snowflake? Patterns like this are the result of *symmetry breaking*.

The symmetry of a uniform gas can be broken by applying a force. Gravity will suffice: in a gravitational field the gas is denser where the field is stronger (closer to the ground). Thus the atmosphere has a density that increases steadily towards ground level. The gas is then no longer homogeneous or isotropic. In this example, the symmetry of the force dictates the symmetry of the distribution of matter that it produces: gravity acts downwards, and it is only in the downwards direction that symmetry is broken. Within horizontal planes (more properly, concentric spherical shells around the Earth) a constant distance from the ground, the atmosphere has a constant density (well, it *would* have if the Earth were a perfect sphere and there were no winds). We might intuitively expect that this will always be so: that the final symmetry of a system will be dictated by that of the symmetry-breaking force that destroys an initially uniform state. In other words, we might expect that matter will rearrange itself only in the direction in which it is pushed or pulled. Within this picture, if you want to pile up sand into mounds arranged in a square, checkerboard array, you will have to apply a force with this 'square' symmetry.

But it is the central surprise of the science of pattern formation that this is not necessarily so. The symmetry of a pattern formed by a symmetry-breaking force does *not* always reflect the symmetry of that force. Of the many examples that I shall describe throughout this book, one will serve here to illustrate what I mean, and why this seems at first sight to be astonishing. If you heat (very carefully—it is not an easy experiment in practice) a shallow pan of oil, it will develop roughly hexagonal circulation cells once the rate of heating exceeds a certain threshold (Plate 1; see also Appendix 1). The system (the fluid) was initially uniform in the plane of the pan, and the symmetry-breaking force (the temperature difference between the top and bottom of the fluid layer) was also applied uniformly in this plane—yet suddenly this uniformity is lost, being replaced by a pattern with hexagonal symmetry. Where has this sixfold pattern come from?

In such cases, one is apparently getting 'order for free'—getting order out without putting order in—although as I say, it is more correct to say that symmetry is being lost rather than gained. The central questions behind many pattern-forming phenomena are: how is it that symmetry can be spontaneously broken? How can the symmetry of the effect differ from that of the cause? And why is symmetry so often broken in similar ways in apparently very different systems? That is to say, why are some patterns universal? These are questions profound enough to last us throughout the rest of the book.

BUBBLES

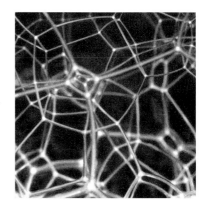

*When I arrived here yesterday Uncle William and Aunt Fanny met me
at the door, Uncle William armed with a vessel of soap and glycerine
prepared for blowing soap bubbles, and a tray with a number of
mathematical figures made of wire. These he dips into the soap mixture
and a film forms or adheres to the wires very beautifully and perfectly
regularly. With some scientific end in view he is studying these films.*

Agnes G. King, niece of Lord Kelvin
1887

I am quite sure that a fascination with patterns in nature is as old as
civilization. When the Egyptians began to keep bees in clay pipes 5000
years ago, they cannot have failed to notice the astonishing hexagonal
pattern of the honeybee's dwelling (Fig. 2.1). Charles Darwin declared it
'absolutely perfect in economising labour and wax', and marvelled at the
bees' instincts for producing such a masterpiece of engineering.

If you want to fill up a plane space with identical, equal-sided and equal-
angled cells, there are only three choices: triangles, squares or hexagons.
Only these regular polygons can be packed together to fill space without
leaving gaps. Pentagons, for example, will not work, and neither will
octagons (Fig. 2.2). Bees making a pentagonal honeycomb would be constantly leaving gaps, and it is not hard to see why these aberrant bees
would not be very successful in the Darwinian struggle for survival. The
same is true for circular cells.

But why do bees not make square or triangular cells
instead of hexagonal? The ancient Greeks suspected that
the bees possessed 'a certain geometrical forethought'
by which they deduced that hexagonal cells could hold
more honey; but the Frenchman R.A.F. de Réaumur

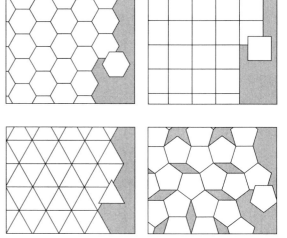

Fig. 2.1 The hexagonal honeycomb of the honey bee was surely one of
the first recognized examples of geometrical pattern in the natural
world. (Photo: Scott Camazine, Pennsylvania State University.)

Fig. 2.2 There are just three types of regular polygon (with equal sides
and angles) that will tile a plane without leaving gaps: equilateral
triangles, squares and hexagons. Pentagons will not fit. But, as we will
see, nature nevertheless has plenty of uses for fivefold symmetry!

proposed in the eighteenth century that it is the area of the walls, not the volume of the cavities, that matters. The total length of the cell walls for hexagonal cells filling a given area is less than that of square or triangular cells enclosing the same area. In other words, it takes less material to make hexagonal walls. It is this drive towards economy that leads bees to make hexagonal honeycombs. Why bees should be economy-conscious was not obvious at that time, however, and Réaumur's contemporaries decided that the bees were guided by mathematical principles according to 'divine guidance and command'. Darwin, of course, removed any residual need for the hand of God in nature's minutiae: he showed that competition and natural selection are the principles that favour organisms who minimize their metabolic costs.

End of story? Hardly. For this was just the kind of Darwinian fable that made D'Arcy Thompson reach for his hammer.

Water's skin

It sounds very neat, but when you start to think about what this explanation requires, it gets uncomfortably elaborate. We must assume that bees and their ancestors have tried out just about every honeycomb pattern a tiler could imagine, before gradually conceding that, yes, hexagons really did leave you less tired and more able to go out foraging. And then they would have had to acquire some kind of sophisticated instinct that allowed them to construct perfect hexagons without the assistance of set-squares, protractors, compasses or any trigonometric know-how.

Why accept this concoction of untested suppositions, asked Thompson, when one could see quite clearly that

Fig. 2.3 A bubble raft of equal-sized bubbles adopts the hexagonal pattern of a honeycomb. Coincidence? (Photo: B.R. Miller.)

the hexagonal honeycomb was an inevitable result of purely physical forces? For everyone knows that a layer of bubbles packs together in just this hexagonal arrangement (Fig. 2.3). If the wax of the comb is made soft enough by the body heat of the bees, suggested Thompson, then it is reasonable to think of the compartments as bubbles surrounded by a sluggish fluid, and so they will be pulled into a perfectly hexagonal array by the same forces of surface tension that organize bubbles into hexagonally packed rafts. In other words, the pattern would form spontaneously, without any great skill on the part of the bees and without the guiding hand of natural selection.

That all sounds plausible enough, perhaps, but it doesn't really explain the hexagonal pattern in any fundamental way—it simply says that the honeycomb is like a bubble-raft, and bubble-rafts make hexagonal arrays. Why hexagons, though? If cellular packings like bubble-rafts and honeycombs really are the product of blind physical forces, why should there be any requirement of equal sides, or of identical shapes, at all? Why not a crazy-paving mosaic of random polygons? At this point we are going to need to know a little more about what a bubble really is, and what controls its shape.

Bubbles are structures made from liquids. We don't often think of liquids as having characteristic shapes—a liquid is fluid, it takes on the shape of the vessel that contains it. But liquids most certainly can have shapes of their own, though these are acutely sensitive to forces such as gravity. In a mist, tiny droplets of water small enough to be buoyed against gravity's tug by the buffeting of air molecules take on the form of near-perfect spheres. Raindrops too take this shape, slightly modified by the frictional forces of their passage through the air and by the urgent pull of gravity.

A spherical droplet provides an illustration of that counter-intuitive aspect of symmetry mentioned in the first chapter: it is generally greatest in the presence of extreme randomness. Unlike crystals, in which the atoms are stacked into regular arrays like eggs in an eggbox, liquids have no ordering of their constituent particles over long distances. The position of one molecule of water bears no relation to the position of another a few millionths of a millimetre away—everything is a jumble. This means that the liquid looks the same in all directions—it is isotropic, and that is reflected in the 'perfect' spherical symmetry of a droplet. But there is something more to the spherical shape, because it is robust: a droplet returns to this shape if momentarily deformed. In other words, there is some factor that *selects* a spherical form. That factor is surface tension.

Liquids and solids are held together by forces of attraction between the constituent molecules, which prevent them from flying apart into vapour. These forces can take many forms. In solids like diamond, strong chemical bonds bind the atoms into structures that can be disrupted only by very energetic processes. In a molecular liquid like water, these same strong bonds hold together two atoms of hydrogen with one of oxygen in each water molecule; but the molecules themselves are bound only by much weaker forces, which give the liquid some cohesion even though the individual molecules are free to move around. These forces of attraction are electrical in origin: regions of the water molecules that bear a slight positive charge (the hydrogen atoms) are electrically attracted to regions on other molecules with a slight negative charge (the oxygen atoms).

Deep within the bulk of the water, a molecule feels attractive forces from all directions. But molecules at the surface are attracted only by the molecules below it, since above is only air (and very diffuse water vapour). There is, therefore, a net inward force on the surface molecules, which we call surface tension. Since the attractive forces have the effect of lowering a molecule's energy (stabilizing the molecule), the surface molecules are more energetic than those deep in the bulk. So there is an excess energy at the surface. Surface tension and surface excess energy are two equivalent manifestations of the fact that surfaces are less stable than the interior of a substance. This means that surfaces cost energy.

As all physical systems like to reach their most energetically stable state (that is, their *equilibrium* state—see Box 2.1), they tend to minimize the area of their surfaces. For a mass of a substance with a certain volume, the shape that has the smallest surface area is a sphere. So a droplet of water forms a sphere to minimize its surface excess energy. It is a statement of the same thing to say that surface tension pulls at the surface of the droplet equally from all directions, so that it acquires spherical symmetry.

I might point out here that surface tension can play a crucial role in determining the forms of solid objects too, in particular those of crystals. Crystals grow by adding atoms to those already packed into regular arrays; but there are several alternatives for where the newly added atoms might sit, and the positioning of these determines the shape of the facetted object. Is it better to add atoms onto the face of an existing layer, or to add them on at the edges of the layer? In other words, which face of a facetted crystal will grow fastest?

Whereas in a liquid droplet the surface tension is the same in all directions, the different faces of a crystal have different surface tensions (because the arrangement of atoms is different on each). The face that grows the fastest will often be that with the greatest surface tension. These considerations determine whether, for example, a crystal like rock salt (sodium chloride) will grow as cubes or as octahedra. Either can be generated from the stacking arrangement of sodium and chlorine atoms, but the cubic shape is selected because of the way that certain facets grow faster than others.

Surface tension controls the shapes that droplets adopt when they sit on surfaces. If a droplet spreads, it increases its surface area and thus its surface excess energy; but on the other hand, it covers the surface below, which also has a surface excess energy. If the total surface excess energy is lower for a fully liquid-covered surface, the droplet will spread into a liquid film; if not, it remains a glistening bead (Fig. 2.4).

Thus, it is not hard to see how surface tension produces the spherical form of liquid droplets. Perhaps more surprisingly, it can also be responsible for regular patterns. In Fig. 2.5a I show a string of pearl-like beads of fly-catching glue attached to the thread of a spider's web. The spider has not painstakingly placed all of these beads at regular intervals along the thread; they have formed spontaneously in a regular pattern through the action of surface tension. A thin, cylindrical column of liquid like the coating of glue on a spider's thread is unstable in the face of tiny disturbances: if the column develops a slight wavy unevenness (Fig. 2.5b), surface tension acts to accentuate the convex curving faces, pulling each undulation into a roughly spherical droplet. This 'pearling'

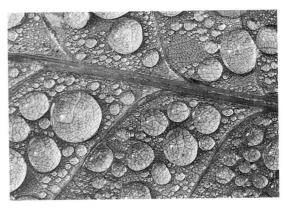

Fig. 2.4 Water droplets will not spread on the waxy surface of a leaf, but instead form an array of beads. (Photo: Christoph Burki, Tony Stone Images.)

phenomenon is called the Rayleigh instability, after Lord Rayleigh who studied it at the end of the nineteenth century. Although the instability acts for perturbations of all sizes, there is a certain wavelength of undulation that is the *most* unstable, and this determines the size and separation of the resulting string of pearl-like droplets. The Rayleigh instability also acts on a thin columnar jet of water, breaking it up into droplets (Fig. 2.5c).

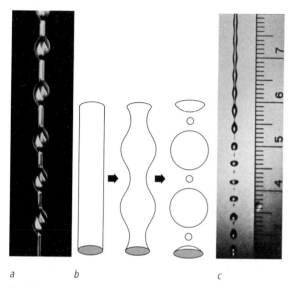

a   *b*   *c*

Fig. 2.5 A coating of glue on the threads of a spider's web breaks up spontaneously into a string of pearl-like beads (*a*). This beading process, called the Rayleigh instability, is a fundamental property of a narrow cylindrical column of liquid, and it selects a certain wavelength (*b*). It can be seen also in the break-up of a narrow jet of water (*c*) (From: Tritton 1988).

We will find throughout this book that pattern-forming processes are often initiated by abrupt instabilities. Generally an instability sets in suddenly when some critical parameter is surpassed. For instance, a person on a bicycle is potentially unstable to falling over (with an equal probability of tipping to the left or the right), but this instability sets in only when the speed falls below a certain threshold. Two common aspects of pattern-forming instabilities are that they involve symmetry-breaking (in the present case, the liquid film is initially uniform (symmetric) along the thread's axis, but the instability breaks this symmetry) and that they have a characteristic *wavelength*, so that the features of the pattern have a specific size.

Balloon games

A bubble seems to defy the exigencies of surface tension. It is spherical, sure enough—but what a surface area! The liquid is stretched into a thin film with a surface area far, far greater that that of a spherical droplet with the same volume of liquid. What has happened?

Everyone knows that, while it is well-nigh impossible to blow bubbles from pure water, they can be made in abundance from water to which a little soap or detergent has been added. Soaps contain molecules called surfactants, which have a tendency to migrate from the bulk of the liquid to the surface, where their presence greatly reduces the surface tension. This means that surfaces cost less, and a larger surface area can be sustained. Notice that, although our intuition tells us that bubbles have a 'stronger skin' than pure water, they can exist at all only because their surface tension is *lower*.

Box 2.1: **Energy and equilibrium**

Energy is a term that is put to many uses, but in science its meaning is precise: a system's energy is its capacity for doing mechanical work, for moving objects against forces. Every process—every movement, every change—in the real world involves a conversion of energy from one form to another. My muscle movements change chemical energy to kinetic energy (the energy of matter in motion), and also to heat. A light bulb changes electrical energy into heat and light energy.

Just about every energy conversion process that we encounter in everyday life produces some quantity of heat, which for our purposes is often 'wasted' energy (I don't need the heat from my ceiling light). With this in mind, there is a maximum amount of useful work that can be extracted from

any system or process, which is less than the total amount of energy converted—some is always squandered. This maximum amount of extractable work is called the *free energy*. The direction of spontaneous change is always that which results in a decrease in free energy. At equilibrium, the free energy is minimized and no further change takes place.

I shall say more about these concepts, which underpin the discipline of thermodynamics, in the next chapter. For now, you might like simply to imagine processes of change as being like a ball rolling down a hill—this entails the lowering of the ball/hill system's free energy. At equilibrium, the ball comes to rest in a valley at the foot of the hill—a static, unchanging state.

Surfactants are molecules that have a double nature: part of them is soluble in water, and part is not. In soaps the surfactants are salts of fatty acids, which have a compact, negatively charged 'head group' attached to a long, fatty tail (Fig. 2.6). The head group can interact strongly with the electrical charges on water molecules, and so is water-soluble. The tails do not interact strongly with water at all, although they do have an affinity for oily hydrocarbon liquids and greases, whose chemical structure resembles theirs. Molecules with this dual nature are called amphiphiles ('liking both'); the term *surfactant* (a condensation of 'surface-active agent') originated in the detergent industry and is often now synonymous with amphiphile, although in fact it has the rather more general meaning of a molecule that mediates surface interactions.

Although soap surfactants will dissolve in water, they prefer to position themselves at the water surface, where the water-insoluble tails can poke above the surface while the water-soluble heads remain in solution (Fig. 2.6). Surfactants will therefore form a film, just one molecule thick, at the surface of water. Because the surface layer of 'unsatisfied' water molecules becomes replaced with a layer of fatty tails that didn't want to be in the water anyway, this film lowers the surface tension.

When you blow a bubble from a soap film, the hollow sphere is filled with air. The pressure inside the bubble is greater than that outside, by an amount that is proportional to the inverse of the bubble's radius: the smaller the bubble, the greater the pressure inside. Thomas Young and Pierre Laplace independently established this relationship in 1805. A bubble's size is determined by a balance between the force of surface tension, which acts to shrink the bubble and decrease its surface area, and the internal pressure, which opposes shrinkage by increasing as the bubble gets smaller. The spherical form, meanwhile, is a consequence of the fact that, of all shapes that can enclose a given volume of space, the sphere has the smallest surface area (and thus the smallest surface excess energy). Mathematically, it is called a minimal surface, about whose properties I shall have more to say later.

This minimization principle determines the shapes of all soap films: when confined between boundaries, the film adopts the shape that has the smallest surface area. Soap films stretched between wire frames take on elegant, smoothly curved shapes that have inspired architects such as the German Frei Otto. From the 1950s, Otto designed lightweight membrane structures in which sheets of translucent material form tent-like shapes whose curvature is calculated to minimize surface area (Fig. 2.7*a*). These structures experience almost exclusively *tensile*, rather than compressive, stresses—just as a soap film is moulded by surface tension. Otto made use of soap films draped across wire frames (Fig. 2.7*b*) to plan the curves of his buildings: these models provide an instant experimental solution to the mathematical problem of how to connect specified boundaries with the minimum of material.

A good head

When bubbles are packed together, the result is a foam. Foams are amongst nature's most complicated architectural structures, and it is safe to say that, while they have been studied for centuries, they are still not fully understood. Nature has learned to make use of foams—the spittle bug, for instance, blows a foamy froth to obscure its larvae on leaves, hiding them from predators. They are of great technological value too: foams are used to fight fires, by smothering them with a light but semi-rigid blanket. They will also damp the power of an explosion, absorbing most of its energy as the bubbles are converted to droplets, which then evaporate. Foams blown in plastics are used as insulation and packaging,

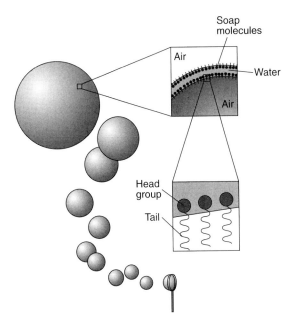

Fig. 2.6 The surface tension of the liquid in a soap bubble's skin is lowered by the presence of the soap molecules at the surface. These molecules, members of a class called surfactants, have a water-soluble head and a water-insoluble tail, which pokes out from the water surface.

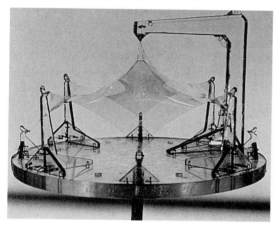

Fig. 2.7 The elegant area-minimizing shapes of soap films have inspired architects such as Frei Otto, whose design for the Olympic Swimming Arena in Munich is shown here (*a*). Otto used soap films stretched across wire frames to plan the curves of his membrane structures (*b*). (Photo (*b*): Michele Emmer, University of Rome 'La Sapienza'.)

while watery foams are used in mineral extraction and metal foams promise strong, lightweight engineering materials. And considerable effort goes into the creation of a good head of beer, although the value is purely aesthetic. So there are plenty of practical as well as academic reasons for wishing to understand the factors that govern foam structure.

But one difficulty is that we're shooting at a moving target. The structure of a foam depends on when you look. A freshly formed foam in water (an aqueous foam), such as that on a newly poured glass of beer, is heavy with water (it is called a wet foam), and the bubbles are mostly spherical (Fig. 2.8*a*). Later the walls become thinner and the bubbles take on a polyhedral shape with more or less flat faces (Fig. 2.8*b*). This is called a dry foam, as much of the liquid has drained from the walls between bubbles. Typically, a foam then begins a process of coarsening, whereby bubbles merge so that their average size increases with time. Eventually, coarsening and evaporation of the liquid leads to collapse.

A wet foam is rather like a box of marbles of different sizes—the spherical bubbles are jumbled together haphazardly, with smaller ones filling the spaces between

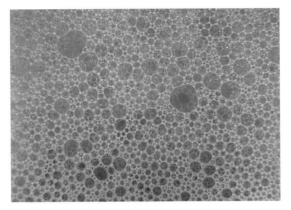

a

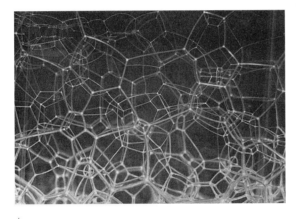

b

Fig. 2.8 A wet foam (*a*) consists of roughly spherical bubbles with water-laden walls. As the water drains away under gravity, the bubbles become more polyhedral and the result is a dry foam (*b*). (Photos: Burkhard Prause, University of Notre Dame, Indiana.)

larger ones. But as gravity sucks out the liquid from the walls and the cells become more like flat-sided polyhedra, the foam starts to take on some very particular geometric features. At first sight, it might seem to be a random mass of polyhedra of all shapes and sizes. At the end of the nineteenth century, however, a Belgian physicist named Joseph Antoine Ferdinand Plateau discerned some rules amongst the chaos.

First, the walls between cells are smooth, but not generally flat—they curve gently one way or another. This curvature indicates that the pressure of the gas inside the two adjacent cells is not equal: it is higher on the concave side of the wall. Smaller cells in a dry foam are the remnants of small bubbles, which (as Young and Laplace showed) have a higher internal pressure than

large bubbles; so where the two meet, the walls of the small cells bulge outwards (you can see this in Fig. 2.8*b*).

Where three walls meet, there is a junction in which the liquid film is slightly thicker than in the walls themselves (Fig. 2.9). Because the walls are necessarily curved at these junctions, the Young–Laplace relationship means that the pressure inside them must be lower than that in the flat walls; as a result, water is squeezed from the walls into the junction region. The consequence is that the junctions, called Plateau borders, contain most of the liquid in the foam.

Where three films meet in a Plateau border, the surface tensions in the films achieve a mechanical balance only if the walls meet at an angle of 120° (Fig. 2.9*a*). Equally, when four films meet, the angles at the junction would have to be 90° to achieve this balance of forces. But Plateau noticed a curious thing: he could find no fourfold junctions in his foams, nor any junctions of still greater numbers of walls. Three was the limit, and always with angles close to 120°.

The explanation for this requires a careful mathematical analysis of the various forces acting on the films, which I won't delve into. Suffice it to say that if four bubble walls *do* meet at a Plateau border, this turns out to be unstable and will rapidly rearrange to two threefold junctions (Fig. 2.9*b*). So here we have an explanation for why a two-dimensional packing of bubbles

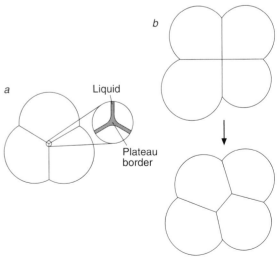

Fig. 2.9 Bubble walls meet at Plateau borders, where the walls are slightly thickened. Three walls will always meet at an angle of 120° at equilibrium (*a*). If, as a foam coarsens, four walls happen to come together at a junction, they will rapidly rearrange into two threefold junctions (*b*).

forms a foam of roughly hexagonal cells—only these satisfy the criterion that the walls always meet in threes with a 120° angle between them. Whether or not D'Arcy Thompson was right to ascribe the origin of the honeycomb's design to this effect, he was right about the way that bubbles pack.

But most foams are three-dimensional, and this means that Plateau borders along the edges of the polyhedral cells converge at their vertices. Here Plateau made another discovery: the number of Plateau borders that meet at a vertex seems always to be four—no more, no less. And they meet at an angle of about 109.5°, the 'tetrahedral angle': the four borders pointed to the vertices of a tetrahedron (Fig. 2.10). Again, this arrangement emerges from the requirements for mechanical stability of the cell walls. These geometrical rules govern the structures that all soap films will form when they meet (Fig. 2.11). They attest to an underlying regularity in the architecture of foams.

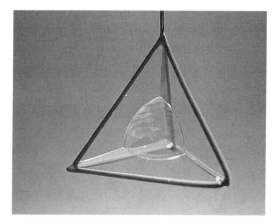

Fig. 2.10 Plateau borders converge at fourfold vertices, where they meet at the tetrahedral angle of about 109.5°. This is beautifully illustrated by soap films formed within a tetrahedral wire frame (see Appendix 1). (Photo: Michele Emmer, University of Rome 'La Sapienza'.)

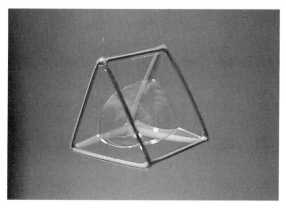

a

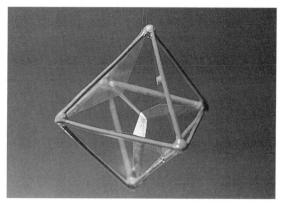

c

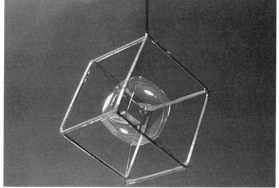

b

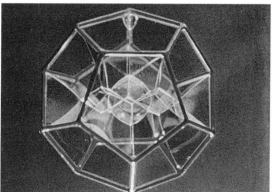

d

Fig. 2.11 The structures taken up by soap films and bubbles held within fixed boundaries are dictated by Plateau's rules. (Photos: Michele Emmer.)

Looked at more closely, however, Plateau's rules run into a problem. While we can understand how each arises in isolated packings of a few bubbles, we then have to ask whether it is in fact possible to fill up space with polyhedra that always conform to the rules. The simplest approach to the problem is to consider every cell to be identical in volume (as they are in a *monodisperse* foam), and to try to find a single polyhedral shape that can be packed together to give a network that obeys the rules governing borders and vertices. As well as satisfying these geometric criteria, the cells in this ideal three-dimensional foam should also minimize their total surface area. Is there a single, well-defined way to partition space so as to both satisfy Plateau's rules and provide the greatest economy in surface area? So far, no unique cellular packing of this sort has been identified.

This problem of cellular packing has a long history. In the eighteenth century, the English clergyman Stephen Hales took an inventive experimental approach, by compressing peas to see what shapes the spheres would take when flattened together. He claimed that the peas were pressed into 'pretty regular Dodecahedra', by which he apparently meant *rhombic* dodecahedra (Fig. 2.12*a*). These experiments were made widely known (though without attribution to Hales) by the French zoologist G.L.L. Buffon in 1753, and for a long time the rhombic dodecahedron was taken to be the best solution to the problem of economy. A rigorous mathematical proof was lacking, however, and in 1887

Lord Kelvin identified a cell shape that did better in terms of minimizing surface area: a 14-sided polyhedron (called a tetrakaidekahedron) with six square and eight hexagonal faces (Fig. 2.12*b*). This object, also known as a truncated octahedron, will pack together to fill space while coming close to satisfying Plateau's rules: at each vertex there are two 120° angles and one 90° angle, but Kelvin showed that only a slight curvature of the hexagonal faces is sufficient to adapt the vertices to the tetrahedral angle of 109.5°. Kelvin was not able to prove, however, that this was the most economical solution of all possible cellular packings, and no such proof has followed subsequently. Nevertheless, some mathematicians (including Hermann Weyl in his famous book *Symmetry*) have long suspected that Kelvin's solution cannot be bettered.

D'Arcy Thompson claimed that if a mass of clay pellets is compressed like Hale's peas, they will form shapes close to rhombic dodecahedra; but if they are first made wet, so that they can slide over one another, they show instead square and hexagonal facets like those of Kelvin's tetrakaidekahedron. So he was happy to conclude that soap bubbles of equal size, which can slide over one another, will form a froth with Kelvin's configuration. All the same, he cautioned that the solution to the packing problem depended in subtle ways on the conditions of packing: he described experiments by J.W. Marvin on compression of lead balls, which apparently formed rhombic dodecahedra if first stacked like a greengrocer's oranges in regular hexagonal layers, but irregular polyhedra with an average of 14 sides if poured into the vessel at random.

Moreover, the *regular* polyhedron (that is, one with identical faces) that comes closest to satisfying Plateau's rules is not the rhombic dodecahedron but the pentagonal dodecahedron, which has 12 pentagonal faces (Fig. 2.12*c*). This object doesn't stack to fill space exactly, and in addition the angles are slightly wrong—116° between faces, 108° between vertices—but it will do the job with a little distortion. Another candidate for the cell shape in a monodisperse foam is an irregular 14-sided polyhedron called a beta-tetrakaidecahedron (Fig. 2.12*d*); but even this needs to be distorted to meet the rules.

So much for the models; what do the cells of real foams look like? The botanist Edwin Matzke conducted a detailed study of the shapes of monodisperse foams in 1946, and found that none of the ideal models provides, by itself, an accurate description of the cellular structure. For one thing, Matzke's foams were far from

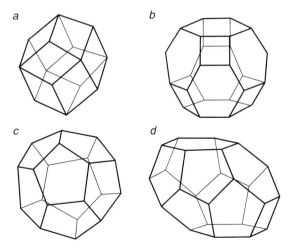

Fig. 2.12 Candidate cell shapes for a 'perfect' foam: (*a*) the rhombic dodecahedron; (*b*) the truncated octahedron promoted by Lord Kelvin; (*c*) the pentagonal dodecahedron; (*d*) the beta-tetrakaidecahedron.

Plate 1 When a liquid is heated uniformly from below, it will spontaneously develop a pattern of hexagonal circulating cells. Here the cells are made visible by metal flakes suspended in the fluid. (Photo: Manuel Velarde, Universidad Complutense, Madrid.)

Plate 2 The rainbow colours of a soap film thinning under gravity. As the liquid in the film gets pulled downwards, the film's thickness varies from top to bottom. Interference between light reflected from the front and back of the film then selects different wavelengths of reflected light for different film thicknesses. The film turns silvery and then black before rupturing. (Photo: Michele Emmer, University of Rome 'La Sapienza'.)

a

b

Plate 3 Two of the simplest periodic minimal surfaces and their repeat units: (*a*) the diamond or D-surface and (*b*) the gyroid or G-surface. The third member of this family is the P-surface, shown in Fig. 2.28. (Images: Alan Mackay, Birkbeck College, London.)

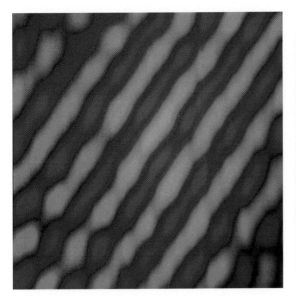
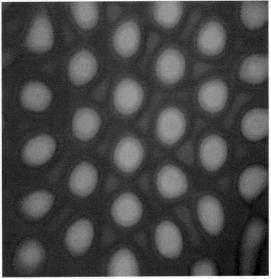

a   b

Plate 4 Turing patterns in a chemical medium. These patterns arise spontaneously through the competition between a localized autocatalytic chemical reaction and the long-ranged diffusion of a substance that inhibits the reaction. The colours denote regions of different chemical composition. The patterns are stationary, but a transition from the stripes to the spots can be induced by changing the composition of the reaction medium. (Images: Harry Swinney, University of Texas at Austin.)

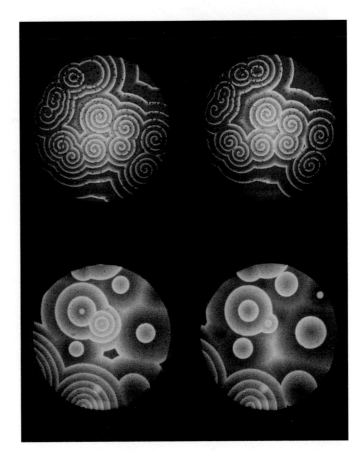

Plate 5 The spectacular patterns of the Belousov–Zhabotinsky reaction consist of target and spiral waves. (Images: Arthur Winfree, University of Arizona.)

Plate 6 The concentric patterns of banded agate are a consequence of a periodic variation in precipitation of the mineral as it forms in the Earth. (Photo: Peter Heaney, Princeton University.)

Plate 7 Spiral waves in the heart, shown here in a simulation of a dog heart modelled as an excitable medium. (Image: James Keener, University of Utah.)

Plate 8 Chevron patterns in a colony of the bacterium *Escherichia coli* grown in a gel. The patterning is a response to adverse conditions (such as scarcity of nutrients), which induces the bacteria to aggregate through chemical signalling (chemotaxis). (Photo: Elena Budrene, Harvard University.)

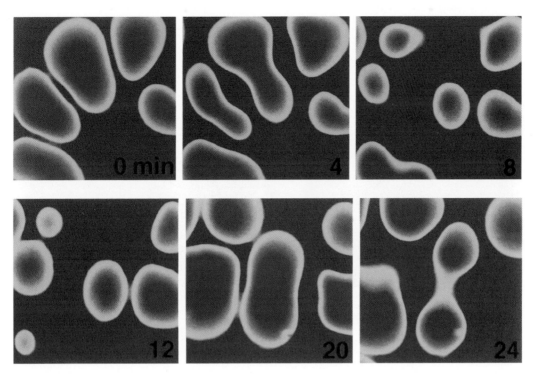

Plate 9 Replicating spots in an activator–inhibitor chemical reaction. These spots grow and divide. When too many are located in one region, they may 'die' through overcrowding. (Photo: Harry Swinney, University of Texas at Austin.)

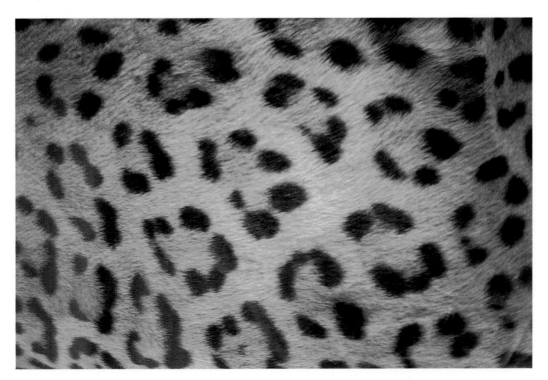

Plate 10 The characteristic markings on the pelts of great cats (here a Paragyuan jaguar) are produced by epidermal cells that generate hair-colouring pigments called melanins.

Plate 11 Some of nature's most spectacular patterns are found in the multicoloured designs of butterfly wings (Photo: Fredrick Nijhout, Duke University.)

a   b

Plate 12 (*a*) The spectacular 'chiral' growth patterns of a bacterial colony of *Bacillus*. (*b*) The 'vortex' growth mode. Here clumps of bacteria swarm in a blob-like vortex at the tip of each branch. (Photos: Eshel Ben-Jacob, Tel Aviv University.)

Plate 13 Rivers carve out fractal networks, and at the same time they shape the terrain into a rough surface of hills, mountains and valleys that also has a fractal structure. (Photo: Jim Kirchner, University of California at Berkeley.)

Plate 14 Fractal geometry can be used to simulate mountainous topography. (Image: John Beale.)

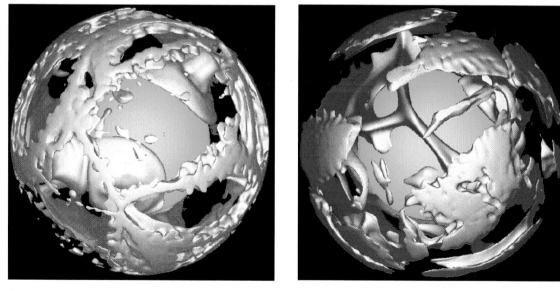

Plate 15 Avalanches in the Earth's mantle are revealed in these fully three-dimensional simulations of mantle convection. The upwelling (red) and downwelling (blue) regions are shown separately, for clarity. The former are plume-like structures which rise through the entire mantle. The latter are sheet-like, and become lodged at the 670-km boundary until catastrophic flushing carries the material through and down to the base of the mantle. (Images: Paul Tackley, California Institute of Technology, Pasadena.)

Plate 16 Freeze-thaw cycles of groundwater at the edge of a lake in Norway produce convection cells that are traced out by stones on the lake bed. (Photo: Bill Krantz, University of Colorado.)

Plate 17 This complex pattern in fluid flow, in which different streamlines (fluid trajectories) are coloured like the contours of a relief map, arises when a two-dimensional flow in the vertical direction is driven by a force with fivefold symmetry throughout the plane. (Image: George Zaslavsky, New York University.)

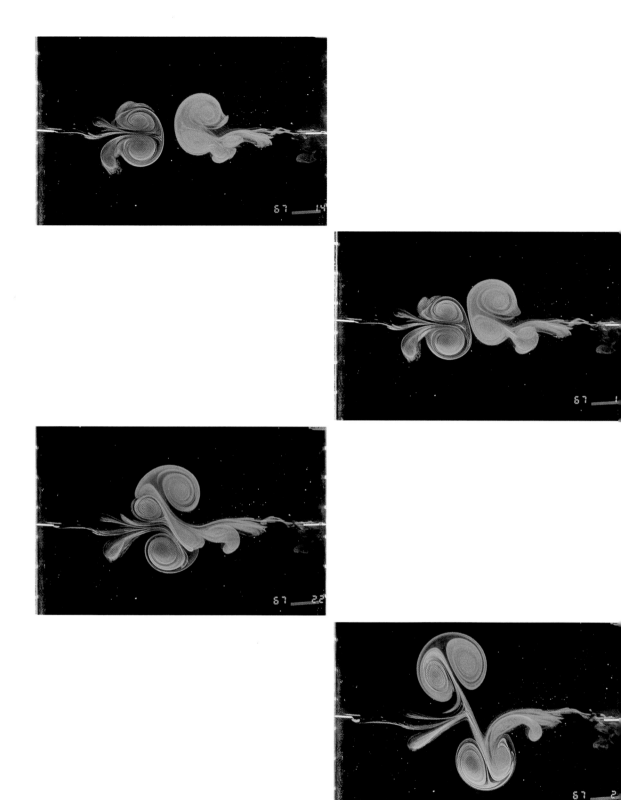

Plate 18 When two dipolar vortices collide, the coherent structures maintain their integrity. The two mushroom-like heads merely exchange vortices and set off on a new course. Here the dipoles have been coloured with dyes to keep them identifiable. (Photo: GertJan van Heijst and Jan-Bert Flór, University of Utrecht.)

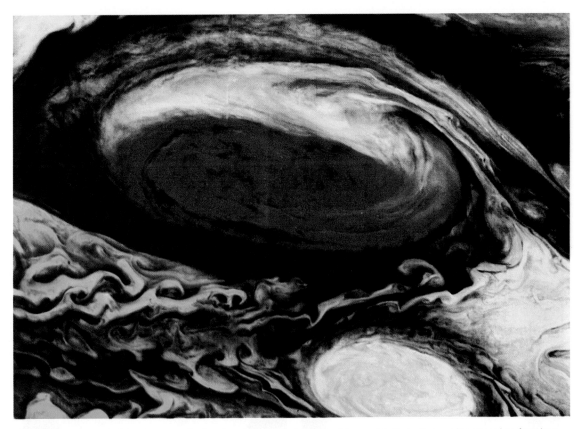

Plate 19 Jupiter's Great Red Spot is an example of a coherent structure in turbulence. It has persisted in Jupiter's swirling atmosphere for at least 300 years. Other, shorter-lived structures also come and go, like the white spot visible to the lower right. (Photo: NASA.)

Plate 20 Jupiter's banded structure can be mimicked in a laboratory model of the planet, in which water laced with a fluorescent dye is spun around between an inner and a clear plastic outer sphere at different temperatures. The bands are due to convective motions in the fluid. (Photo: Peter Olson, Johns Hopkins University.)

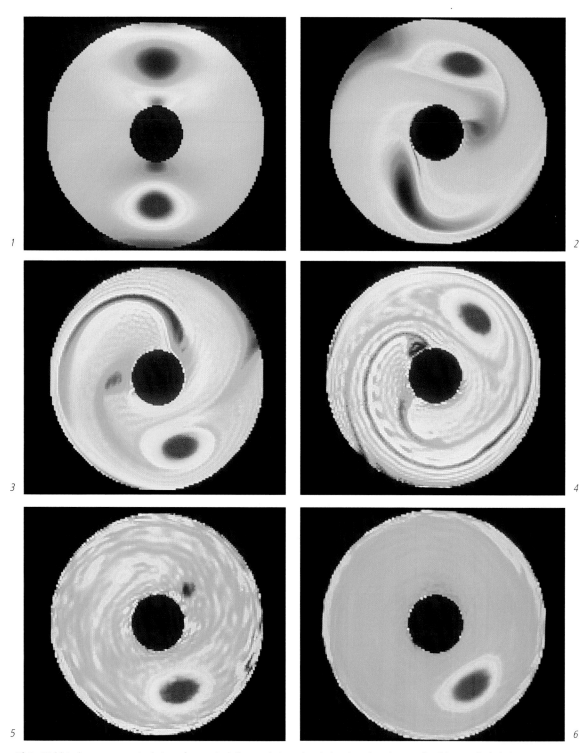

Plate 21 (*a*) In these computer simulations of atmospheric flow on Jupiter, a hemisphere is projected onto a flat disk around which the gases rotate. Two vortices are superimposed on the flow (frame 1): one (*red*) rotating in the same direction as the mean flow, and the other (*blue*) rotating in the opposite direction. The colours roughly track the trajectories that a coloured dye would follow if injected into the vortices. The vortex rotating in the same direction as the flow remains stable, even though the Reynolds number of the flow is in the turbulent regime. The counter-rotating vortex, however, is rapidly pulled apart into a spiral, which breaks up via the Kelvin–Helmholtz instability into a mass of tiny whorls (you can see some indication of the periodicity of this instability in frame 4). The single remaining vortex, like Jupiter's Gread Red Spot, swallows up these little whirlpools, purging them from the rest of the flow (frame 6).

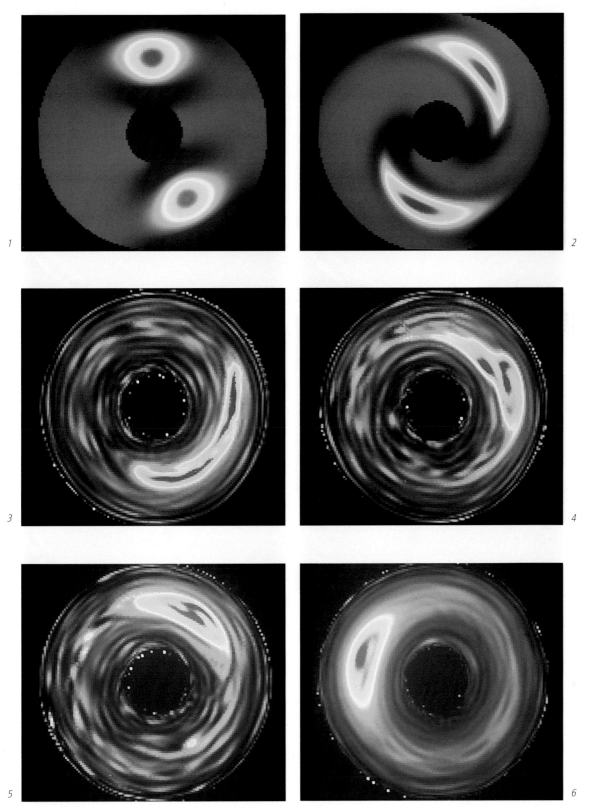

(*b*) Two large vortices, both with the same sense of rotation as the mean flow (frame 1), merge into one (frame 6), implying that a flow containing a single large vortex represents a stable 'attractor' for this system.

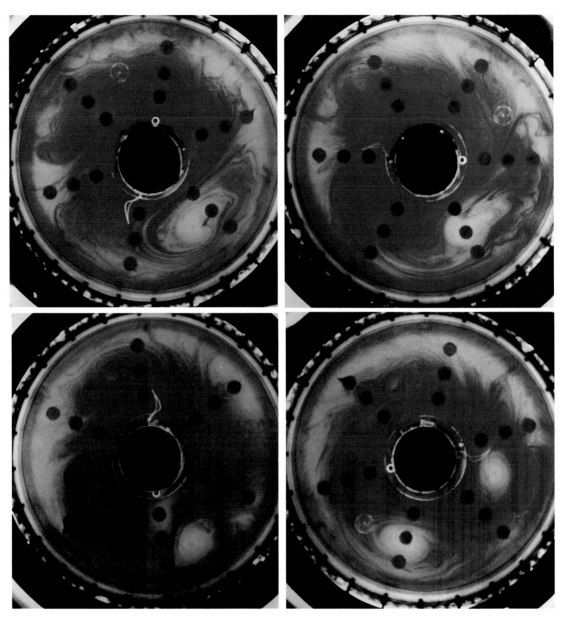

Plate 22 An experimental model of Jupiter's flow in a rotating tank produces a model 'Red Spot' (revealed here by injecting dye into the fluid), which remains stable despite the turbulence of the flow as a whole. (Photo: Harry Swinney, University of Texas at Austin.)

Plate 23 Sand dunes are an example of a self-organized pattern on a grand scale. (Photo: Jackie Cohen.)

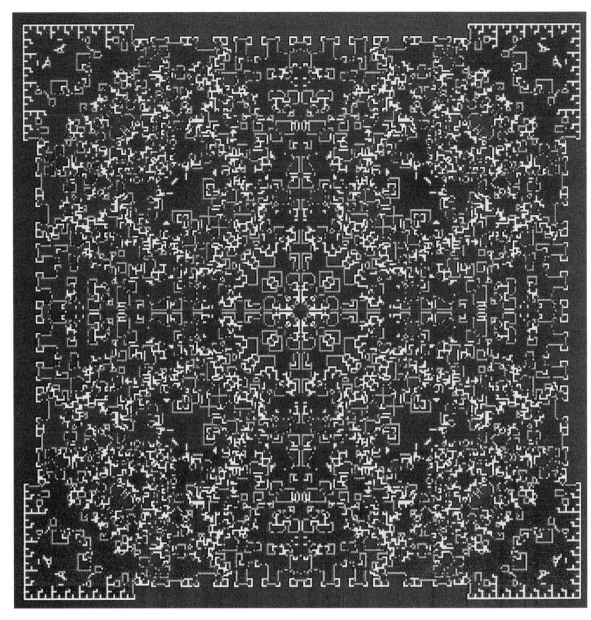

Plate 24 The game of cooperation and defection devised by Nowak and May can produce beautiful, symmetric and highly complex patterns in the two populations of cooperators and defectors. Blue sites represent cooperators that were cooperators in the last round too, and red sites are defectors that were defectors. Yellow and green sites denote, respectively, defectors that were cooperators and vice versa; so these sites show where the community boundaries are changing. These patterns are constantly shifting in a kaleidoscopic manner. (From Nowak and May 1993).

regular—they contained cells of many different shapes, so that the structure could be described only in statistical terms. He observed that about 8% of the cells had roughly the shape of a pentagonal dodecahedron, although over half of the faces had five sides. Cells approximating Kelvin's truncated octahedra were even rarer—only 10% of the faces were four-sided, and Matzke found no cells resembling Kelvin's overall. Most of the cells tended instead to be rather like Marvin's squashed lead pellets, averaging about 14 sides each but with irregular shapes that might be best approximated by the beta-tetrakaidecahedron (Fig. 2.12d). Matzke's experiments suggested that the packing problem was purely academic, since perfectly regular foams are a Platonic ideal with no relevance to the real world.

But recently, physicists Dennis Weaire and Robert Phelan at Trinity College, Dublin, have questioned this conclusion. In 1993 they discovered a new type of cell shape for regular foams that finally deposed Kelvin's solution—after over a hundred years of supremacy—as the most economical solution to the packing problem. Their solution is less elegant than Kelvin's. Rather than a single cell type with faces that are regular polygons, the foam described by Weaire and Phelan has a repeat unit built up from eight cells, six of which have 14 faces and two of which have 12 (Fig. 2.13). The latter are pen-

tagonal dodecahedra, while the former have two hexagonal faces and 12 pentagons. But only the hexagonal faces are regular (with equal sides and angles); the pentagons in these cells have sides of differing lengths and corners of differing angles. All the same, this unit can be stacked together to give a regularly repeating foam structure whose surface area is about 0.3% less than that of a Kelvin-type structure of the same volume, while still maintaining Plateau's rules if the faces are almost imperceptibly curved.

Having identified this improved solution to the packing problem, Weaire and Phelan wanted to see if they could see it in real foams. So they decided to conduct a survey like Matzke's. But whereas Matzke had specified a highly complicated procedure for making mono-disperse foams by adding bubbles one at a time, Weaire and Phelan found that they could produce these foams simply by using the 'drinking straw' technique of blowing bubbles underwater in a cylinder of liquid. They found first of all that the foams produced this way were not necessarily totally irregular and disordered, like Matzke's, but could contain regions in which regular cells were packed together. In parts of the foam close to the cylinder walls they often observed cells with square and hexagonal faces like those proposed by Kelvin (Fig. 2.14a); but these cell shapes seldom persisted beyond the first three or four layers. Within the bulk of the foam, meanwhile, they spotted regions where the cells had pentagonal and hexagonal faces, fitting together into structures very much like the one they had put forward as an improvement on Kelvin's (Fig. 2.14b,c). So it seems that after all, foams can be more geometrically precise—and more adept at the economical filling of space—than has long been believed.

Face to face

The problem of how to fill space with identical polyhedral cells, subject to a minimization principle for surface areas, is one that bees face too. The major part of the honeycomb problem is *two*-dimensional, because the cells are just prisms that are uniform along their length. What matters in this case is the cross-sectional shape of the cells, and the optimal solution in this regard is clear: hexagonal cells minimize the cross-sectional perimeter of the cell walls and so cost the bees less wax. But in the honeycomb, *two* such layers of cells are placed back to back, and the bees must then find the best way of marrying the two layers. The problem becomes three-dimensional, and so more complex, at the interface of the layers of cells.

Fig. 2.13 A better foam? This cellular structure, proposed by Weaire and Phelan, has a slightly smaller surface area than that made of Kelvin's cells, for the same enclosed volume. The repeat unit consists of eight slightly irregular cells. (Image: Dennis Weaire and Robert Phelan, Trinity College, Dublin.)

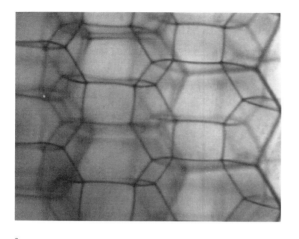

a

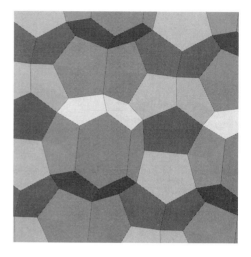

c

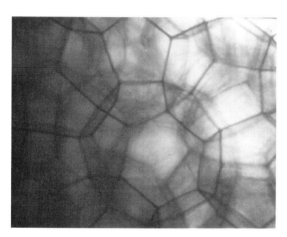

b

Fig. 2.14 What does a real dry 'ideal' foam look like? At its boundaries are regions containing cells like Kelvin's (*a*), but deeper inside (*b*) are regions with cells like those of the 'minimal foam' of Weaire and Phelan (*c*). (Images: Dennis Weaire and Robert Phelan.)

This packing problem is entirely equivalent to that of filling space with polyhedral cells, except that it is confined to a single layer. In a real honeycomb each cell ends in three rhombic (four-sided) faces (Fig. 2.15*a*), which together constitute one fragment of the rhombic dodecahedron (Fig. 2.12*a*)—this relationship to the polyhedron seems to have been first identified by the sixteenth-century German astronomer Johannes Kepler. Back-to-back cells with these end caps marry perfectly, and in cross-section the interface has a zigzag structure (Fig. 2.15*b*). Is this the most economic solution to the problem?

Réaumur concluded in the eighteenth century that it was. He considered the case of two arrays of hexagonal cells meeting such that their end caps consist of three identical and equal-edged rhombuses, and asked the Swiss mathematician Samuel Koenig to find the shape of the rhombuses that minimized the surface area. Koenig showed that the angles of each rhombic face should be about 109.5° and 70.5°, which are those in the regular rhombic dodecahedron—and also those observed in real honeycombs. It was this finding that led the secretary of the French Academy, Fontenelle, to issue the pronouncement on the divine guidance of bees quoted on page 17. To reach his solution, Koenig had had to employ the methods of differential calculus introduced less than half-a-century previously by Isaac Newton and Gottfried Wilhelm Leibniz, and it was too much for Fontenelle to suppose that the bees could possess this knowledge that surpassed 'the forces of common geometry'—for that would surely mean that 'in the end these Bees would know too much, and their

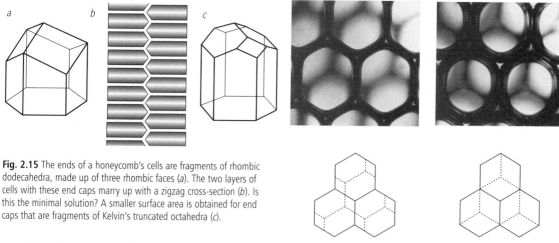

Fig. 2.15 The ends of a honeycomb's cells are fragments of rhombic dodecahedra, made up of three rhombic faces (*a*). The two layers of cells with these end caps marry up with a zigzag cross-section (*b*). Is this the minimal solution? A smaller surface area is obtained for end caps that are fragments of Kelvin's truncated octahedra (*c*).

Fig. 2.16 Tóth's structures can be seen at the interface of a double layer of hexagonal bubbles (*a*). But if the bubbles contain more liquid in their walls, the faces at the interface change to rhombuses (*b*), giving a junction like that in real honeycombs (Fig. 2.15*a*, *b*). (Photos: Dennis Weaire and Robert Phelan.)

exceeding glory would be their own ruin'. Evidently the geometric excellence was that of God, not of mere creatures.

But in posing the problem, Réaumur had imposed constraints (the requirement of three identical rhombuses) that left doubt as to whether the bees have truly found the optimal answer. In 1964 the Hungarian mathematician L. Fejes Tóth pondered on the economy of the honeycomb in a lecture entitled 'What the bees know and what they do not know'. He showed that a better solution exists in which the cells' end caps are more elaborate—a combination of squares and hexagons (Fig. 2.15*c*). This structure represents a total saving of a tiny fraction of a percent of each cell's surface area. Just as the rhombic cap is related to the rhombic dodecahedron, so Tóth's cell is closely related to the truncated octahedron (Fig. 2.12*b*) that Kelvin showed to be more economical in three dimensions. Tóth emphasized that, while his was mathematically a superior solution, there was no guarantee that it was biologically better—for the bees might have to expend more effort in making the more elaborate end-caps.

Weaire and Phelan have used their foam-blowing technique to put Tóth's idea to an experimental test. They looked at the cell structures in a thin foam—two layers of bubbles—constrained between glass plates. The bubbles adopt hexagonal faces at the interface with the glass, so that the foam is a precise analogue of the honeycomb. They found that the interface between the two layers of bubbles does adopt Tóth's structure (Fig. 2.16*a*), which can be identified by the distinctive pattern made by the junctions of bubbles in projection. But if Weaire and Phelan thickened the bubble walls by adding more liquid (creating a wet foam), they found

something unexpected: as the bubbles become more rounded, there is a point at which the interface suddenly switches to the three-rhombus configuration found in the real honeycomb (Fig. 2.16*b*). The thickening of the walls and curving of the bubble sides apparently changes the balance in surface energies so that this structure becomes more stable instead. So in thicker-walled honeycombs, maybe the bees *do* have the best solution. Do they know more than we thought? I return to this question at the end of the chapter.

Curved spaces

Cells, starfish and doughnuts

Soap bubbles and foams do not last for ever, and I suppose that is part of their appeal: fragile beauty, gone in a moment. The collapse of foams is brought about partly by the drainage of the films, under the influence of gravity and capillary forces, until they become too thin to resist the slightest disturbance—a vibration or a breath of air. But in their passing, soap films can treat us to a wonderful display. Held vertically on a wire frame, a thinning soap film becomes striated with bands of rainbow colours that pass from top to bottom (Plate 2). Finally the top becomes silvery and then black; and the blackness, like a premonition of the film's demise,

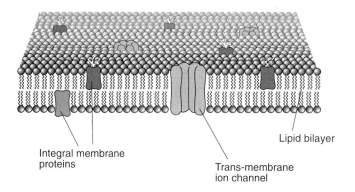

Fig. 2.17 Cell membranes are made from double layers of surfactants called phospholipids. These *bilayers* are studded with other membrane components, such as protein molecules, and are sometimes strengthened with a protein web called the cytoskeleton.

Lipid bilayer

Integral membrane proteins

Trans-membrane ion channel

moves over the entire surface. Once it is black, the film is doomed to burst at the merest perturbation.

These colours are the result of interference between light reflected from the front and the back of the film. Interference takes place when the distance between the front and back becomes comparable to the wavelength of light (a few hundred millionths of a millimetre), and as this distance changes, so too does the wavelength (that is, the colour) of the light affected by interference. When the film is black, all reflected visible light cancels itself out by interference. The film is by that stage only about four-and-a-half millionths of a millimetre thick—about the same thickness as a double layer of soap molecules. The two films at the surfaces have almost met back to back.

This back-to-back arrangement of surfactant molecules has some similarities to the wall of a living cell. Cell membranes (Fig. 2.17) are composed of amphiphilic molecules—biological surfactants, if you like—called phospholipids, or just lipids. A double layer of lipids, called a bilayer, is one of the fundamental architectural features of living organisms, providing the housing in which nature's chemistry takes place. Lipid bilayers also divide up cells of multicelled organisms (like ourselves) into several compartments, each of which acts as the location for specific biological processes. One critical difference between a black soap film and a lipid bilayer, however, is that in the former the surfactants meet head to head and in the latter they meet tail to tail. Thus lipid bilayers present a horde of water-soluble head groups at their surface, and the water-insoluble tails are buried within, where they are shielded from water. In a loose sense, cell membranes can be considered to be microscopic, inside-out bubbles, afloat in a watery sea. Of course, real cells are anything but 'hollow'—their insides are filled with biological hardware, including the DNA that allows

the cells to generate copies of themselves. But in the 1960s, researchers at Cambridge University found that phospholipids would come together spontaneously in solution to form empty cell-like structures called vesicles, when the solution was jiggled by sound waves. This self-assembly of vesicles is driven by the tendency of lipids to form bilayers in order to bury their insoluble tails.

A lot of work has been devoted to studying the shapes that lipid bilayer vesicles can adopt, because these can provide clues about the factors that control the shapes of real cells. The range of shapes is far more varied and interesting than those of soap bubbles: vesicles can be spherical, but they can also take on other stable shapes too. In the broadest sense, these shapes are determined by the same driving force that dictates the shapes of soap films: the tendency to minimize the total (free) energy. The principal contribution to the energy of a soap film comes from the surface tension, so the film adopts a shape that minimizes this by finding the smallest surface area. But for a vesicle, the surface area is essentially fixed: once a vesicle is formed, the number of surfactant molecules in its wall stays pretty much the same, and each molecule occupies a fixed area on the bilayer surface. This means that another factor is able to exercise a dominant influence on the energy: the surface *curvature*. The way that shape affects the curvature energy is rather subtle, and it may turn out that the lowest-energy shape is not that with constant mean curvature—a sphere—but some other, more complex shape. This balance can be shifted by changing the nature of the vesicle's environment—for example, by warming it up—and so the vesicle may undergo changes in shape as the temperature is changed.

The German biophysicist Erich Sackmann and co-workers have shown that under certain conditions, the most stable shape of a vesicle is that of a disk with dim-

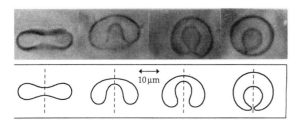

a

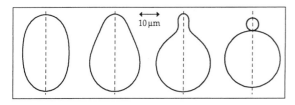

b

Fig. 2.18 Vesicles are closed, cell-like bilayer membranes. They adopt a range of different shapes at different temperatures, which are determined by the subtle influences of elastic and curvature energy. In (*a*) a flattened vesicle with a shape like a red blood cell develops a concavity which becomes a separate internal vesicle. In (*b*) an elongated vesicle develops a bud, which eventually separates from the main body. Both of these sequences, seen experimentally under a microscope (*top frames*), can be reproduced by calculations of the equilibrium shape that minimizes the total energy (*lower frames*). (Images: Reinhard Lipowsky, Max Planck Institute for Colloid Science, Teltow-Seehof, Germany.)

and budding off of a small interior vesicle—happen in real cells, where they are called exocytosis and endocytosis. The former process allows cells or interior subcompartments of cells called organelles to send out little chemical messages—a package of protein molecules, perhaps—in soft wrappers, while the latter enables a cell to ingest material. In cells these processes are controlled by protein molecules embedded in the cell membranes, but we can see that they can also come about through nothing more than the 'blind' physical forces that determine a membrane's geometry.

Udo Seifert and co-workers at the Max-Planck Institute for Colloid Science in Teltow-Seehof, Germany, have found that under some conditions the driving force to minimize curvature energy can push vesicles through extremely bizarre contortions. Under the microscope they saw multi-armed vesicles that looked like starfish or ink blots (Fig. 2.19*a*). If these were living amoeba dragging themselves around by extending pseudopodia, we might not consider the shapes surprising; but they are merely empty sacs whose shapes are the product of a mathematically well-defined minimization principle! Seifert and colleagues showed that they could reproduce the shapes theoretically by minimizing the curvature energy of the bilayers subject to the constraints of fixed surface area and enclosed volume (Fig. 2.19*b*). This 'mathematics of blobs' appears to hold some symmetry principles: the researchers could

ples in the top and bottom (Fig. 2.18*a*), which is precisely the shape that a red blood cell adopts. They saw these vesicles change shape to become bowl-like entities as the temperature was increased (Fig. 2.18*a*), and were able to show theoretically that these shape changes are to be expected because of the changing balance in energies. The bowl-like shape, called a stomatocyte, may eventually curl up on itself to generate a small, spherical vesicle inside a larger one, connected via a narrow neck which eventually became pinched off. Under different conditions, a vesicle can become elongated from an egg-like shape into a pear shape, ultimately pinching off a little bud at the thin end (Fig. 2.18*b*). Both of these processes—the budding and expulsion of a small vesicle from the outside of a cell membrane and the engulfing

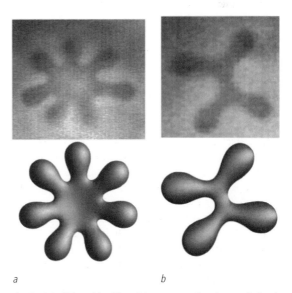

a   *b*

Fig. 2.19 Starfish vesicles (*a*), and the corresponding shapes calculated with an energy-minimization model (*b*). (Images: Udo Seifert, Max Planck Institute for Colloid Science, Teltow-Seehof, Germany.)

not, for example, find starfish vesicles with four symmetrical arms either experimentally or theoretically.

A part of the curvature energy of a vesicle arises not from the size or shape of the surface but from its topology—the overall 'connectedness' between different parts of the membrane. Two shapes are topologically equivalent if one can be converted into the other without any tearing or puncturing. For example, a spherical vesicle is topologically equivalent to all of the disk- and bowl-like vesicles in Fig. 2.18a, because they can be made just by flattening and bending the sphere. However, the shape on the far right of Fig. 2.18b is topologically *non*-equivalent to a sphere: when the small vesicle is pinched off at the neck, so that it can float free from the larger one, the topology is altered because the membrane has to be ruptured to create this arrangement.

Another shape that is topologically different from the spherical vesicle is the doughnut, technically called a torus. Vesicles with this shape have been seen by David Bensimon and co-workers at the Ecole Normale Supérieure in Paris (Fig. 2.20a). Bensimon's team showed that these shapes can become the most energetically favourable under some circumstances. You might notice that they can be generated from an extreme version of the disk-like shapes on the left of Fig. 2.18a, when the two dimples touch each other in the middle. At that stage the upper and lower membranes may merge and a hole open up in the middle—the topology is then abruptly transformed. Bensimon's group have reported even more topologically complex vesicle shapes, such as double toruses (Fig. 2.20b), which are topologically distinct from the single toruses. The point to bear in mind here is that even these apparently complicated shapes are selected according to relatively simple physical principles that minimize the vesicle's energy.

Bubbles in flatland

Vesicles are rarely formed in solutions of surfactants or lipids unless given some encouragement, in the form of sonic vibration for instance. Left to their own devices, surfactants display a gallery of other aggregate structures with their own propensity for pattern formation. Imagine gradually adding soap molecules to water. The first thing they'll do is gather at the water surface, where the insoluble, hydrophobic tails can poke out into the air. The water surface becomes gradually covered with a molecular film just one molecule thick. Benjamin Franklin was captivated by these thin films in the

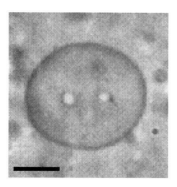

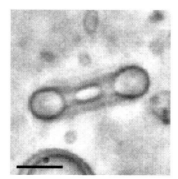

Fig. 2.20 Vesicles with holes: a doughnut or torus (a, showing top and side views and a double torus (b). Even these topologically complex shapes correspond to equilibrium structures that represent energetic minima. The scale bar indicates 10 micrometres in all frames. (Photos: Xavier Michalet and David Bensimon, Ecole Normale Supérieure, Paris.)

a

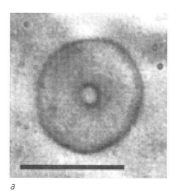

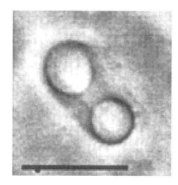

b

eighteenth century, which he observed by gently pouring an oil onto the surface of a pond. He took to carrying oil in a little vial in his walking stick, and would merrily create a miniature oil slick on every pond he encountered, particularly that on London's Clapham Common. What amused him was that just the tiny volume of oil that he carried would spread across the entire pond, and as it did so it would lower the surface tension of the water surface and leave it smooth as a mirror. I don't recommend trying this, however, unless you fancy you can explain to a park attendant that you are reproducing a historical experiment by Ben Franklin.

The study of surfactant films (particularly those of the soap-like molecules called fatty acids) on the surface of water was pioneered by Lord Rayleigh at the end of the nineteenth century and by the American chemist Irving Langmuir and his students at the beginning of the twentieth century. These films now bear the name Langmuir films, and they exhibit an astonishing range of pattern-forming behaviour. Langmuir created them in a shallow trough in which a movable barrier skimming the water surface allowed him to marshal the surfactant molecules into an ever smaller area of water surface and so control their density—which is to say, the average surface area commanded by each. As this density increases, a Langmuir film can undergo abrupt changes that are two-dimensional 'flatland' versions of the transformations from gas to liquid to solid that a material in three dimensions will undergo as it is compressed. But these films have an extra state: there are *two* kinds of 'flat' liquid, in both of which the molecules are mobile and disordered but which have distinctly differ-

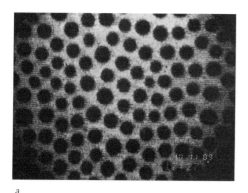

a

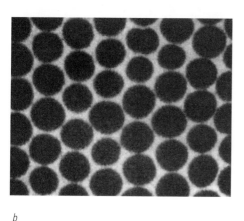

b

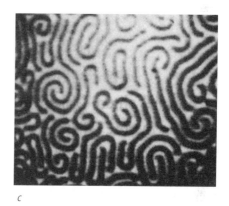

c

d

Fig. 2.21 Surfactants at the water surface will form a variety of different states when the surface layer is compressed. (*a*) A dense, disordered liquid-like state (called the LC state, *dark patches*) grows within a less-dense state (LE) that contains a fluorescent dye (*light regions*). (*b*) As the LC domains grow, they become ordered in a hexagonal pattern. (*c*) Eventually the LC domains become squeezed into worm-like shapes by their mutual repulsion. (*d*) The stripe phase of surfactant films is analogous to the striped arrangement of magnetic domains in thin films of garnet. Here too the stripes arise from mutual repulsion of the domains. (Photos: (*a*) S. Akamatsu and E. To, University of Paris IV; (*b*) Helmut Möhwald, Max Planck Institute for Colloid and Interface Science, Berlin; (*c*) Charles Knobler, University of California at Los Angeles; (*d*) Michael Seul, BioArray Solutions, Fanwood, New Jersey.)

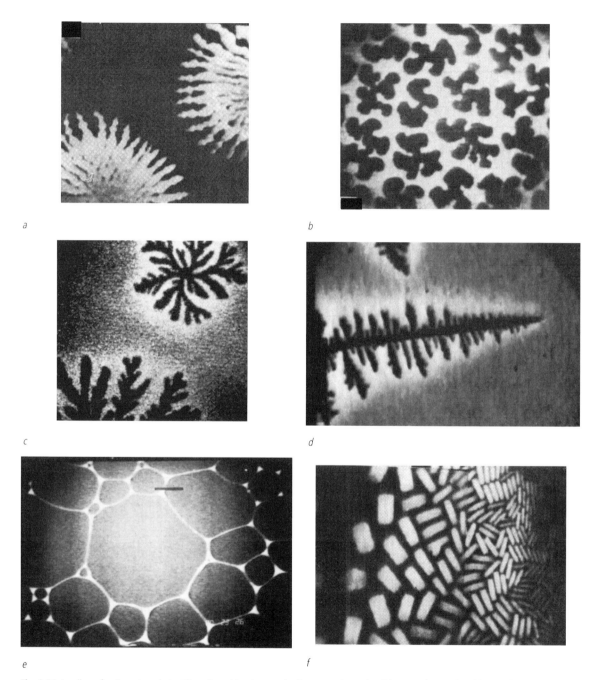

a

b

c

d

e

f

Fig. 2.22 A gallery of patterns in surfactant films, formed by the growth of one state in another. (Photos: Helmut Möhwald, Max Planck Institute for Colloid and Interface Science, Berlin, except for (*e*) from: Prost and Rondelez 1991.)

ent densities. By adding to the film a fluorescent dye that dissolves more readily in the less dense liquid (the liquid-expanded (LE) phase) than in the more dense liquid (the liquid-condensed (LC) phase), we can 'light up' the LE phase and watch darker 'droplets' of the LC phase coalesce and grow within it (Fig. 2.21*a*).

Aside from the extra phase, all of this is not so different to the condensing and freezing of a normal, three-dimensional liquid. But there is one other important difference: all of the molecules in these denser phases of Langmuir films point in the same direction, with their tails more or less perpendicular to the water surface. Because they also have an imbalance of electrical charge owing to the charged head group at one end, an electric field is set up around each molecule that repels those around neighbouring molecules, in much the same way as the *magnetic* fields of bar magnets will repel one another when they point in the same direction.

This means that each of the bubble-like domains of an LC phase condensing within an LE phase acts like an electrically charged bubble that repels the other domains. The result is that the domains tend to organize themselves so as to keep roughly the same distance between each, and the film becomes organized into a peculiar kind of 'crystal' in which the domains are packed together in a regular manner (Fig. 2.21b). Unlike a normal molecular crystal, however, in which the size of the characteristic repeating unit is of the same order as the size of the constituent molecules, here the scale of the pattern bears no direct relationship to the scale of the component parts from which it is made.

As these domains are squeezed ever more closely together, something even more dramatic can happen: the strength of the electrostatic repulsion between domains makes them deform into elongated shapes, which can eventually fuse together to form worm-like structures (Fig. 2.21c). This pattern is called a stripe phase, and it is also found in other two-dimensional systems containing domains of different structures that repel one another. For example, within a thin film of a magnetic material, domains may appear in which the direction of the magnetic field points in different directions; these too can adopt a stripe phase (Fig. 2.21d).

A subtle interplay between packing effects of the surfactant molecules, electrostatic repulsion of domains and dynamic (time-dependent) effects arising from the diffusion of impurities from one phase to another can give rise to all manner of strange patterns in Langmuir films, and I don't have space here to do much more than show a selective gallery (Fig. 2.22). Let me just point out, however, the similarity between one of these shapes (Fig. 2.22d) and those discussed in Chapter 5 (see p. 123)—this is a generic pattern called a dendrite, most familiar to us from the ornate branched arms of a snowflake. Notice too how a Langmuir film will form a two-dimensional foam (Fig. 2.22e) in which Plateau's rules, particularly that regarding 120° junctions between walls, can be seen to be obeyed.

The plumber's nightmare

What happens if we go on adding surfactant molecules to a solution? Only so many can accumulate at the surface; after the surface is full, they are forced to remain in solution. There is then the unfortunate fact that most of the molecule is a fatty, water-insoluble tail, and the surfactants have to do something about it. What they do is to aggregate together into a bewildering number of different structures, which can impose a regular pattern on the whole system.

The simplest aggregates are just blobs of surfactants, containing typically a few hundred molecules. These blobs, called micelles, have an internal logic: all the surfactants are arranged with their head groups pointing outwards onto the micelle surface, while the tails are buried in the interior (Fig. 2.23a). In this way, the molecules hide their tails from the water, and show only their water-soluble heads. G.S. Hartley proposed in the 1930s that micelles are roughly spherical, and experiments in the 1950s showed this to be the case. They are formed when the concentration of surfactant in solution is increased above a certain critical level, called the critical micelle concentration.

a

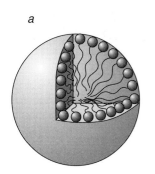

b

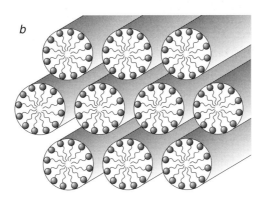

Fig. 2.23 (*a*) A micelle. (*b*) Cylindrical micelles packed together in the hexagonal phase.

In the early 1950s, before the shape of micelles was finally established, the physicist Paul Debye proposed that they might be rod-shaped instead. In Debye's model, the tails are still buried in the interior of the aggregates, but the blobs are extended along one direction into cylinders. Although experiments and theoretical studies later seemed to favour Hartley's spherical model, it gradually became clear that sausage-shaped micelles like Debye's *can* be formed too. As the concentration of surfactant is increased, it becomes more favourable for the new molecules to attach themselves to existing micelles, extending them into worm-like structures, than to form new micelles. If their concentration is not too great, the cylindrical micelles are entangled like spaghetti; but at higher concentrations they are compelled to pack together into a regular array, like a stack of logs. Each micelle is surrounded by six others in a so-called hexagonal phase, reminiscent of the arrangement in a layer of bubbles (Fig. 2.23b).

In 1992 researchers from the Mobil Corporation's laboratories in Princeton showed that the hexagonal phase of surfactants can be used to make patterned solid materials, by precipitating a hard mineral around the soft organic material. The regularly packed columns of surfactants act as a mould, imprinting the solid with an array of regular channels. The Mobil team allowed silica—the stuff of sand, window glass and quartz—to solidify from a solution of silicate ions, and then expelled the surfactants by heating the material. This left behind a porous form of silica with a honeycomb of channels about 10 to 100 millionths of a millimetre

wide (Fig. 2.24). (I should say that the process may not be *quite* as simple as taking a silica cast of the preformed surfactant pattern, because it appears that the structure of the surfactant aggregates might change over time as the silica precipitates.) The Mobil discovery created tremendous excitement amongst materials scientists, because there are many possible technological uses of solids patterned on these scales—as ultrafine sieves, for example, or as chemical catalysts. Notice that the characteristic length scale of the pattern in these materials is much larger than the characteristic size of the component parts—the silicate ions or surfactant molecules. So you would never guess that the system has the potential for forming such a pattern by looking at these components individually. The hexagonal pattern is the result of a *self-organizing* process.

Another structure that surfactants may form spontaneously in a sufficiently concentrated solution is made up of flat bilayer sheets, like the walls of vesicles, stacked on top of each other. This is called the lamellar phase (Fig. 2.25). These assemblies too can be 'fossilized' by precipitating silica around them; shortly after the Mobil discovery, researchers at the University of California in Santa Barbara made a layered silica material this way.

For a separation between lamellar bilayers greater than a certain threshold, it can become energetically favourable for two adjacent layers to fuse together at one point around a hole or pore (Fig. 2.26). You can see that this introduces bent regions in the bilayers, and we saw earlier that curvature in bilayers costs energy. But this energetic deficit can be more than compensated by the favourable increase in *entropy*—in disorder—that arises from these disturbances to the regular stacking of sheets.

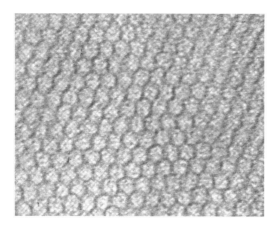

Fig. 2.24 A mixture of surfactants and silicate ions will cooperate to form a patterned solid, in which the silica walls are cast around an ordered surfactant template. The resulting material acquires a honeycomb pattern of pores. (Photo: Charles Kresge, Mobil Research Laboratories, Princeton.)

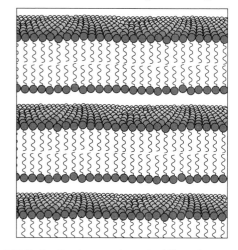

Fig. 2.25 The lamellar phase contains stacks of bilayer sheets.

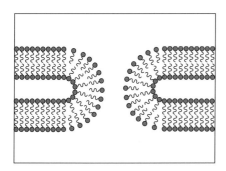

Fig. 2.26 Pores can form spontaneously between adjacent sheets in a lamellar phase, thereby providing channels of communication between previously separated regions of space.

Under these conditions, pores can proliferate between sheets, and the stacked layers break up into a web of tunnels that divides the system into two distinct subspaces. A molecular-scale fish could swim throughout all parts of one subspace or the other, depending on where it began, but it could not pass from one subspace to the other without punching through a bilayer (Fig. 2.27). A structure like this is said to be bicontinuous, because it consists of two continuously connected but independent networks of channels, each intimately woven through the other.

The channels of a bicontinuous phase of surfactant bilayers may be arranged in a haphazard way, in which case the system has the random, perforated structure characteristic of a sponge (and is indeed called the sponge

phase—or more figuratively, the plumber's nightmare). But more interesting from the perspective of pattern formation is the alternative in which the pores are positioned in a regular, orderly manner. Why should the pores be ordered? Because they have a tendency to repel one another: if two pores get too close together, they create very pronounced curvature of the bilayers in their vicinity, and this costs energy. So when there are many pores, they tend to sit at an optimal distance from each other on a regular lattice. The surfactant structure then becomes a kind of 'tubular crystal'. The sponge phase is really a 'melted', disordered version of this curious crystal.

The most common of the ordered bicontinuous surfactant phases are the cubic phases (Fig. 2.28), so-called because the symmetry properties of the labyrinth are the same as those of a cube. They are examples of what mathematicians call a periodic minimal surface: one that encloses two distinct volumes of space such that the area of their interface is as small as possible (minimizing surface tension), while maintaining an equal pressure on both sides. This latter constraint means that the surface must have zero *mean* curvature everywhere on the surface, since, as I mentioned earlier, curvature arises from an imbalance in pressure. It may seem odd to suggest that a surface like that in Fig. 2.28 can have zero mean curvature—it is obviously *highly* curved. But curvature is defined mathematically to be positive or negative depending on which way it bends—a concave surface (like the inside of a bowl) has negative curvature, while a convex surface (like the outer surface of a balloon) is positively curved. If the surface bends one way in one direction and the other way in the perpendicular direction, like the middle of a saddle, then the positive and negative curvatures can cancel out at that point. On surfaces of zero mean curvature, they cancel out everywhere.

Fig. 2.27 In a bicontinuous phase, the surfactant bilayers divide three-dimensional space into two distinct networks that interpenetrate without being interconnected. Here I show a slice through a simple bicontinuous phase made up from pores connecting adjacent bilayers. The two subspaces are denoted by the different shades of grey.

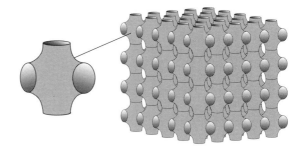

Fig. 2.28 The cubic P-phase, one of the ordered bicontinuous phases that may be adopted by surfactant bilayers in water. It is made up of identical units with the symmetry of a cube (*left*), and is an example of a periodic minimal surface.

We saw earlier that although the spherical shape of a bubble gives it minimal surface area, the curvature of the surface means that the pressure inside is greater than that outside. The soap bubble has a surface of *constant* (but non-zero) mean curvature. It is possible to construct periodic bicontinuous surfaces that also maintain a *constant* pressure difference across the interface—they are then periodic surfaces of constant mean curvature, of which those with strictly *zero* mean curvature (periodic minimal surfaces) are a special case.

Making the least of things

Periodic minimal surfaces have a distinguished history that starts with Joseph Louis Lagrange in 1762. Lagrange took advantage of the newly invented technique of variational calculus to explore the shapes of surfaces that have a minimal surface area for a given perimeter shape. In 1744 the Swiss mathematician Leonhard Euler discovered the *catenoid*, the minimal surface bounded by two coaxial circles (Fig. 2.29). You can easily make a catenoid from a soap film (I explain how in Appendix 1). In 1776 J. Meusnier identified the crucial property of a true minimal surface: more fundamental than the minimal surface area is the fact that the mean curvature is zero everywhere on the surface. Nearly sixty years later, the first *periodic* minimal surface—one made from building blocks that repeat regularly through space—was discovered by H.F. Scherk. Arguably our understanding of these curious structures owes most, however, to the German mathematician Hermann Schwarz, who explored their properties in the late nineteenth century. He was interested in deducing the shape that a soap film will take when stretched between the edges of a geometric figure like a tetrahedron, a problem posed by Joseph Plateau. How could the film touch all four vertices while maintaining a minimal surface area? Schwarz found that the solution takes the form of a saddle-like surface (Fig. 2.30). He realized

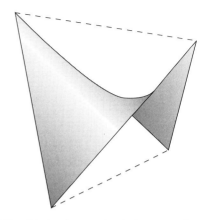

Fig. 2.30 The minimal surface spanning the four corners of a tetrahedron. This surface, first discovered by Schwarz in the nineteenth century, can be used as the building block of several periodic minimal surfaces.

that these tetrahedron-spanning films can be stacked together to form bicontinuous, periodic labyrinths whose mean curvature is zero. One of these is the so-called P-surface shown in Fig. 2.28; another is the D-surface shown in Plate 3*a*. A third simple periodic minimal surface, the gyroid or G-surface (Plate 3*b*), was discovered by Alan Schoen in the 1960s. All of these surfaces can be constructed by stacking a different kind of 'unit cell' containing a minimal surface. Examples of the P-, D- and G-surfaces have all been identified in the cubic phases of surfactants in water.

Cells get cubic

When the Italian chemist Vittorio Luzzati first discovered surfactant cubic phases in the 1960s, he wondered whether they might be more than a laboratory curiosity—whether they might, in fact, be found in living cells. After all, cell membranes are made up of bilayers too. Might these membranes curl up under the right conditions to make such labyrinthine tunnels? Perhaps the bicontinuous networks might even serve a useful purpose as cellular plumbing systems? Certainly it was known already that cell membranes do form neck-like pores—these are found in profusion, for example, in the bilayers that constitute the nuclear envelope around the DNA-containing nucleus of our own cells. And a complex tangle of bilayer channels called the smooth endoplasmic reticulum, which forms part of the cell's transport system and is where lipids and some proteins are manufactured, is nature's version of the disordered sponge phase (Fig. 2.31). Are there ordered networks in cells too?

Fig. 2.29 The catenoid, a minimal surface bounded by two coaxial circles.

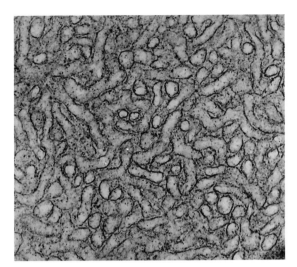

Fig. 2.31 The cell's smooth endoplasmic reticulum is a disordered 'sponge' of natural membranes. (Photo: Don Fawcett.)

There most certainly are! Membranes with regular channel structures akin to periodic minimal surfaces have now been identified in the cells of countless organisms ranging from bacteria to plants to rats. Kåre Larssen, Tomas Landh and colleagues at Lund University in Sweden have shown that the biological literature is replete with images of ordered membrane networks (Fig. 2.32), many of them apparently corresponding to periodic minimal surfaces or surfaces of constant mean curvature. They had not previously been recognized as such, says Landh, because cell biologists,

unfamiliar with these mathematical abstractions, had been unable to interpret what they saw.

The first pictures of such structures were presented in 1965 by B.E. Gunning, who observed them in electron microscope images of plant cells. These are much like the pictures that one can see through a light microscope—lighter where there is less dense matter and darker where the density is greater—but because the images are formed by the scattering of electrons rather than light, they have a higher resolution: smaller features can be seen. (The limit on the size of the objects a microscope can resolve is set by the wavelength of the imaging beam, and a beam of electrons typically has a shorter wavelength than visible light.) The complication, however, is that these electron micrographs show projections—two-dimensional 'shadows' of the three-dimensional structure. This can make it very hard to decide exactly what kind of three-dimensional pattern is being imaged, and in general researchers have to rely on comparisons between the real images and simulated images calculated by assuming a particular 3D structure (Fig. 2.33).

Gunning saw a hexagonal pattern in micrographs of leaf cells, and he proposed that it was the projected image of a regular network formed from the biological membranes. But it was not until 1980 that Kåre Larssen and his co-workers at Lund made the connection between these pictures and minimal surfaces. They suggested that Gunning's model might correspond to the D-surface (Plate 3a). Soon other structures began to come to light in the organelles—the functional compartments—of many other cells. They are particularly common in the endoplasmic reticulum, but are also

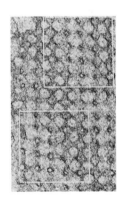

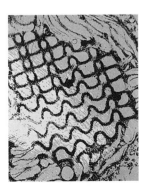

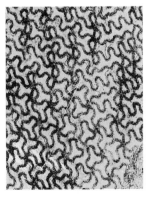

a   b   c

Fig. 2.32 Periodic membrane structures are common in living cells. Many of these appear to be related to periodic minimal surfaces: (a) the D-surface in leaf membranes; (b) the P-surface in algae; (c) the G-surface in lamprey epithelial cells. (Photos: Tomas Landh, Lund University.)

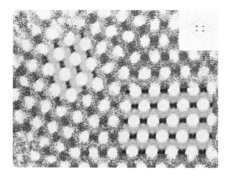

a

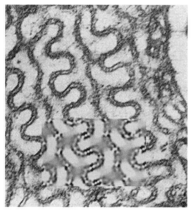

b

Fig. 2.33 The three-dimensional structures of cubic membranes can be deduced by comparing the two-dimensional projections in electron micrographs with those calculated on the assumption that the structures are periodic minimal surfaces. Here I show calculated images superimposed on the corresponding micrographs. (*a*) The D-surface (diamond) in leaf membranes (simulations on *top left* and *lower right*). (*b*) The G-surface (gyroid) (simulation *lower centre*). (Photos: Tomas Landh, Lund University.)

found in the membranes of the mitochondria (the cell's metabolic powerhouses) and the lysosomes (compartments that break down proteins and lipids), and in the cell's outer membranes. Whether or not cells contain these ordered networks seems to depend on their age—they are more common in older, mature cells.

Are these biological structures equivalent to the cubic phases of surfactants in water (Fig. 2.28)? That's still an open question. Both are made from bilayer membranes, but the cellular membranes also contain embedded protein molecules, which might affect their propensity to curve. And the repeat units in the cell structures are commonly larger than those in surfactant cubic phases by a factor of five or more. Perhaps most significantly

of all, the cell structures are formed under non-equilibrium conditions, whereas the cubic phases represent equilibrium structures. So the relationship between the two is not clear, and although the ordered cell structures are called *cubic membranes*, this does not mean that they are strictly equivalent to the cubic phases—rather, it simply denotes that they too have regular cubic structures similar to those of periodic minimal surfaces.

But what is clear is that nature has found a use for these spontaneously formed patterns. Their ubiquity implies that these are no freaks—in biology we expect patterns to serve a purpose. No one yet knows, however, what this purpose might be. It could well be connected with organization: cubic membranes divide up regions of the cell into neatly organized compartments, like the rooms of a house, in which tasks can be apportioned. The fact that they are common in organelles such as the endoplasmic reticulum, where repetitive operations like protein synthesis take place, suggests that a regularly organized space might optimize the efficiency of this sort of assembly line. Making cubic membranes is also a good way of creating a lot of surface in a small volume, which might be useful when the task being performed requires a surface to work on—the synthetic processes that take place within the endoplasmic reticulum are like this. And the intricate interweaving of two distinct subspaces might enhance the communication and transport between them.

Well, the idea that the organization of the cell might be guided in part by geometry and physics is unpalatable to some biologists, and not everyone has been ready to accept these interpretations of the electron micrographs, let alone speculations on their implications. 'Biologists tend to believe everything is controlled by proteins', says George Oster, a biophysicist from the University of California at Berkeley. Sure, cubic membranes *could* form through physics alone, but 'cells tend to be dirtier and messier than that'. This is true enough—but might not biology sometimes choose to do things the easy way?

Fossil foams

Whether or not complex, regular membrane patterns play a role in the biology of the cell, one thing is for sure: many organisms use membranes as scaffolds for erecting stronger, more rigid superstructures with fantastic architectures. Bone, for example, is a mixture of a mineral (hydroxyapatite) and proteins (largely collagen) that is wrought into an intricate, porous network by

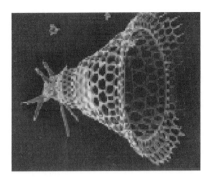

a

b

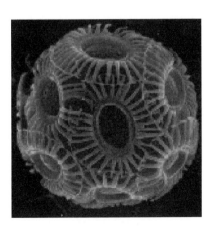

c

Fig. 2.34 The skeletons of radiolarians (*a*) and diatoms (*b*) are thought to be the mineralized casts of a froth of vesicles. Coccolithophores (*c*) also have delicately patterned plates shaped by organic tissues. (Photos: (*a*) The Museum of Science, Boston; (*b*) Dee Breger, Lamont-Doherty Earth Observatory; (*c*) Jeremy Young, Natural History Museum, London.)

cells called osteoblasts, which deposit the hard stuff amongst the membranous web of organic tissues. Far more dramatically, the shells of marine organisms such as radiolarians and diatoms are the casts of patterns formed by ephemeral membranes and vesicles packed into foams (Fig. 2.34).

To scientists interested in pattern formation, these microscopic follies have surely been the most inspirational of life's constructions. And no wonder—for in both their beauty and their diversity, they are the biological equivalent of snowflakes. But as the biologist Karl von Frisch points out, nature is indifferent to aesthetics. 'I do not want to wax philosophical about so much "useless" beauty scattered over the oceans', he says, 'Nature is prodigal.'

The structures are not, strictly speaking, shells at all, but rather exoskeletons—external skeletons that enclose the soft, organic tissues of their architects. Several classes of marine organisms construct exoskeletons. Radiolarians are tiny, single-celled animals (protozoans) whose exoskeletons are made of silica. Diatoms, dinoflagellates and coccolithophores, on the other hand, are members of the class of microscopic plants called phytoplankton. Diatoms and dinoflagellates live mainly in coastal and polar waters, and their exoskeletons are also made of silica; coccolithophores are more abundant in warmer, tropical seas, and they make their elaborate cages from calcium carbonate, the fabric of chalk and marble.

When Christian Gottfried Ehrenberg made the first recorded observation of coccolithophores in 1836 while inspecting chalk from an island in the Baltic Sea, he thought that they must be inorganic mineral formations of some kind. All Ehrenberg saw were the 'bones'—oval-shaped platelets of the hard exoskeletons of these creatures preserved in the rock, their organic tissues having long since decayed. Ironically, while today those searching for ancient forms of fossilized microorganisms run the risk of being misled by complex mineral formations of organic appearance formed without the aid of living creatures (Chapter 1), Ehrenberg was initially deceived in the other direction: he could not imagine that the elaborate carbonate structures he found could have anything to do with life, and decided instead that they must be related to previously known spherical crystals called spherulites. Ehrenberg spent 14 years recording thousands of different forms of coccolith skeletons in meticulous drawings, all the time under the impression that he was drawing curious crystals.

In 1857 the biologist Thomas Huxley observed similar 'rounded bodies' in the muddy sediment pulled

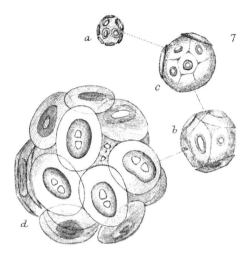

Fig. 2.35 Thomas Huxley sketched many coccoliths in 1868, but believed them to be inorganic formations. (Image: Jeremy Young, Natural History Museum, London.)

armour of protoplasmic organisms—coccolithophores—that dwell within.

Much of what was known at this time about coccoliths and radiolarians came from the sediment samples collected by the British research vessel *HMS Challenger*, which from 1872 to 1876 embarked on a cruise to probe the secrets of the abyssal ooze. Ernst Haeckel was captivated by the geometric wonders of *Challenger*'s bounty, and he catalogued hundreds of radiolarian exoskeletons in a vast Atlas (Fig. 2.36). Whereas coccolithophore shells are generally composed of overlapping, disk-shaped platelets, radiolarian exoskeletons are typically an ornate latticework of geometric polygons, with hexagons being particularly prominent. Haeckel's drawing of the organism *Aulonia hexagona* (Fig. 2.37a) showed a perfect sphere traced out in a web of hexagonal cells. But when, around the beginning of this century, D'Arcy Thompson came to exercise his awesome interpretive faculties on Haeckel's atlas, he noted something important: 'No system of hexagons

up from the deep North Atlantic Ocean. Although he noted that they looked 'Somewhat like single cells of the plant *Protococcus*', he too decided that they must be inorganic in origin, and he called them coccoliths (from the Greek *lithos*, stone) (Fig. 2.35). But in 1861, G.C. Wallich found these same oval-shaped platelets in seafloor sediments, and noticed that sometimes they were stuck together in spherical aggregates like those that Huxley sketched. These 'coccospheres' were often associated with plankton called foraminifers, and so Wallich decided that they were probably of biological origin. At the same time, the Englishman Henry Clifton Sorby came to the same conclusion after studying coccoliths in chalk. When Wallich and Sorby published their findings, most biologists, including Huxley, came to accept the biological origin of coccoliths. But not Ehrenberg, who resolutely maintained that they were inorganic until his death in 1876.

Huxley took a close look at his coccolith samples under a microscope, and observed that many were embedded in a transparent jelly-like slime, a 'protoplasm' of the sort identified a few years earlier by the German biologist Ernst Haeckel. He decided that the coccospheres were skeletal structures that helped to support this slime. Although it later became clear that the jelly was simply a product of chemical reactions between the sea water and alcohol used to preserve the specimens, in 1898 George Murray and V.H. Blackman proposed that the coccospheres are the protective

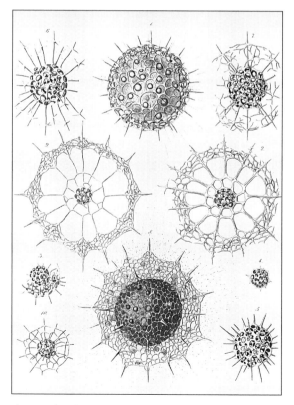

Fig. 2.36 The Atlas prepared by Ernst Haeckel depicts a vast selection of beautiful radiolarian skeletons. (Image: Scott Camazine, Pennsylvania State University.)

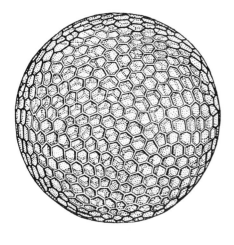

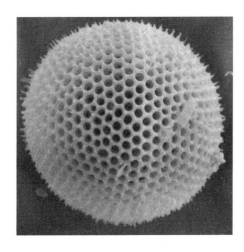

Fig. 2.37 The radiolarian *Aulonia hexagona* as drawn by Haeckel (*a*) and as it appears in the electron microscope (*b*). The shell is a closed sphere of primarily hexagons, but pentagons are also needed for closure. A few of these can be discerned in both images. (Images: (*a*) from Thompson 1961; (*b*) Tibor Tarnai, Technical University of Budapest.)

can enclose space… the array of hexagons may be extended as far as you please, … but it never closes in'. This, Thompson pointed out, was a consequence of a relationship deduced by mathematician Leonhard Euler between the number of faces, vertices and edges of a polyhedron. Euler's formula tells us that such a polyhedron cannot be made of hexagons alone. Instead, Thompson realized, there must be pentagonal or square facets in such a polyhedron to allow it to form a closed shell. Precisely 12 pentagons will suffice to close a polyhedral shell whose other faces are all hexagons, no matter how big the shell is. And indeed, said Thompson, Haeckel did allude to the presence of some pentagonal and square cells in the framework of the *Aulonia*

Fig. 2.38 Richard Buckminster Fuller used hexagonal and pentagonal elements to construct his geodesic domes, most notably that used in the US exhibit for Expo '67 in Montreal. (Photo: Copyright 1967 Allegra Fuller Snyder, courtesy of the Buckminster Fuller Institute, Santa Barbara.)

exoskeletons. This need for pentagons was understood by the American architect Richard Buckminster Fuller when he designed his distinctive geodesic domes (Fig. 2.38), and it became gradually evident to the discoverers of the carbon-60 molecule, a new form of carbon dubbed 'buckminsterfullerene', in 1985 as they struggled to understand how hexagonal sheets of carbon atoms like those in graphite could be induced to curl up into a closed spherical cage.

How on earth does a lump of protoplasmic jelly put together an edifice as fantastic as these? For Haeckel's radiolarians, D'Arcy Thompson felt that nothing could be simpler. The organism blows bubbles, and then sets them in stone. That is to say, the organism surrounds itself with a foam of vesicle-like bubbles (called vacuoles) and allows the mineral to precipitate from the solution held in the Plateau borders where the bubble walls meet. The foam is a two-dimensional hexagonal layer of bubbles curved into a spherical shell around the organism itself, and so the Plateau borders meet at angles of 120°. Take away the soft, organic material of the vesicles and you are left with a lattice of hexagons: a foam preserved in mineral form.

Is nature really this simple? We now know that D'Arcy Thompson's notion was extraordinarily prescient. The silica lattices of diatoms and radiolarians are formed around close-packed arrays of large 'areolar vesicles' secreted from and attached to the organism's membrane wall (called the plasmalemma). Between the areolar vesicles is assembled a system of thin, tubular vesicles, and silica is deposited within them. The result is a hexagonal mesh of silica (Fig. 2.39). So while there is a considerable degree of orchestration in this process, particularly to transport and confine the inorganic

material and prevent it from precipitating willy-nilly around the 'bubble-raft' template, the basic elements of shell formation are indeed those suggested by Thompson.

Clearly, however, many marine microorganisms produce exoskeletons that look considerably more complex than the mineral replica of a foam. Diatoms, for instance, commonly sport delicate patterns within patterns, such as fine perforations of a larger-scale mesh. Here there seems to be a hierarchy of patterning processes: the large areolar vesicles may become detached from the plasmalemma once the basic silica mesh is in place, and the intervening space becomes infiltrated with smaller vesicles, some that deposit silica and others that do not. By again packing into regular arrays, these can impose a finer mesh on the larger one. In sponges, meanwhile, clusters of just a few vesicles can provide the template for star-like mineral structures called spicules formed at the vertices of converging Plateau borders (Fig. 2.40a), while for so-called silico-flagellates, similar clusters generate a small fragment of a foam framework in which Plateau's rules and the curving of Plateau borders owing to pressure differences (page 22) are clearly visible (Fig. 2.40b).

Coccoliths pose another kind of puzzle: their platelets, with shapes that are typically bowl- or mushroom-like, are not obviously akin to the shapes made by packing bubbles. All the same, we know that these structures are moulded by organic membranes within the single-celled organisms, although the forces that shape them are surely more complex than surface tension alone. These platelets are usually cast in vesicular compartments within the cells before being transported to the cell surface. Nonetheless, the coccolith-forming

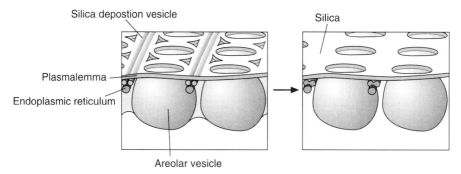

Fig. 2.39 The formation of exoskeletons of diatoms and radiolarians is a highly orchestrated process. A froth of areolar vesicles is attached to the outer membrane wall (the plasmalemma), and a scaffolding of tubular vesicles is constructed in the gaps between areolar vesicles. The tubular vesicles secrete silica, which forms a geometric mesh around the froth. (After: Mann and Ozin 1996).

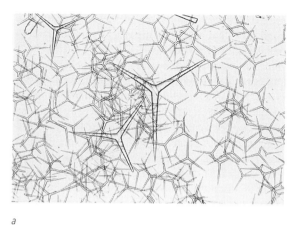

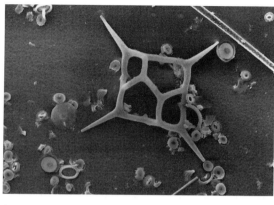

a   *b*

Fig. 2.40 (*a*) The spicules of sponges appear to represent the casts of Plateau borders between a few vesicles. (*b*) Plateau junctions are clearly evident in the exoskeleton of this silicoflagellate (compare Fig. 2.9). (Photos: (*a*) Michelle Kelly-Borges, Natural History Museum, London; (*b*) Stephen Mann, University of Bath.)

vesicles sometimes themselves become patterned with fine ornamentation that is transferred to the mineral platelet: a mesh of pores, presumably from the packing of smaller vesicles, is quite common on coccoliths.

We can see a particularly striking example of bio-mineral patterning in the skeletons of the sea-urchin *Cidaris rugosa*. The skeleton is a regular mesh of calcite (Fig. 2.41), which bears a remarkable resemblance to

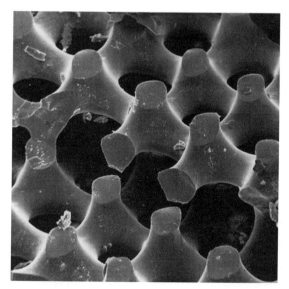

Fig. 2.41 The calcite skeleton of the sea urchin *Cidaris rugosa* appears to be a mineralized cast of a periodic minimal surface, the P-surface. (Photo: Hans-Udde Nissen, kindly supplied by Michele Emmer.)

the cubic P-surface (see Fig. 2.28). It seems most likely that the organic tissues within which the mineral is originally deposited have conspired to adopt a structure very much like this periodic minimal surface, which acts as a template for skeleton formation. The smooth, continuous curvature of the mineral means that it can distribute loads evenly and is not liable to split along the atomic planes of the crystal. As a consequence, skeletons like these can attain strengths greater than that of reinforced concrete. So there are clearly practical benefits to these complex patterns.

Test-tube skeletons

As I indicated earlier, there can be practical value, as well as aesthetic pleasure, in patterned materials. There is now a whole battery of sophisticated techniques that materials scientists have at their disposal for imposing a pattern on a substance, and armed with electron and ion beams they can carve the most intricate circuitry into a silicon chip or etch semiconductor films into a microscopic mesh (Fig. 2.42). But these are extremely costly and labour-intensive methods, so the products do not come cheap. Nature, meanwhile, forms her patterns in very impure, messy chemical mixtures under the mildest of conditions and with profligate abundance. How much cheaper and easier it would be if we could learn a few tricks from her so as to effect the kind of patterning shown in Fig. 2.42 by throwing together a few chemical reagents in a bucket.

But if the delicate filigree of radiolarians and diatoms were the product of some complicated biological

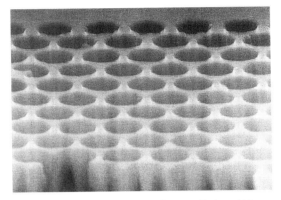

Fig. 2.42 A film of the semiconductor gallium arsenide into which a honeycomb array of holes, just a micrometre or so wide, has been etched with an ion beam. Patterned materials like this might have important technological applications as filters or as optical 'waveguides'. But they are expensive to produce this way. (Photo: J.R. Wendt and G.A. Vanter, Sandia National Laboratories, New Mexico.)

needed to make a radiolarian skeleton. If that is truly so, might one adapt nature's strategies to make artificial radiolarians in a test tube?

A question like this was very much in the mind of the Dutch zoologist and microscopist Pieter Harting when in the 1870s—four decades before the publication of *On Growth and Form*!—he conducted experiments to discover whether by chemical means alone he could produce 'calcareous formations' like the patterned shells that Huxley, Haeckel and others had found in the sediments of the deep sea. In 1872 Harting published a paper in which he summarized the fruits of his efforts to mimic nature with chemistry. His concoctions were Shakespearean: he attempted to crystallize calcium carbonate and phosphate in liquid mixtures that included 'Blood, bile, mucus … and the liquor obtained by triturating chopped-up oysters in a mortar'. Out of this witches' brew came forth 'A considerable number of forms … which are, for the most part, found in nature'.

In particular, Harting found many patterned spherical deposits of calcium carbonate, like the spherulites familiar to Ehrenberg, which he called calcospherites. In his drawings (Fig. 2.43*a*) these look remarkably similar to coccoliths both in size and form, as Harting himself remarked. Sometimes they aggregated into columns or fused into polyhedral plates (Fig. 2.43*b*), resembling the

process requiring (as most biological construction does) a detailed DNA blueprint and protein work force to put the pieces into place, there would not be much chance that one could reproduce patterns like this synthetically. D'Arcy Thompson's speculations give materials scientists the hope, however, that nothing more than simple physical and chemical forces are

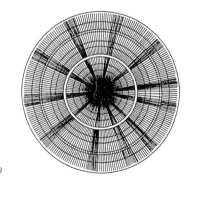

a

c

Fig. 2.43 The drawings of Pieter Harting bear witness to the extraordinary outcome of his experiments on 'artificial biomineralization'. He saw patterns plates like those of coccolithophores (*a*), polyhedral plates (*b*), fibrous bands (*c*) and 'warty' growths reminiscent of spicules (*d*). (After: Harting 1872.)

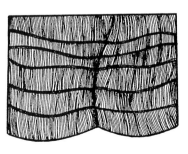

b

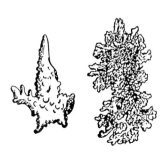

d

shells of marine gastropods. Harting observed that the calcospherites were not simply mineral, but contained organic matter too—just like the shells of organisms. From mixtures of calcium phosphate and carbonate Harting obtained 'plates, which sometimes attain a considerable size, and are more or less curved' and laced with 'fine fibres' (Fig. 2.43c). In other experiments he saw 'warty' and branched forms (Fig. 2.43d) that reminded him of spicules.

It is hard now to know what processes might have been going on in Harting's reaction vessels, or to tell whether these bore any resemblance to the ways in which patterned biological minerals are formed. But his use of gel-like substances like albumin and gelatin is interesting, because it meant that the mixtures could not have been well-mixed—the dissolved chemical components would have been able to diffuse only rather slowly through the jelly-like medium. Thus Harting may have created what is now known as a reaction-diffusion system, in which the rate at which the chemical components can react (or in this case, precipitate) is limited by the rate at which they diffuse. I will show in later chapters how such systems can give rise to a wide range of complex patterns.

Harting's work had none of the impact of D'Arcy Thompson's, and although Thompson mentioned it in his *magnum opus*, it later fell into neglect. But in 1995 Geoffrey Ozin of the University of Toronto realized that it was a precursor to his own efforts at making patterned materials. While tinkering with synthetic techniques similar to those developed by the Mobil researchers in 1992, Ozin found that he could make patterns at much larger scales than those of the Mobil team. Moreover, these patterns were considerably more varied and complex: whereas the Mobil group created hexagonal honeycombs in silica (Fig. 2.24), Ozin's patterns, fashioned instead from an aluminophosphate mineral, showed features over a wide range of length scales, from a few millionths of a millimetre to little less than a millimetre (Fig. 2.44). In other words, he had extended the scale of patterning to that comparable with diatom and radiolarian shells.

How did he do it? Ozin and his co-workers believe that these patterns are the imprints of vesicles, which act as templates for the patterned mineral just as the areolar vesicles do for the shells of diatoms. They mixed together alumina (aluminium oxide) and phosphate ions (which together precipitate to form the mineral), along with surfactants like those used by the Mobil group, in a solvent of the organic compound tetraethylene glycol

(TEG). Ozin showed that the surfactants and phosphate ions together form a layered material akin to the lamellar phase of simple surfactant solutions. But he suggested that the TEG solvent molecules worm their way into the layers and force them to curve, eventually triggering the formation of closed vesicles. The solid formed by precipitation of the phosphate and alumina then bears the imprint of these packed vesicles.

In some parts of the resulting material the pattern is a regular honeycomb (Fig. 2.44a), in others an irregular two-dimensional foam (Fig. 2.44b). Both are consistent with the idea of patterning by packed bubble-like vesicles. Elsewhere the structures are more complex (Fig. 2.44c–e), exhibiting patterns within patterns—like those seen in some diatoms and coccoliths. Ozin suggests that here a hierarchy of templating structures is formed from the organic components of the system—the surfactants and TEG molecules.

Given the sheer diversity in type and size of patterns found in even this relatively simple chemical system, we cannot any longer be too surprised that living organisms, which undoubtedly contain still richer mixtures of chemical components, provide us with a seemingly limitless gallery. But behind all of this artistry we can now begin to see that there are probably some rather simple governing processes, chief amongst which is the formation of templates from a foam of organic bubbles whose architecture is dictated by D'Arcy Thompson's 'physical forces'.

A poor mix

The ordered patterns formed by surfactants in water are rather fluid affairs. Individual surfactant molecules can shuffle around within the aggregates or sheets, or can defect into solution. The sheets of the lamellar phase are distinctly floppy, and the cylindrical micelles of the hexagonal phase are wobbly columns. But many of these same structures can be found congealed into a more robust state in materials known as block co-polymers, which give rise to microscopically patterned plastics.

Polymers are molecules in which many small molecular units, called monomers, are linked together into large 'macromolecules' that can contain thousands of atoms. Proteins are natural polymers, made up of amino-acid units, while synthetic polymers like polyethylene and polystyrene give us the ubiquitous plastics of modern culture. Most polymers are simple straight chains of interlinked monomers. Commodity plastics like polystyrene contain chains made up of just one

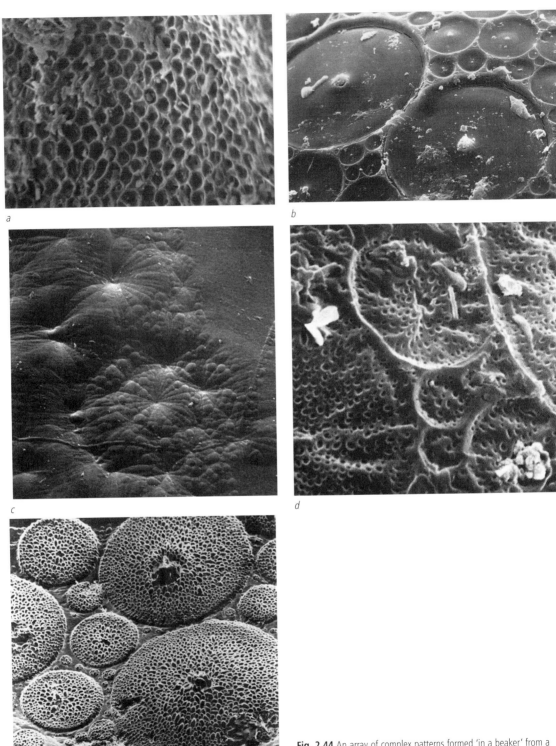

a

b

c

d

Fig. 2.44 An array of complex patterns formed 'in a beaker' from a mixture of organic surfactants and inorganic ions. (Photos: Scott Oliver and Geoffrey Ozin, University of Toronto.)

e

kind of monomer (styrene in this case). Co-polymers, meanwhile, contain more than one sort of monomeric unit. In block co-polymers these different units alternate in blocks along the chain—there will be a block of perhaps a few dozen or a few hundred monomers of one type followed by so many of another.

What gives block co-polymers their propensity to form patterns is the fact that one type of polymer does not mix well with another. Like oil and water, two liquid polymers will separate out into distinct layers (the more dense on the bottom). But block co-polymers are essentially different polymers linked together by chemical bonds, so the different blocks cannot simply separate out into two distinct layers. Like Siamese twins, they are inextricably joined together.

What this means is that block co-polymers have to reach some sort of compromise between mixing and separating. Blocks of the same constitution gather together in domains whose width is determined by the block length. This can lead to the formation of roughly spherical domains of one polymer type immersed in a sea of the other (Fig. 2.45). A co-polymer containing a block of polybutadiene sandwiched between two blocks of polystyrene, which has this kind of structure, forms a rubbery material that, unlike real rubber, can be melted and reset. This 'thermosetting elastomer' is used in the soles of training shoes.

You might notice that the domains in Fig. 2.45 are somewhat akin to micelles, in that they are roughly spherical and that the molecules hide a part of themselves from an incompatible medium all around. But there is still an interface between the two immiscible substances at the surface of the spheres, and the surface tension of this interface imposes an energetic penalty. Spherical domains are formed from a co-polymer con-

taining two blocks (a diblock co-polymer) in which one block (that which forms the spheres) is much shorter than the other. These domains can be distributed at random in some co-polymers, but it can become favourable under some conditions for the domains to be ordered (Fig. 2.46a).

If the domain-forming blocks are a little longer, they may form long cylindrical domains instead of spheres (Fig. 2.46b)—just as surfactants form cylindrical micelles at higher concentrations. When the blocks are of roughly equal length, the co-polymer separates into flat lamellae, analogous to the bilayer sheets of surfactants. But in between the lamellar and the cylindrical phases, some co-polymers adopt the bicontinuous gyroid phase (Plate 3b). Electron micrographs of the gyroid structure show a projection with a characteristic six-armed star shape (Fig. 2.46c). Triblock co-polymers meanwhile (which have three different blocks in each chain) can form even more complex patterns (Fig. 2.46d).

Edwin Thomas and co-workers at the University of Massachusetts at Amherst observed bicontinuous structures in diblock copolymers in 1988, and suggested that they represent an attempt to minimize the total area of the interface between blocks subject to the constraints imposed by the block lengths. Because the two blocks of a gyroid-forming co-polymer are not equivalent (they are of unequal length, for one thing), the two sides of the bicontinuous surfaces are not equivalent either— the networks are formed from the shorter blocks, surrounded by the 'medium' of the longer blocks. So the structures do not correspond exactly to Schwarz's periodic minimal surfaces; but Thomas and colleagues suggested that they do represent area-minimizing surfaces of *constant* mean curvature.

But Sol Gruner of Princeton University and co-workers have pointed out that other factors are also at play in the patterns adopted by these complex materials. In particular, if the polymer chains have to pack in such a way as to form a surface of perfectly constant mean curvature, this imposes severe restrictions on their freedom to crumple up into random tangles—in effect, it means that some of the chains get stretched. This stretching has its own energetic cost, which may act to deform the bicontinuous structure somewhat. The equilibrium structure therefore represents a balance between these opposing tendencies (minimization of surface area and minimization of chain stretching), and in general it deviates from a mathematically ideal surface of constant mean curvature.

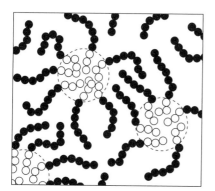

Fig. 2.45 Block co-polymers contain molecular chains with unlike segments, which separate out into distinct domains.

Fig. 2.46 The domains in block co-polymers can form ordered patterns: (*a*) a hexagonal array of spheres; (*b*, here seen partly edge-on and partly face-on) a hexagonal array of cylinders; (*c*) the gyroid phase; and (*d*) a complex morphology seen in a three-block co-polymer. (Photos: Edwin Thomas, Massachusetts Institute of Technology.)

As we will see, complex patterns are very often born of such compromises. They emerge spontaneously from a delicate interplay of forces, and can often be altered in scale or in structure by a small shift in the balance of this interplay. They cannot be predicted by simply considering how the building blocks might be stacked together, but are *emergent* properties of the system as a whole. This is a theme that will recur throughout the book.

What *do* the bees know?

Was D'Arcy Thompson right, then, to see in the astonishing symmetry of a honeycomb nothing more remarkable than the inexorable pull of surface tension and the natural hexagonal symmetry of packed bubbles? Was he justified in asserting that 'The bee makes no economies; and whatever economies lie in the theoretical construction, the bee's handiwork is not fine nor accurate enough to take advantage of them'? On the contrary, I'm afraid that he was deeply underestimating the honeybee, and this example provides us with a cautionary lesson about making too much of physics and too little of evolutionary biology. For the bee is a master craftswoman, and does the job the hard way.

I do mean crafts*woman*, for in the hive I fear it is the females who do all the work. They are the ones who collect the pollen that is used to make honey, and it is they who build the house. The male bees, meanwhile (the drones), do nothing of note except mate with the queen at the appropriate time. But lest you should consider this an indecent inequality on nature's behalf, bear in mind that the males are often driven away from the hive by the females once their use as a source of sperm is expended.

The workers build the hexagonal cells of the hive from wax secreted from wax glands, during a particular stage in their development when these glands mature to a functional form. They set about the job much as we might build an intricate wooden trellis: putting it in place piece by piece, making constant measurements with accurate tools. The activity within the hive during cell building maintains a constant temperature of 35°C to keep the wax malleable, and the workers push each fleck of wax into place in the walls, which they arrange at the correct 120° angle with respect to one another.

Exactly how these angles are measured remains uncertain, but we know at least that the bee has a reference direction defined by the downward pull of gravity. The combs are generally oriented such that two parallel sides of each cell are vertical. The direction of gravity is determined by an organ on the bee's neck, which allows its head to serve as a kind of plumb line for the vertical direction. This organ can sense whether the head is being held up against gravity (in which case it tilts towards the body) or hanging like a plummet below the body.

As well as setting the correct angle between the cell walls, the bees are able to measure the tilt of the cell along its axis. The cells are tilted at an angle of about 13° with respect to the horizontal, which is just enough to prevent the honey from running out.

The thickness of the cell walls, meanwhile, is machined to the incredibly fine tolerance of two-thousandths of a millimetre. To achieve this, the bee has a set of tools for measuring the wall's resilience. She pushes on the wall with her mandibles, creating an indentation, and then uses tactile organs on the tips of her antennae to monitor the flexure of the wax sheet. This flexure depends sensitively on the wall's thickness, and the bee shaves off surplus wax or adds it where necessary to obtain the required elastic behaviour.

Finally, I might add that the bees are able to standardize the alignment of a comb as a whole by orienting it with respect to the Earth's magnetic field. In this way the full, many-layered comb, produced by thousands of bees working simultaneously and in succession, sometimes in total darkness, ends up as a well ordered affair rather than a chaotic assembly of uncorrelated contributions.

So, then, we should not too readily assume that any regularity in nature is the result of symmetric physical forces acting in complete isolation from any genetic programming. For honeybees are very clearly given a set of genetic instructions, and a set of genetically constructed tools, for making their elaborate store rooms. But even if in this instance D'Arcy Thompson was guilty of an overenthusiastic application of physical arguments, there is nevertheless a sense in which geometric imperatives have directed the outcome of nature's processes. For, armed with this impressive array of devices, there is no reason why bees could not produce cells of other equally regular geometries. Yet, as we saw at the beginning of this chapter, the hexagonal honeycomb is the most economic in wax and thus in labour.

Of course, the bees do not *know* the geometric principle that makes this so, and neither does each generation have to learn it by harsh trial and error. Rather, it is genetically hardwired into their instincts. (I suppose it is here that Fontenelle would have to search for the handiwork of his God; but evolution will suffice.) But the point that is relevant for the present argument is that there is no way we could detect the 'presence' of this geometric principle by decoding the genetic information in the bee's DNA. Rather, we could see that they make use of it only by watching the organism as a whole go about her job. Sometimes, to see more you need not to peer more closely but to take a step back. I feel sure D'Arcy Thompson would have no argument with that.

WAVES

A wave is but a single thing, we're told; but from its hue you'd think it was a mixture—flowers and snow!

Ki no Tsurayuki
The Toas Diary

Amongst the sprawling modern myth of Thomas Pynchon's *Gravity's Rainbow* is a fleeting reference to a man who tried to make patterned paint. The reference is clearly meant to document a quixotic, absurd ambition, for we all know that paint does not unmix into the separate pigments that went into its making.

But life is always stranger than we think. Take a look at Plate 4. Pynchon's ill-fated entrepreneur would have done well to follow the recipe that produced these blue and yellow stripes; for this is indeed a stable pattern that emerges spontaneously from a mixture of chemical compounds.

This brew of chemicals is just one of many that have been found to generate spatial patterns. Some of these, like that in Plate 4, are stationary; others are dynamic, releasing waves of colour to a pulse as regular as a clock's tick. Their crucial characteristic, however, is that the elements of the pattern—its symmetry, its length scale, its rhythms—are set not by any external agency but by the internal dynamics of the chemical system, by the rates at which the molecules react with one another and travel through the medium in which they are dissolved.

If you are surprised that a mixture of chemical compounds can form stripes or beat out a pulse, you are not being naive. Indeed, when rhythmic behaviour was observed in a chemical system in the 1920s, most chemists dismissed the observations out of hand as impossible. The idea seemed to contravene a physical law that scientists believed to be unassailable. I will relate in this chapter the tale of how chemical pattern formation has made the difficult journey from a contemptuously dismissed oddity to an exciting new field of research. It is a shame that the Russian biochemist Boris Pavlovitch Belousov did not live to see this story unravel. For initiating the field of dynamical pattern formation in chemistry, all Belousov got for his troubles during his lifetime was the derision of his peers.

Travelling waves

Off balance

Belousov had not intended to create such peculiar and controversial effects in his chemical reaction; rather, he had devised the mixture with the intention of mimicking some of the aspects of the metabolic biochemical process called glycolysis, in which the energy in glucose is liberated as enzymes break it down. He found, however, that this mix of compounds did not seem to settle down into a steady, equilibrium state. Instead, it kept changing colour with uncanny regularity, from clear to yellow and back again. I don't know if that will sound terribly surprising to anyone who has not studied chemistry, but you might get some indication of how this would once have seemed to a chemist if you imagine pouring cream into your coffee and finding it repeatedly dispersing to a uniform brown and then separating out again into a white swirl in the black liquid. Not just once (which would be odd enough), but again and again, at regular intervals.

There is a scientific way to express our intuition about such seemingly one-way processes, and it was the apparent violation of this principle that upset Belousov's contemporaries so much. All physical processes have a preferred direction, be they the way an apple falls from a tree to the ground, the way iron goes rusty or, indeed, the way cream disperses in coffee. The arrow of time is directed by the condition that the total entropy of the Universe—crudely speaking, its total amount of disorder—must increase. This is the second law of thermodynamics, which is the science of change.*

You can probably see that a uniform dispersion of particles of cream throughout a cup of coffee corresponds to a more disordered state than one in which some well-defined pattern of cream and coffee persists in the cup—the positions of the cream and the coffee are more randomized when the two are fully mixed. For the rusting of iron, or for any other chemical reaction, the direction of entropy increase is not so obvious, but it can be deduced readily enough by making measurements of the heat flow and pressure or volume changes during the reaction. The main point is that there *is* a preferred direction, a preferred equilibrium state. Reactions cannot go first towards one end-point and then back again, because in only one of these directions can entropy increase.

Or so the scientific establishment thought in 1951, when Belousov tried to publish his finding. After being snubbed by the journals to which he sent his papers, the Russian biochemist was forced eventually to publish the work in an obscure conference proceedings devoted to another topic entirely. Outside the Soviet Union, it remained virtually unknown.

Had Belousov only known about it, he could have taken some solace from the fact that others before him had anticipated, and even seen, oscillating chemical reactions—and been met with similar indifference or disbelief. In 1910 the mathematician Alfred Lotka published a paper describing a theoretical chemical reaction that underwent damped oscillations—the direction

changed back and forth in a periodic manner, but these changes gradually died out and the system settled into a steady state. In 1920 he showed that a related hypothetical reaction could sustain oscillations indefinitely. The Italian biologist Vito Volterra showed in the 1930s that Lotka's scheme could be used to model fluctuations in fish populations, since it turns out that the same equations that describe reacting chemicals can provide a crude description of interactions between a predator population and its multiplying prey population. I will return to this in Chapter 9.

Lotka's work made little impact on the chemistry community at the time. One of the few to appreciate its significance was William Bray of the University of California at Berkeley, who found in 1921 that a chemical reaction between hydrogen peroxide and iodate ions, which generates oxygen and molecular iodine, exhibits oscillations in the amount of these products generated over time. Even though Bray referred to Lotka's work in his own report, he was told the same as Belousov would be 30 years later—your claims violate the second law of thermodynamics, so they must be the result of poor experimental technique.

But during the 1960s, biochemist Anatoly Zhabotinsky, then still a graduate student of Moscow State University, began to take Belousov's results seriously. His careful experiments finally persuaded others that the effect was real. Zhabotinsky found a combination of compounds that generated a more pronounced colour change, from red to blue, by adding an indicator whose colour depends on the relative concentrations of metal ions involved in the reaction. In Appendix 2 I have given recipes both for this version of the reaction, which has become known as the Belousov–Zhabotinsky (BZ) reaction, and for a related oscillating reaction in which a colour change from yellow to blue is induced by the presence of starch, which turns blue when iodine is produced in the reaction. These are now called clock reactions, for obvious reasons (but don't set your watch by them).

By the end of the 1970s, the BZ reaction was an accepted and at least partly understood part of textbook chemistry, and in 1980 Belousov and Zhabotinsky (together with colleagues Albert Zaikin, Valentin Krinsky and Genrik Ivanitsky) were awarded the Lenin Prize for their discovery. But Belousov had died 10 years earlier, and so never saw his work reach wide acceptance and recognition.

How, though, can we reconcile the oscillatory behaviour with the second law of thermodynamics? Broadly

---

* In the previous chapter I suggested that the direction of change is determined by free energy, which must always decrease. This is entirely equivalent to the condition that entropy must *increase*, but is simply more convenient: free energy (the amount of mechanical work that can be extracted from a system) can be readily measured, whereas the total entropy of the Universe can't. The definition of free energy is in fact chosen precisely so as to make thermodynamics an experimentally accessible enterprise.

speaking, we just need a little patience—for the oscillations do not last for ever. If we leave it for long enough, a colour-changing beaker of the BZ mixture will at last settle down into a uniform and unchanging steady state. There *is* after all a stable equilibrium for the reaction, and it is one in which the overall entropy of the beaker's contents and their environment has increased, just as the second law requires. It is just that the mixture takes a strange and circuitous route in getting there.

The fact is that the second law, like all of the thermodynamics developed during the late nineteenth and early twentieth centuries, pertains only to equilibrium states. It speaks only of end-points. And those were, for a long time, all chemists were really interested in. From a certain perspective, this is understandable. If you are interested in making a particular chemical compound, the first thing to ask is whether it is thermodynamically possible—whether entropy will increase in transforming the starting materials to the end product. If the answer is no, you can forget about it.

But equilibrium is a dull place to be. Nothing happens there. If the Universe were itself at thermodynamic equilibrium, it would be a lifeless place pervaded by a uniform, dim glow of just a few degrees above absolute zero. Just about every phenomenon that interests us is an out-of-equilibrium process—life, to mention one. All human activity, from thinking to shopping to sleeping, takes place in a state that is far from thermodynamic equilibrium. However we may think we hanker after equilibrium in our lives, we tend to do all we can to avoid it in the truest sense, since genuine equilibrium is death.

Our planet is itself far from equilibrium. Why else would the weather be so unpredictable? The seas, the skies and the ground beneath us are all in constant motion, in a manner discussed further in Chapter 7. The atmosphere is a complex chemical brew whose relative constancy of composition is not at all an indication of a true equilibrium state. Rather, this composition is maintained *actively*, by cycling of carbon, oxygen and nitrogen and other elements between the air, living organisms and the geological environment. It is only because of the presence of life on Earth that our planet has the composition that it does: the non-equilibrium rhythms of life maintain an oxygen content that would be extremely peculiar on a dead planet. If aliens were ever inclined to monitor the Earth for signs of life, they'd have no need to look for cities, roads or radio transmissions—one glance at the composition of the atmosphere (something that can in principle be done light years away) would give the game away.

The theme of rich behaviour in systems out of equilibrium is one that will recur many times throughout this book—it is one of the unifying themes of pattern formation, and has been developed into a formal and exact science. For now, I wish to make a crucial point about such systems: they do not come for free, but need a supply of energy. Without this, they will decay—be it slowly or quickly—to a bland equilibrium.

All the same, there is a kind of magic in this transaction. You put in featureless, indiscriminate energy, and the out-of-equilibrium system uses it to organize itself into patterns that can astonish. This is not quite form for free, but it is nevertheless form from formlessness.

To maintain the out-of-equilibrium processes of life and of our planet's shifting meteorology and climate, the energy comes almost entirely from the Sun, in the form of heat and light. A little comes from the planet's hot interior, energy from radioactive decay or left over from the fiery process of planet formation four-and-a-half billion years ago. To sustain indefinitely the oscillations of the BZ reaction, we also need a source of energy. In practice this supply can take the form of a constant throughflow of reactants and products: the reaction can be conducted in a stirred vessel in which fresh reactants are constantly supplied and end products withdrawn. Such vessels are called continuous stirred-tank reactors (CSTRs). Living organisms can be considered as approximations to CSTRs insofar as they (we) continually (though perhaps not continuously) ingest food (fresh material for metabolism) and excrete waste products. In this way we sustain our out-of-equilibrium (and sometimes oscillatory) biochemistry.

There is, then, no reason to fear for thermodynamics in the BZ reaction. But how is it that the reaction keeps changing its mind while it remains far from equilibrium? To answer that, we need to look at a bit of real chemistry.

The chemical seesaw

The key to the BZ reaction is the chemical process known as catalysis, in which some chemical compound speeds up the rate of a chemical reaction without itself being changed by the process. The majority of industrial chemical processes use some kind of catalyst to accelerate product formation—otherwise, the reactions would be too slow to be economically viable. And almost all biochemical reactions in the body are mediated by natural protein catalysts called enzymes.

A catalyst interacts with the reacting molecules so as to help them become transformed. When special techniques are used to see the intermediate steps in a catalytic reaction, we find that the catalyst is not at all aloof from the action—while the reaction remains out of equilibrium, it might take on a new, ephemeral structure, but once the reaction has gone to completion, the catalyst is reformed as if untouched.

What makes the BZ reaction different from most catalytic reactions is that it makes its own catalyst. That is to say, one of the product molecules acts as a catalyst to speed up the formation of more product. This is an example of a positive feedback process, and left to its own devices it will simply makes things go faster and faster. A nuclear explosion is also an example of positive feedback, as indeed are most chemical combustion and explosion processes, and all demonstrate that this kind of self-catalysis, or *autocatalysis*, literally blows up out of control.

Clearly the BZ reaction does not blow up out of control, and the reason for that is that there is a competing process that kicks in to stop the autocatalytic reaction from going to completion (that is, reaching equilibrium). Because of the positive feedback, an autocatalytic process tends to use up its supplies very quickly—even in a CSTR, it consumes the reactants faster than they can be provided. So the concentration of reactants in the vessel plummets, and the concentration of products surges. Since one of these products is coloured, the mixture of chemicals takes on that colour.

Once the concentration of reactants gets low enough, the autocatalytic process runs out of steam, and this allows the competing reaction (which is not autocatalytic) to take over, and the mixture starts to generate a different product, with a different colour. In time, this process too consumes nearly all of its starting materials, and the autocatalytic process kicks back in. Crucially, each of these two processes regenerates some of the compounds needed to get the other started. So while one reaction holds sway, it is paving the way for the other to take over.

Deducing the various steps of the BZ reaction proved to be a complicated and difficult task, since they involve at least 30 different chemical species. Most Western scientists first learned about the BZ reaction in 1968 at an international symposium in Prague, and by 1972 Richard Field, Endre Körös and Richard Noyes at the University of Oregon had put together a somewhat simplified model for the reaction mechanism, which still retains its essential characteristics. Two years later, Field and Noyes pared down this model to an even simpler one, dubbed the Oregonator, in which there are just five

distinct steps and six different chemical species. I outline this scheme in Box 3.1; I don't think you'll find that any great knowledge of chemistry is required to follow it, but because I know how readily some people's eyes glaze over at the sight of chemical formulae of any sort, I shan't inflict it on the main text. For what comes later, you'll simply need to know the following.

The reaction has two branches, A and B: on each branch, a chemical reaction converts certain electrically charged chemical species (ions) to others. The change in colour between blue and red signifies a switch from one branch to the other—a change-over in predominance of the chemical processes taking place. The 'input' to the reaction (the raw material, as it were) is bromate ions, denoted BrO_3^-. The 'output' (the end product) is BrO^- ions. So to sustain the oscillations indefinitely, we need to keep feeding the system with BrO_3^- and removing BrO^-.

We can see the oscillations visibly because of the colour change; but it is often more useful to depict them in terms of the changes in concentration of the various intermediate ions in the reaction—the ones that come and go as BrO_3^- gets consumed and BrO^- produced. One such is the bromide ion, denoted Br^-. Initially, the concentration of bromide in the mixture is high. As Branch A proceeds, the concentration falls, and it is this decline that eventually allows Branch B to take over. But subsequently, another reaction (eqn 3.5) boosts the concentration of Br^- back up again. Plotted as a graph, we see these variations as a series of regular oscillations (Fig. 3.1). Because of this periodicity in time, Arthur Winfree of the University of Arizona calls the BZ reaction a 'time crystal'. (Note, however, that there is an initial 'induction period' before the mixture settles down to regular oscillations.)

Meanwhile, the concentration of another intermediate ion, BrO_2^-, also rises and falls, with the same periodicity as the concentration of Br^- but with the

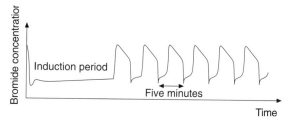

Fig. 3.1 The oscillations of the Belousov–Zhabotinsky reaction can be revealed by monitoring the concentration of bromide ions in the mixture, which rises and falls periodically through time.

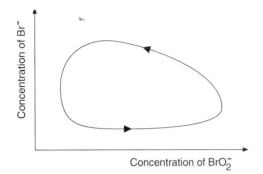

Fig. 3.2 Another way to portray the chemical oscillations is to plot changes in the concentration of one substance (say bromide Br⁻) against another (BrO₂⁻). This plot traces out a closed loop, called a limit cycle.

peaks out of step. This is because the production of BrO_2^- is greatest while Branch B holds sway, whereas the concentration of Br⁻ is greatest just before the switch back to Branch A. One way of depicting both of these rising and falling concentrations at the same time is to plot one against the other (Fig. 3.2). What we see is a closed loop: as time progresses, the concentrations trace out a repetitive circuit of the loop. The reaction will typically be initiated from a point on the graph that lies off the loop, but is quickly drawn onto it—this transient pathway onto the closed loop corresponds to the induction period in Fig. 3.1.

The loop is called a limit cycle, and is a robust characteristic of the reaction. That is to say, for a given set of conditions (temperature and flow rate in the CSTR), any initial point in the plot of the two concentrations will be drawn onto the limit cycle, and the subsequent evolution of the system will be restricted to an endless circulation of the loop, so long as the conditions remain unaltered. You could, if you like, regard the limit cycle as a kind of robust 'form' that the BZ reaction takes in this abstract mathematical space. These stable forms are called *attractors*.

Going places

The oscillating BZ reaction thus has a characteristic pattern of sorts, as depicted by the oscillations in Fig. 3.1 or the limit cycle of Fig. 3.2. But these are patterns in time, whereas on the whole I shall be talking in this book about patterns in space—the kind of pattern that you can see at a glance. The BZ reaction can generate these too. The intricate spatial patterning of the BZ reaction was the second shock it held in store for theorists; says Arthur Winfree, 'Its antics turn out to resemble nothing

foreseen in the thirty years devoted to the subject by theoretical chemists and biologists'.

Above I have considered only the case of a well-mixed reaction, in which an automatic stirring device ensures that the concentrations of the component species, while varying in time, remain uniform at any instant throughout the reaction vessel. But if the reaction is carried out without stirring, there will inevitably be small variations in concentrations from place to place. This is true for any unstirred chemical reaction, but in general it doesn't lead to anything remarkable—the rate of the reaction (which usually depends on the concentrations of some or all of the reactants) then simply varies slightly from place to place too.

For the BZ reaction, however, small variations can make a big difference. This is because the reaction has an autocatalytic component: the positive feedback inherent in this process has the effect of blowing up minor differences into major ones. It means, in particular, that some regions of the mixture can be flipped onto one branch while others remain on the other branch. We then find a mixture in which the colour varies from place to place.

These colour variations do not take the form of a random patchwork of red and blue. Rather, we see complex patterns of astonishing beauty: in a shallow dish of the BZ mixture, concentric rings or twisting spirals of red and blue are produced, which radiate outwards like ripples. The chemical oscillations take the form of travelling chemical waves (Fig. 3.3 and Plate 5).

These patterns were first described by the German scientist H.G. Busse in 1969, although their true nature as chemical waves was perceived the following year by Zhabotinsky and his colleague Albert Zaikin. The pat-

Fig. 3.3 Spiral waves in the unmixed BZ reaction. The waves continuously expand and collide. (Photo: Art Winfree, University of Arizona.)

Box 3.1: The Oregonator

If the BZ reaction is allowed to go to completion (to reach thermodynamic equilibrium), the overall process is one in which an organic compound, malonic acid, is converted to a bromine-containing variant, bromomalonic acid, by reaction with bromate ions, BrO_3^-. (Here the chemical formula connotes that the ion—a negatively charged molecule—contains one bromine atom and three oxygen atoms.) This process is catalysed by certain metal ions, either doubly charged iron (Fe^{2+}) or triply charged cerium (Ce^{3+}).

But curiously, the Oregonator model does *not* trouble itself at all with the conversion of malonic acid—this compound does not feature amongst the six involved in the reaction scheme. This is because the Oregonator is a description of the non-equilibrium states of the reaction, those states that flash alternately red and blue, whereas the malonic acid enters into the scheme only once all the interesting autocatalysis has gone through its paces, and is spat out of the end (in brominated form) as a dull product of the overall equilibrium.

This transformation is the end product of the *non-autocatalytic* branch of the scheme. The Oregonator includes this branch only up to an earlier stage in the full sequence of transformations. The initial reactants on this branch are BrO_3^- ions and bromide (Br⁻) ions, both present in the general recipe (Appendix 2). These ions react to generate the ions BrO_2^- and BrO^-:

$$BrO_3^- + Br^- \rightarrow BrO_2^- + BrO^-. \qquad (3.1)$$

If the chemical formulae look daunting here, just bear in mind that all equation 3.1 is showing is the transfer of an oxygen atom from BrO_3^- to Br⁻. The next step on this branch involves a reaction between the BrO_2^- produced in equation 3.1 and more Br⁻:

$$BrO_2^- + Br^- \rightarrow 2BrO^-. \qquad (3.2)$$

Again, just think of this as the transfer of an oxygen atom from BrO_2^- to Br⁻. It's the BrO⁻ produced in these steps that ultimately goes on to convert malonic acid to bromomalonic acid, but we don't need to worry about that.

Now, the rate at which these two transformations occur depends on the concentration of the reactants (the compounds on the left-hand side of the arrow). Simply speaking, the more there are of these around, the better are their chances of encountering one another and reacting. So initially, when there is a lot of Br⁻ in the mixture, the rate is fast. But as the Br⁻ gets consumed, the rate starts to decline.

That is when the second, autocatalytic branch of the Oregonator comes into play. For it happens that BrO_2^- ions, produced in equation 3.1, can react with BrO_3^- (one of the initial reactants, remember) to generate *two* molecules of BrO_2^-. This reaction can take place only in the presence of the metal ions (Fe^{2+} or Ce^{3+})—the function of these ions is to donate an electron (a negatively charged particle) to the reaction, which leaves them with an extra positive charge. So the metal ions get converted to Fe^{3+} or Ce^{4+}, respectively:

$$BrO_2^- + BrO_3^- + Fe^{2+} \rightarrow 2BrO_2^- + Fe^{3+}. \qquad (3.3)$$

You might notice that this reaction doesn't quite add up—there are five oxygen atoms on the left-hand side but just four on the right. It is bad practice to write down 'unbalanced' equations like this, but I hope you'll excuse it—I've done so to keep things simple. In practice the extra oxygen is taken up by hydrogen ions (H^+) to make water (H_2O). The important thing to notice is that two BrO_2^- ions are produced from one. This is what gives rise to the autocatalysis, because the rate of the reaction depends on the concentration of BrO_2^- (it appears as a reactant on the left-hand side). As the reaction progresses, more and more BrO_2^- is produced, so the reaction goes faster and faster.

The BrO_2^- produced in equation 3.3 goes on, however, to react further—two of these ions exchange an oxygen atom:

$$2BrO_2^- \rightarrow BrO_3^- + BrO^-. \qquad (3.4)$$

Now we can see how, out of equations 3.1 to 3.4, oscillations in the concentration of chemical species in the mixture arise. Equations 3.1 and 3.2 comprise the first branch of the Oregonator scheme—call it Branch A. Initially, this reaction proceeds through to the completion of equation 3.2, which ultimately results in the formation of bromomalonic acid. The cocktail of BZ reagents includes an indicator that turns red in the presence of Fe^{2+} ions, and because these are added to the initial mixture (although they do not partake in Branch A), the mixture starts off red.

You can see that equation 3.1 produces BrO_2^-, which could in principle react with the BrO_3^- as in equation 3.3. But it turns out that BrO_3^- reacts faster with Br⁻ (eqn 3.1) than with BrO_2^- (eqn 3.2), so the former dominates. Eventually, however, the concentration of Br⁻ falls so far—because it is consumed in equations 3.1 and 3.2—that the rate of equation 3.1 slows down significantly. Then equation 3.3 has a chance to take over. Once it does so, Branch B (eqns 3.3 and 3.4) rapidly come to dominate, because equation 3.3 is autocatalytic. Moreover, it converts Fe^{2+} to Fe^{3+}, and the indicator turns blue in the presence of the latter.

If this were all there is to the Oregonator, we'd get just one oscillation—a switch from blue to red. But there is a final step, which switches the conditions back to those that favour the dominance of Branch A. In this step, a rather complex set of reactions between Fe^{3+} and other bromine-containing compounds results in the formation of both Fe^{2+} (turning the indicator red again) and plenty of Br⁻ (allowing equation 3.1 to reassert itself):

$$Fe^{3+} + bromine\ compounds \rightarrow Fe^{2+} + Br^-. \qquad (3.5)$$

(I've left this equation so vague that you can't tell if it's balanced or not, again for the sake of simplicity.)

Equation 3.5 completes the Oregonator, and allows the reaction to flip back from Branch B to Branch A. Then the whole cycle repeats itself. To sustain the oscillations in a CSTR, we need to keep supplying BrO_3^- and removing BrO⁻—you'll notice that the latter appears only as a product on the right-hand side of the reactions, whereas all the other species appear as both reactants and products.

terns are, as we shall see, not unique to the BZ reaction but are generic to a whole class of non-equilibrium systems in chemistry and beyond. They can be captured most easily in a BZ mixture by infusing it into a gel, which slows down the rate at which the chemical species can diffuse through the medium and so stabilizes the chemical waves. The patterns are clearly an example of symmetry breaking: the uniform reaction medium breaks up into complex structures that manifest a degree of organization in space and time. I explain how to create them in Appendix 3.

We can imagine readily enough that a fluctuation in the relative concentrations of the reacting species might shift the reaction from Branch A (red) to Branch B (blue). But why then does this disturbance radiate outwards as a wave with a specific period?

As the autocatalytic cycle of Branch B takes hold, its influence spreads into the surrounding medium and the blue region expands from its origin. But as the wavefront advances, the cycle is played out behind it: the Fe^{3+} produced by equation 3.3 partakes in equation 3.5, regenerating the Fe^{2+} and Br^- needed for Branch A. In effect, the red regions to either side of the wavefront are then no longer equivalent: beyond the wavefront, the medium is ripe for 'colonization' by Branch B, whereas behind the wavefront this branch has run itself to exhaustion and Branch A has begun a new cycle, which is completed when the next wavefront arrives.

This non-uniform BZ mixture is what physicists call an *excitable medium*. Such a medium can change its state locally—switching from the red Branch A to the blue Branch B, for instance—when some stimulus (here the concentrations of chemical species conducive to sparking off Branch B) reaches a certain threshold. But, crucially for formation of these complex patterns, the medium goes through a 'refractory' period once it has been excited. During this time, it cannot be excited again. It is this refractory period that enables steady, periodic oscillations to be set up, giving rise to intricate spatio-temporal patterns.

This mechanism accounts for the target patterns; but what is happening with the spirals? These are basically a mutation of the targets, generated by a perturbation to the expanding circular waves. Such perturbations can happen by accident if, for example, the wavefront encounters some obstacle in the reaction medium; or they can be introduced on purpose, for instance by blowing air onto the wavefront through a pipette. Disturbances like this might break apart the circular wavefront, and the fragmented ends become 'rotors'—

rotating centres of excitation from which the arm of the spiral wave emanates (Fig. 3.4a). Because a broken wavefront must have two ends, spiral waves are commonly formed in pairs that rotate in opposite directions. German chemists Stefan Müller and O. Steinbock have shown that a laser beam can be used to marshal several rotors together to create multi-armed spirals in a modified BZ reaction mixture that is sensitive to light (Fig. 3.4b).

I explained in the first chapter that one of the goals of studies in pattern formation is to reduce a pattern-forming system to its barest essentials, so that we can start to see which patterns are universal—characteristic of certain classes of systems rather than dependent on the fine details. For the BZ reaction, even the simplified

a

Disruption

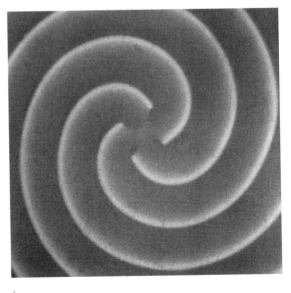

b

Fig. 3.4 (*a*) Spiral waves develop in the BZ reaction from a disturbance that breaks up a circular wavefront of the target pattern. The two broken ends curl into counter-rotating spirals. (*b*) Multi-armed spirals can be created by bringing several rotors together. (Photo: Stefan Müller, University of Magdeburg.)

Oregonator scheme is quite a complicated one to deal with mathematically, although Richard Field and co-workers have subsequently managed to concoct a theoretical scheme that includes just three independent variables rather than the six concentrations of chemical species in the original Oregonator, while still capturing most of the essential behaviour. But to model a non-uniform BZ medium using one of these chemical schemes, with their menagerie of mutually interacting compounds, is a highly computer-intensive task that would not tell us much about the *universality* of the patterns formed.

It is possible to reproduce these patterns, however, in a model that includes no chemistry whatsoever. In this model the flat reaction medium is represented by a two-dimensional checkerboard lattice of little compartments or cells, each of which interacts with those around it. To model the properties of an excitable medium, each cell can exist in three states: receptive (meaning that it is liable to become excited), excited and refractory (which means that it is recovering from a period of excitation). When in the excited state, the cells deliver a stimulus to those around it. If any receptive cell receives a sufficiently large stimulus from its neighbours (equivalent, in the BZ reaction, to receiving a certain influx of diffusing chemical species of a certain type), it too becomes excited. But once excited, a cell eventually enters the refractory state, during which time it remains unresponsive to stimuli regardless of what its neighbours are up to.

This kind of model is called a cellular automaton, reflecting the fact that the cells are mindless and respond to stimuli in a kind of automatic, knee-jerk way. Cellular automata were devised in the 1960s by mathematicians John von Neumann and Stanislaw Ulam, who were interested in modelling self-reproducing entities; I will say more about them in Chapter 9. For now it is enough to say that they represent a very general, computationally tractable way to model complex interacting systems. The behaviour of the system as a whole depends on the rules that govern the interactions between neighbouring cells. In the case of the cellular excitable medium, travelling spiral and target patterns arise when excitations are initiated at a few points (Fig. 3.5). Moreover, the wavefronts annihilate each other in just the same way as they do in the BZ reaction. Because these patterns are formed without including any ingredients of the specific chemical reactions taking place, we should expect them to be characteristic of *any* medium that is excitable. Notice

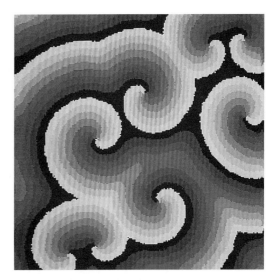

Fig. 3.5 The wave patterns of the BZ reaction can be mimicked by a model that takes no account of the chemical specifics, but simply presents the reaction medium as a lattice of cells that can be excited by receiving stimuli from their neighbours. (Images: Mario Markus and Benno Hess, Max Planck Institute for Molecular Physiology, Dortmund.)

too that the cellular automaton model does not include any description of the *oscillatory* nature of the BZ reaction; rather, it is the *excitable* nature of the reaction medium that creates the spatial patterns.

I've talked so far about BZ mixtures in thin layers, which are essentially two-dimensional. In three dimensions the patterns become more complex. The simplest is a scroll wave, in which a two-dimensional spiral wave is drawn out into a kind of curled-up scroll (Fig. 3.6a). The question naturally arises of what happens at the ends of a scroll, and the answer is that commonly they join up to form a *scroll ring*. Cross-sections of a scroll ring in the plane of the ring look like concentric circles—target patterns—while those perpendicular to the ring appear as spiral waves curling in opposite directions. Scroll rings have been seen in the laboratory in BZ media since the 1970s. Arthur Winfree and co-workers have made theoretical investigations of more complicated scroll rings in which the scroll acquires a twist around the ring: this leads to complex patterns of wavefront collision and annihilation (Fig. 3.6b).

Frontal assault

There's an important refinement to be made in the relation between a chemical *travelling wave* and the pulsating target and spiral patterns of the BZ reaction.

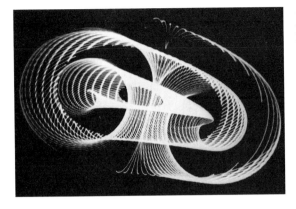

a

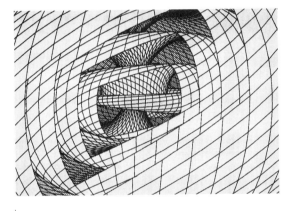

b

Fig. 3.6 The patterns in a three-dimensional BZ mixture are more complex. One of the simplest, the analogue of a two-dimensional spiral, is the scroll wave (*a*), which can curl up on itself to form a scroll ring. Twisted scroll rings have more complicated structures (*b*). (Images: Arthur Winfree, University of Arizona.)

chemical waves in a reaction between oxalic acid and permanganate ions, projected onto a screen in front of the audience.

Luther suggested that the waves arose from a competition between an autocatalytic reaction and the process of diffusion that transports the chemical reagents through the reaction medium. Diffusion is a random process—molecules of the reacting molecules are buffeted from all directions by collisions with molecules of the surrounding solvent (generally water), and as a result they execute a convoluted, meandering path often likened to a drunkard's walk. Despite this randomness, the molecules do actually get somewhere rather than just meandering a little around their initial positions—but the direction they take is random, and the distance travelled from some initial location increases only rather slowly as time progresses. (Whereas the distance covered by walking along a straight path at constant speed increases in direct proportion to the time elapsed, the distance travelled by a random walker is proportional to the square root of the elapsed time.) Random walks owing to diffusion were much studied at the beginning of the century, notably by Albert Einstein.

When a chemical reaction is conducted under conditions where the concentrations are not maintained uniformly throughout the medium by vigorous mixing, diffusion becomes important, since it limits the rate at which a reagent that has become used up in one region can be replenished from elsewhere to sustain further reaction. This is particularly important for autocatalytic reactions, since they can use up a reagent locally at an extremely rapid rate. If diffusion cannot keep pace with this, the reaction runs into problems. This is precisely the situation that I described earlier—although not quite in these terms—in the vicinity of a wavefront in the BZ reaction. The inadequacies of diffusional transport create the refractory period in the medium just behind an advancing wavefront, where the reaction has exhausted itself but has not yet been replenished with fresh reagents. The poorly mixed BZ reaction is thus an example of a so-called *reaction–diffusion* system, which is now clearly recognized as one of the most fertile generic pattern-forming systems that we know of.

After Luther's pioneering studies, the theory of reaction–diffusion systems was placed on a firm mathematical footing by the eminent population biologist Ronald Fisher and by the Russian mathematician Andrei Kolmogoroff and co-workers, both of whom published seminal works in 1937. Fisher was interested

Periodic pulsations can arise only in a medium that is *excitable*, but a propagating wavefront is a rather more common beast, something that could arise for instance from a single, one-off disturbance.

The idea that chemical reactions can develop travelling waves goes back a long way—before, even, the theory of oscillating reactions (which, as we saw, started with Lotka in 1910). At a meeting of German chemists in Dresden in 1906, Robert Luther, director of the Physical Chemistry Laboratory in Leipzig, presented a paper on the discovery and analysis of propagating chemical wavefronts in autocatalytic reactions. Sceptics were apparently quelled by Luther's demonstration of the phenomenon before their very eyes—he showed

in reaction–diffusion processes for modelling the spread of an advantageous gene in a population, not with their manifestation in chemistry—a curious repetition of Volterra's assimilation of Lotka's ideas on oscillating chemical reactions into mathematical biology earlier in the century. It is almost as if chemists were for decades unwilling to face up to the existence of these complex and surprising phenomena in their own field!

All the same, studies of waves in chemical media were conducted in parallel with, but independently from, work on oscillatory reactions since the beginning of the century. In 1900 the German physical chemist Wilhelm Ostwald described travelling pulses in an electrochemical system. When he used a zinc needle to prick the dark coating of oxidized iron on the surface of an iron wire immersed in acid, Ostwald saw a colour change that propagated away from the point of contact at high speeds. From the 1920s onwards, many researchers studied this simple system as an analogue of nerve impulses (which are also propagating electrochemical waves), and in the early 1960s Jin-Ichi Nagumo and co-workers in Tokyo observed spiral waves on the surface of a two-dimensional grid of iron wire subjected to this treatment. But this work, published in Japanese, met the fate so common for studies that are not reported in the English language—it was ignored in the West, until Zhabotinsky's efforts had established the significance of this sort of wave activity.

The ripples spread

The BZ reaction is by no means unique: several other chemical mixtures share the same general features of autocatalysis, feedback and competing reactions that lead to excitable and oscillatory behaviour. It has been seen too in many biochemical processes, including, rather pleasingly, the glycolytic cycle of metabolism that Belousov had first set out to emulate. Similar effects crop up in some corrosion and combustion reactions. When these processes take place in poorly mixed conditions, spatio-temporal patterns can arise whose forms are attractively diverse.

One of the functions of a catalytic converter in automobiles is to reduce emissions of carbon monoxide (CO), a poisonous gas, in the exhaust fumes. This is done by combining CO with oxygen gas in the converter to create carbon dioxide (CO_2), a reaction that is

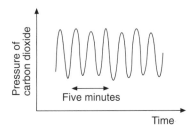

Fig. 3.7 Oscillations in the reaction of carbon monoxide and oxygen on a platinum surface. The reaction produces carbon dioxide.

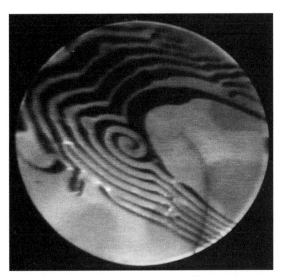

Fig. 3.8 Target (*a*) and spiral (*b*) waves in the reaction of carbon monoxide and oxygen on platinum. The images are all several tenths of a millimetre across. (Photos: Gerhard Ertl, Fritz Haber Institute, Berlin.)

a   *b*

speeded up by the use of a metal catalyst consisting of a mixture of rhodium and platinum. The reaction takes place on the metal surface, where the chemical bonds in the reactant molecules are broken or loosened up. So the reaction between CO and oxygen on a platinum surface is of considerable technological interest. There is no obvious mechanism for *auto*catalysis here, however—the product is simply CO_2, which is not then involved in subsequent reactions.

So it was a surprise to Gerhard Ertl and colleagues at the Fritz Haber Institute in Berlin when they found oscillatory behaviour in the rate of this reaction in 1985 (Fig. 3.7). And when in the early 1990s the Berlin group developed a new kind of microscope to look at the way that the CO and oxygen were distributed on the surface, they saw spiral and target patterns just like those of the BZ reaction, albeit just a fraction of a millimetre across (Fig. 3.8). The bright regions in this figure correspond to parts of the metal surface covered with CO molecules, and the dark regions are richer in oxygen atoms. Ertl's team deduced that the molecules of CO that became stuck to the metal surface were altering its structure, and thereby its catalytic behaviour, in a way that introduces feedback into this apparently simple reaction.

Platinum metal is a crystal: its atoms are packed together in a regular array like oranges on a fruit stall. On a clean platinum surface exposed by cutting through the metal, the arrangement of atoms depends on the angle at which the cut is made; for one particular cleavage plane, the surface looks like that in Fig. 3.9a. This is called the {110} surface, and the arrangement of surface atoms is termed the (1×1) phase. In a vacuum, the topmost atoms of a freshly exposed platinum (1×1) surface will spontaneously shift their positions to create a different surface structure with a lower surface energy. This is called the (1×2) phase, and has a 'missing' row of surface atoms (Fig. 3.9b). The rearrangement process is called a surface reconstruction.

If CO molecules become attached to the reconstructed (1×2) surface of platinum, the balance of energies gets shifted around, and the original (1×1) phase becomes more favourable. This means that, as the reaction between CO and oxygen atoms on the platinum {110} surface proceeds, the surface does not remain passive but shifts its structure between the (1×2) and (1×1) phases, depending on the amount of CO on the surface.

Now the point is that these two surface phases have different catalytic abilities: the (1×1) phase is considerably better at speeding up the reaction with oxygen than is the (1×2) phase. We can now see the possibility of some subtle and complex interactions, which can give rise to feedback. The more the bare (1×2) surface becomes covered in CO, the greater the extent of reconstruction to the (1×1) phase and the more the catalytic potential of the metal is enhanced. But as the reaction proceeds, the CO gets converted to CO_2, which departs from the surface and leaves behind a bare (1×1) surface. On its own, this prefers to revert to the reconstructed (1×2) phase.

Gerhard Ertl, David King at Cambridge University, and their co-workers have devised a six-step reaction scheme that is akin to the Oregonator of the BZ reaction, which incorporates these various processes for reactions on platinum surfaces. It includes an autocatalytic process in which the reaction between CO and oxygen on the (1×1) surface creates new 'bare' catalytic sites. They have found that this scheme produces oscillatory behaviour of the various reaction parameters, such as the rate of CO_2 formation or the surface coverage of CO (Fig. 3.10). Like the Oregonator, the process jumps between two branches—essentially a low-reactivity branch involving the (1×2) surface and a high-reactivity branch involving the (1×1) surface—with the autocatalytic steps providing a mechanism for rapid switching between the branches. It is easy to see that sites of non-uniformity in these surface reactions can act as the centres for the formation of travelling waves like those shown in Fig. 3.8.

Several other metal-catalysed surface reactions are now known to show oscillatory behaviour. One difference between these essentially two-dimensional processes and those in flat dishes of the BZ mixture is that for the latter the medium is isotropic: it looks the same in all directions. For surface reactions taking place on metal crystals, on the other hand, all directions are not the same, because the metal atoms are lined up in a regular checkerboard-like array. This means that the

a　　　　　　　*b*

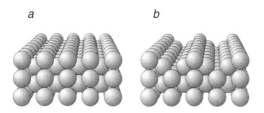

Fig. 3.9 (*a*) The atomic structure of the 1 × 1 surface phase of platinum. (*b*) In a vacuum, this surface will rearrange itself to the 1 × 2 reconstruction.

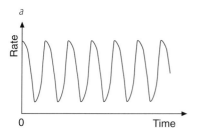

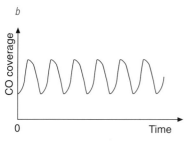

Fig. 3.10 The oscillations in the surface reaction of CO and oxygen can be reproduced by a theoretical model that includes the autocatalytic processes. Oscillations are seen in both the rate of reaction (*a*) and the amount of carbon monoxide on the surface (*b*).

ability of the reacting molecules to move about can be similarly anisotropic (direction-dependent). It is for this reason that the target and spiral patterns in Fig. 3.8 are elliptical rather than circular—the speed of the chemical wave fronts differs in different directions. In extreme cases, this anisotropy means that the symmetry of the underlying metal crystal surface can leave itself

Fig. 3.11 The spiral waves of the oscillatory reaction of nitric oxide and hydrogen on a rhodium surface have a square appearance, which derives from the square symmetry of the underlying atomic lattice. (Photo: Ronald Imbihl, Fritz Haber Institute, Berlin.)

imprinted on the spatial patterns that arise. For example, Ertl's colleague Ronald Imbihl has seen square travelling waves in the reaction of nitric oxide and hydrogen on a rhodium surface, an echo of the square symmetry of the metal crystal surface (Fig. 3.11).

Rock art

If you are a rock collector, the target patterns in Plate 5 may look familiar. They are reminiscent of the stunning concentric bands displayed by agates (Plate 6). Agates are formed when water from rain or snow permeates through fissures in cooling basaltic lava, dissolving metal ions as it goes. Once the body of lava has cooled sufficiently, the ions precipitate out of the mineral-rich solution as agates. This is a process of crystallization occurring far from equilibrium, and so we should perhaps not be too surprised that it can lead to pattern formation.

Periodic patterns due to non-equilibrium crystallization and precipitation have a history that predates the discovery of oscillating chemical reactions. In 1896, the German chemist Raphael Eduard Liesegang performed experiments in which he reacted silver nitrate with potassium chromate in a gelatin gel. This reaction generates insoluble silver chromate, which precipitates as a dark deposit. In solution, this precipitate would all be flushed out at once, as the two salts would mix very quickly. But in a gel, the mixing is much slower, limited by the slow diffusion of the ions. Liesegang saturated the gel with potassium chromate, and then allowed a drop of silver nitrate solution to diffuse through it. He found that the dark precipitate appears in a series of rings behind a reaction front that advances through the reaction vessel. Many other chemical reactions that generate an insoluble compound show the same behaviour when limited by diffusion through a gel (Fig. 3.12), and you can try it for yourself using the recipe in Appendix 4.

Liesegang's experiments are not nearly so obtuse as they might sound. The precipitation of silver metal and salts in gelatin gels became a subject of intense interest in the late nineteenth century owing to its relevance to photography: black-and-white photographic emulsion is essentially a gel containing a silver salt, which is converted to a dark, fine precipitate of silver metal on exposure to light. Indeed, Liesegang's father and grandfather were both early pioneers of photography. Raphael Liesegang himself was by all accounts a remarkable, not to say eccentric, character, with interests every bit as catholic as D'Arcy Thompson's. He

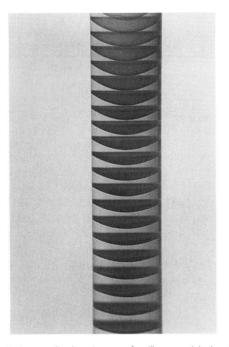

Fig. 3.12 Liesegang bands, a signature of oscillatory precipitation at an advancing diffusion front. Here the bands are produced by cobalt hydroxide as hydroxide ions diffuse down a column of cobalt-laden gelatin. (Photo: R. Sultan, American University of Beirut.)

wrote about the possibility of television in 1891 and, as well as his work on photography, he pursued research on bacteriology, chromosomes, plant physiology, neurology, anaesthesia and the disease of silicosis.

Liesegang's rings (only later was the reaction performed in cylindrical test tubes, so that the precipitation fronts appeared instead as a series of band-like disks) captured the imagination of many of the leading scientists of the time, including Lord Rayleigh, J.J.Thompson and Wilhelm Ostwald. Some early enthusiasts around the turn of the century suggested that in the bands and rings one might be seeing a simplified version of the stripes of tigers and zebras or the patterns on butterfly wings. In this, remarked one critic in 1931, 'enthusiasm has been carried beyond the bounds of prudence'. But as we will see in the next chapter, on one level at least such scepticism is misplaced (although given what was known at the time about chemical pattern formation—next to nothing—we can't really regard these speculations as anything more than a lucky guess).

As the gel medium of the Liesegang process evidently makes diffusion a critical aspect, it's not hard to guess

from the preceding discussion that a reaction–diffusion process lies behind the pattern formation. But while this is no doubt the case, the phenomenon is not fully understood even today. One idea, which was first proposed by Ostwald a year after Liesegang published his findings, is based on the proposition that the reaction product does not precipitate until the solution becomes supersaturated above some critical threshold concentration. Precipitation can potentially occur as soon as the concentration of the reaction product becomes too high for the solution to bear—as soon as it becomes supersaturated. But in practice, particles of the insoluble product will grow large enough to precipitate only after they have first attained a certain critical size. This is one of the basic tenets of the theory of crystal growth, which Ostwald helped to establish. If the product molecules cannot cluster into these 'critical nuclei', the solution can become highly supersaturated.

Ostwald suggested that in Liesegang's experiments, formation of the critical nuclei was slowed down by the fact that the reaction product diffuses only slowly through the gel. The reaction is all the while increasing the degree of supersaturation, however, and once this exceeds a threshold, the concentration of the product is at last great enough everywhere for nucleation to occur. Then the nuclei grow rapidly, accreting the reaction product from the solution around it and precipitating as a dark band. Precipitation leaves the reaction front depleted in the product, and so precipitation stops—and it takes some time for it to build up again to the critical threshold, by which time the front has moved forward. This cycle of nucleation-precipitation-depletion dumps a train of bands in the wake of the front.

Ostwald's theory was refined in 1923 by K. Jablczynski, who showed that it could be used to predict the spacing between successive bands. Jablczynski's spacing law states that the ratio of the positions of two consecutive bands (defined relative to, say, the first band) approaches a constant value as the number of bands gets larger. The theory was further refined by S. Prager in 1956, who turned it into a well-defined mathematical model; but unfortunately Prager's model predicted that the bands will be infinitely narrow, which is certainly not what is observed. Peter Ortoleva at the University of Indiana and co-workers made further improvements to the theory in the 1980s to overcome this shortcoming. More recently, Bastien Chopard from the University of Geneva and colleagues have devised a cellular-automaton model which takes into account some of

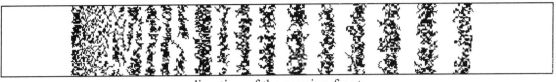

direction of the moving front

Fig. 3.13 Liesegang bands generated in a cellular automaton model of a precipitation-diffusion process. (Image: Bastien Chopard, University of Geneva.)

the microscopic processes that control the diffusion, nucleation and precipitation of the reacting species in Liesegang systems. Their model is able to produce precipitation bands (Fig. 3.13) which obey Jablczynski's spacing law.

But the trouble is that the band spacings in real experiments don't by any means always observe this law: several different relationships have been reported, and there seems to be no general law that applies to all Liesegang-type experiments. Other explanations for the banding have been put forward, many of which generally involve processes that take place *after* nucleation has occurred. But with such a diversity of observations, it isn't hard to find results that will fit most models, while preventing unambiguous discrimination between them.

Liesegang realized that the banded patterns he saw were similar to those found in certain rocks. There is now good reason to suppose that many banded rock formations do indeed arise from cyclic precipitation as mineral-rich water infiltrates a porous rock and reacts to form an insoluble product. Amongst the mineral patterns that have been attributed to Liesegang-type processes are the bands seen in some iron oxide minerals, the wood-grain texture of cherts, the striations of a mineral called zebrastone, and perhaps most familiarly of all, the bands of agates. And Ostwald's idea is just one of a whole class of models involving particle transport, nucleation and precipitation that have been put forward to explain such formations. To take just one example: geologists Peter Heaney from Princeton University and Andrew Davis from the University of Chicago showed in 1995 that Liesegang precipitation–diffusion cycles can account for the iridescence of iris agates. Whereas the colour banding shown in Plate 6 is perhaps the most spectacular feature of these and other agates, the iridescence comes from a periodic banded structure on a scale too small to see by eye. The bands vary in width from about a tenth of a micrometre to several micrometres

a

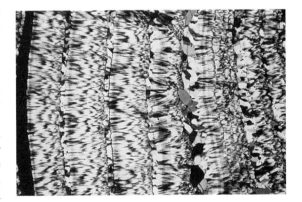

b

Fig. 3.14 Liesegang banding at very small length scales in iris quartz gives it an iridescent appearance. (*a*) The bands here are about seven micrometres apart, and are caused by periodic differences in the concentration of defects in the crystal structure. (*b*) At a larger scale, thin bands of quartz alternate with thicker bands of chalcedony. The bands run from top to bottom; the horizontal striations have a different origin, caused by the fibrous structure of the mineral. The image here is about 2.5 mm across. Banding is also evident on scales of about a centimetre or so (Plate 6). (Photos: Peter Heaney, Princeton University.)

(Fig. 3.14*a*), and this banded 'grating' scatters visible light (because the light has wavelengths of comparable dimensions), creating the iridescent effect. Heaney and Davis showed that these bands correspond to differences in the crystal structure of the mineral: regions of highly crystalline quartz alternate with regions in which a high concentration of defects disrupt the regularity of the crystalline lattice. They postulated that the defective regions formed by the initial linking together of soluble silicate ions into long chains, which precipitate to give a poorly crystalline form of the mineral chalcedony. The highly crystalline regions, meanwhile, are formed by precipitation of individual silicate ions as quartz. This latter is possible when the concentration of silicate ions at the crystallization front is low. But because quartz precipitates slowly, silicate ions diffuse towards the front more rapidly than they are removed by precipitation, and eventually the concentration builds up to a degree that allows their linking into chains. Then the more rapid precipitation of chalcedony takes precedence, until this depletes the silicate solution once again.

Heaney and Davis pointed out that, while this mechanism could account for the iris banding, the agates are in fact patterned on several length scales. There are also oscillations between defect-rich chalcedony and defect-poor quartz with wavelengths of several hundred micrometres (Fig. 3.14*b*) and of a centimetre or so (Plate 6), suggesting that there are several *hierarchical* mechanisms for oscillatory patterning at play here. This kind of hierarchical repetition of pattern over several length scales is a feature of some of the patterns that we will encounter in later chapters.

Burn up

I have already mentioned that combustion processes are autocatalytic; but normally this doesn't produce anything more interesting than a big bang, because there is nothing to keep the process in check. When, however, an explosive combustion process such as the burning of hydrogen in air is conducted under experimentally well controlled conditions, oscillations in the reaction rate can arise. The overall reaction looks simple enough: two molecules of hydrogen combine with one of oxygen to form two molecules of water:

$$2H_2 + O_2 \rightarrow 2H_2O. \qquad (3.6)$$

But the detailed evolution of this reaction is rather complicated, involving short-lived, reactive intermediate species such as lone hydrogen and oxygen

atoms and the hydroxyl free radical, OH. In an autocatalytic process, *three* molecules of hydrogen and one of oxygen can react with a lone hydrogen atom to produce two molecules of water and *three* hydrogen atoms—thus the products of this process represent a multiplication of the reactants. This autocatalytic process arises because hydrogen atoms are less reactive, and so hang around for longer, than oxygen or hydroxyl radicals. When the reaction of hydrogen and oxygen is carried out in a stream of flowing gases in a CSTR, the result of these autocatalytic processes is an oscillatory variation in the burning rate, which shows up as a rise and fall of the temperature generated in the combustion flame (Fig. 3.15). In effect, the mixture of gases repeatedly ignites and then subsides into an unreactive state. The reaction between carbon monoxide and oxygen—the same reaction as that studied by Ertl's group, but this time carried out by burning the free gases rather than bringing them together on a catalytic metal surface—also shows oscillatory behaviour in a CSTR.

Can these combustion processes also generate spatial patterns if they are not well mixed? Well, it has been known for a long time that when a hydrocarbon fuel such as butane is burnt in a flame under carefully controlled conditions, the flame can become very non-uniform, separating into a number of distinct cells. The cells are bright regions, separated by darker boundaries where the temperature is lower and there is less emission of light by the excited gas molecules. These cellular flames were first reported by A. Smithells and H. Ingel in 1892, who described a flame that separated into petal-like segments that rotated around the flame's axis. George Markstein made a careful study of such flames in the 1950s, and in 1977 G.I. Sivashinsky showed theoretically that the cell patterns could be the result of

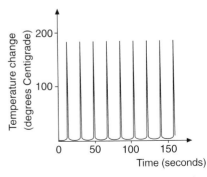

Fig. 3.15 Oscillations in the combustion of hydrogen in a flow reactor, revealed by variations in the temperature. (After: Scott 1992.)

a reaction–diffusion process. A burning flame requires both fuel (generally a hydrocarbon like butane) and oxygen—the former cannot burn without the latter. But molecules of these two compounds travel (diffuse) through the gas mixture at different rates—the oxygen molecules are lighter and so diffuse more rapidly. The combustion reaction can be sustained only as long as the rates of diffusion of both species are sufficient to

Fig. 3.16 Cellular patterns in a shallow cylindrical flame seen from above. The temperature of the flame is lower in the dark regions. (These dark regions are not truly dark to the eye; they are simply a result of the limited dynamic range of the video tape on which the images were recorded.) The cellular flames adopt ordered states. Here I show a sequence of ordered states of increasing complexity as the rate of gas flow in the flame is increased. (Photos: Michael Gorman, University of Houston, Texas.)

feed it. Spatial irregularities can arise when the oxygen in one region burns up the fuel more rapidly than it is replenished by diffusion, and combustion cannot then continue until more fuel diffuses there. Another theory was developed around much the same time, however, which held that flow effects in the gas streams, not molecular diffusion, were responsible for cellular flames.

Sivashinsky showed that the diffusive mechanism could in principle produce ordered hexagonal arrangements of cells. But the cells that Markstein and others had reported were irregularly shaped and were in constant, disordered motion, breaking up and coalescing. It was not until 1994 that Michael Gorman and Mohamed El-Hamdi of the University of Houston in Texas and Kay Robbins of the University of Texas at San Antonio found the first clear examples of regular patterns in cellular flames.

They studied flames of a butane–oxygen mixture passing through a stainless steel plug in a cylindrical chamber. The flame appeared as a luminous disk just half a millimetre thick. As the researchers increased the rate of gas flow through the disk, they saw an initially uniform disk-shaped flame break up into a ring of cells (Fig. 3.16). When the cells first appeared, there were four of them; but as the flow rate was increased, a fifth and then a sixth cell appeared. Increasing the rate still further created a new inner ring of cells, which multiplied from one to six and then spawned a third ring (Fig. 3.16*j*). An hexagonal array of cells appeared as the last ordered structure; for higher flow rates the cells began to exhibit rapid, chaotic motion in which no recognizable pattern was apparent.

The overall number of cells in each concentric ring stays steady for a given flow rate, but their positions keep altering. In some experiments, each ring of cells seemed to rotate independently in a kind of hopping motion: the cells would stay put for a while, then the whole ring would abruptly rotate like a gear wheel shifting position. Under other conditions the motion was more chaotic: the ordered states might disappear intermittently into randomly shaped cells, and then reappear (Fig. 3.17). Sometimes the innermost cells took on a spiral shape, which circulated like a rotating yin-yang symbol.

The Texas researchers concluded that Sivashinsky's diffusion mechanism, not gas-flow effects, lies behind the ordered patterns that they saw—mainly because the latter mechanism was expected to produce bigger cells than those observed. These ordered cellular flames rep-

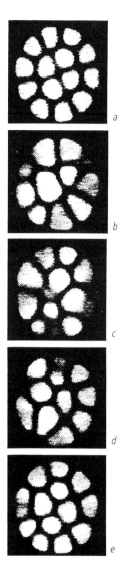

Fig. 3.17 Under some conditions, ordering in the cellular flames is intermittent, being interrupted sporadically by the appearance of more-random cell arrangements. Time advances here from (*a*) to (*e*). (Photos: Michael Gorman.)

resent an unusual kind of dynamic reaction–diffusion pattern, however, because they are all unstable, undergoing intermittent rearrangements via more chaotic states. On the other hand, Howard Pearlman of NASA's Lewis Research Center and Paul Ronney of the University of Southern California have observed flame patterns that look very much like the spiral waves of the BZ reaction, reinforcing the idea that they are the products of a reaction–diffusion process.

a

b

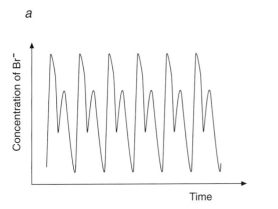

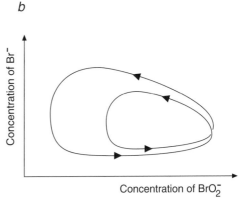

Fig. 3.18 As the flow rate of a BZ mixture in a continuous stirred-tank reactor is increased, the oscillations double up, a phenomenon called periodic doubling (*a*). The limit cycle of the period-doubled oscillations develops two loops (*b*).

Going wild

Another thing that this combustion process and the BZ reaction share in common is that gradual changes in the flow rate of the reacting molecules do not induce correspondingly gradual changes in behaviour; rather, there are abrupt jumps to a new mode of behaviour when a certain threshold is reached. In the well-mixed BZ reaction conducted in a CSTR, I indicated that the oscillatory behaviour defines a certain limit cycle that remained robust in the face of changes in, say, flow rate or initial concentrations of the reagents. But this is true only up to a point. If the flow rate is increased far enough, the colour-changing mixture suddenly starts to exhibit a new temporal pattern. It alternates between blue and red, sure enough, but if we were to time the colour changes—or better still, to measure the rise and fall in concentration of one of the intermediates such as the bromide ion—we would find that something new has happened. The switching now appears to have a double pulse (Fig. 3.18*a*). Plotted as a limit cycle, this behaviour manifests itself as a double loop (Fig. 3.18*b*). The system has to traverse both lobes before it repeats itself.

This is called a period-doubling bifurcation, and it was observed in the BZ reaction in the 1980s by J.C. Roux and co-workers. 'Period-doubling' is obvious enough—the system now has two stable pulses or periods. 'Bifurcation' simply means that the stable, oscillating state of the system has forked into two, with each state corresponding to a loop of the limit cycle.

I should point out that the initial oscillatory state of the BZ mixture in a CSTR is itself the product of a bifurcation, because the flow rate has to reach a certain threshold before the indefinitely oscillating colour change is stable at all; below this flow rate, the reaction will simply go through a series of transient colour changes before settling down into an unchanging, uniform state. (Admittedly, this may take some time, which is why 'clock' reactions like this still look temporarily like regular oscillators even in a closed vessel.) This kind of abrupt transition from a stable, steady state to an oscillatory one was first identified mathematically by the German Eberhard Hopf, long before anyone knew about chemical oscillators. It is therefore called a Hopf bifurcation. Hopf bifurcations are a common source of periodic motion from initially steady motion—the mathematicians Ian Stewart and Martin Golubitsky describe them appealingly as the onset of a wobble.

It does not yet seem to be clear whether the transitions seen in cellular flames are indeed Hopf bifurcations or some other kind of bifurcation. All the same, both these and the period-doubling jumps seen in the BZ mixture in a CSTR share the characteristic that they just keep on coming as the flow rate is increased, giving patterns of ever more complexity. In the latter case, a further increase in flow rate induces another bifurcation into a limit cycle with four lobes, and then eight, and so forth. In cellular flames, each jump adds more cells or even a new ring of cells. With each jump, the amount that the flow rate has to be *further* increased to induce another bifurcation decreases: the jumps get closer and closer together and the patterns become more and more

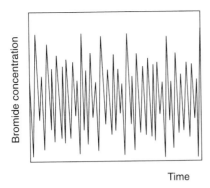

Fig. 3.19 At high flow rates, the oscillations of the BZ reaction become apparently random—the system becomes chaotic.

many diverse systems that exhibit chaos, including lasers and populations of predators and their prey (Chapter 9). Very clear period-doubling bifurcations leading to chaotic oscillations have been seen in the combustion of carbon monoxide and oxygen gases in a CSTR (Fig. 3.20). There are other ways for a chemical system to become chaotic too, and Jack Hudson and colleagues at the University of Virginia identified one such in the late 1970s. In studies of the BZ reaction in a CSTR at high flow rates, they saw 'mixed-mode' oscillations in which a single cycle involves both small- and large-amplitude oscillations (Fig. 3.21a). Typically, each large-amplitude oscillation (in the concentration of, say, bromide) is accompanied by a little train of small oscillations—perhaps just one, perhaps more. Under some conditions these mixed sequences keep repeating regularly, but in other cases different mixed modes may alternate with no apparent periodicity (Fig. 3.21b). Although these non-periodic states satisfy all of the mathematical criteria for chaos (which distinguish them from purely random processes), there was much debate initially about whether they were genuine examples of 'chemical chaos' rather than effects induced by poor mixing in the experiments. But it is now clear that theoretical models of oscillatory reactions (which don't have to suffer any experimental deficiencies) can gener-

complex (which is to say, of lower and lower symmetry). There eventually comes a point at which all pretence of pattern is thrown to the winds and the system descends into chaos. For the BZ reactor, this means that the oscillations in concentrations no longer show any sign of periodicity at all—they appear to be random (Fig. 3.19). The cellular flames, on the other hand, dissolve into a random pattern of irregular cells that is forever shifting.

The route to chaotic behaviour through a series of period-doubling bifurcations is a common one, seen in

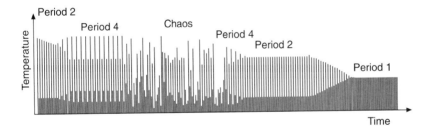

Fig. 3.20 Period doubling and the transition to chaos in the reaction of carbon monoxide and oxygen in a flow reactor. (After: Scott 1992.)

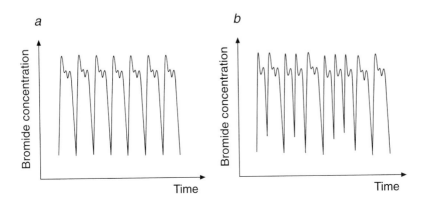

Fig. 3.21 Mixed-mode oscillations in the BZ reaction consist of a mixture of large- and small-amplitude oscillations (a). Mixed modes may alternate apparently at random (chaotically) when the flow rate is high (b).

ate chaotic mixed modes, so this seems to be a *bona fide* example of a new route to chaos.

Rhythms of life

Chemical waves are not merely a curiosity conjured up in laboratories under highly specialized conditions. For many living organisms, ourselves included, they are a matter of life and death.

Heart attacks are the leading cause of death in industrialized nations. The majority of these result from a pathological condition of the heart called ventricular fibrillation—a medical term which, roughly translated, means that the heart forgets how to beat. Instead of acting in a coordinated manner to generate a regular pumping motion, the tissues of a heart that has entered into ventricular fibrillation lose their ability to execute large-scale coordinated contractions, and the heart appears to flutter feebly to no great effect, like a frightened bird. During the onset of ventricular fibrillation, the heart enters a kind of behaviour called cardiac arrhythmia, which, despite the name, actually denotes a new rhythmic activity in which the regular beats of about one per second give way to rapid pulsations about five times faster. These eventually dissolve into uncoordinated fibrillation, leading to sudden cardiac death. That eventual heart stoppage in such cases is preceded by this frenzied activity was recognized as early as 1888, when J.A. MacWilliam described the fateful events in colourful terms: 'The cardiac pump is thrown out of gear, and the last of its vital energy is dissipated in a violent and prolonged turmoil of fruitless activity in the ventricular wall'. These changes in heart activity can be seen in electrocardiograms, which record the change in electrical voltage in a region of the heart tissue (Fig. 3.22).

How do cells in a healthy heart act in synchrony in the first place? Each heartbeat corresponds to a travelling wave of electrical activity, which begins at a pacemaker region of the heart called the sinoatrial node and travels throughout the heart tissue. At the front of this travelling wave, the electrical voltage across the cell walls alters as electrically charged ions move from one side to the other. The wave of electrical activity (which is akin to a nerve impulse) induces muscle contraction, causing the heart to pump blood. Once the wavefront has passed, the cells become refractory (immune to a further pulse of electrical activity), while they 'reset' their across-membrane voltages by redistributing the ions. Thus, heart tissue is an excitable medium, and the heartbeat is induced by a spatio-temporal pattern—a travelling wave—very much akin to that of the BZ reaction. This is one of the major reasons why the BZ reaction has attracted such interest: scientists are interested in the patterns not only for their own sake (pretty as they are) but because they might provide us with a model to help understand some aspects of heart behaviour.

It now seems clear that the fatal condition of ventricular fibrillation is associated with the initiation of spiral waves in the heart. You can see from Plate 5 that spiral travelling waves in an excitable medium tend to have a shorter periodicity than target waves (adjacent wavefronts are closer together). The consequence of this is that, once they are created, spiral waves come to dominate over target waves, because they 'jump in' to excite the medium more quickly. A clue to the role of spiral waves in ventricular fibrillation (VF) is given by the fact that, when cardiac arrhythmia is initiated, the frequency of the heart's oscillations increases.

It is now possible to see these lethal spiral waves directly in beating hearts. Early experiments involved

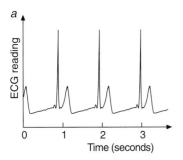 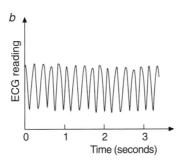 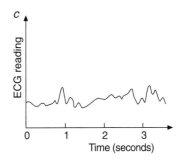

Fig. 3.22 The onset of ventricular fibrillation is evident in electrocardiograms as a transition from regular heartbeats a few seconds apart (*a*), to rapid oscillations that flutter several times a second (*b*), a condition called cardiac arrhythmia. Finally this regular behaviour dissolves into the feeble, uncoordinated behaviour of ventricular fibrillation (*c*), which is quickly fatal if not arrested.

studies of fragments of heart tissue, kept 'active' by mimicking physiological conditions: suspending the tissue in a salt solution and supplying nutrients (glucose) and oxygen. In such an experiment, researchers from the University of Amsterdam were able in 1972 to see evidence of a wave of electrical activity spinning like a turbine blade, at ten revolutions per second, in a piece of rabbit heart tissue. Subsequent studies showed these pirouettes ever more clearly, but it was not until 1995 that they were revealed in whole hearts. Richard Gray and co-workers at the Health Science Center of the State University of New York showed that a single rotating spiral wave could give rise to VF in whole rabbit and sheep hearts sustained in a culture medium. They followed the patterns of electrical activity propagating through the hearts by adding to the artificial 'blood' supply a dye that emitted fluorescent light whose brightness was a measure of the local voltage. The researchers saw spiral waves of electrical activity whose centres meandered over the surface of the heart (Fig. 3.23a). The electrocardiograms associated with this behaviour showed the uncoordinated oscillations characteristic of VF (Fig. 3.23b).

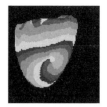

a

b

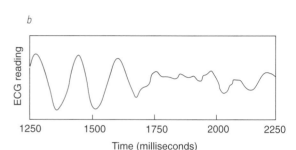

Fig. 3.24 A spiral wave in a numerical model of the heart (a), and the associated electrocardiogram trace (b). (Image: Richard Gray.)

The rotating spiral waves in these experiments are not easy to see in the spatial maps of electrical activity, at least without a trained eye. But Gray and his colleague José Jalife carried out simulations of the heart behaviour on a computer, using equations that were known to describe the basic properties of heart activity. In their simulations, the spirals were very clear (Fig. 3.24a and Plate 7), and when the model was adjusted to allow the tips of these waves to meander, the simulated electrocardiograms were very similar to those seen in the real sheep hearts (Fig. 3.24b).

How might spiral waves arise out of the regular travelling waves found in healthy hearts? It seems likely that obstacles of 'inert' tissue, such as damaged tissue caused by blood clots, can act as the initiating points for such waves. Travelling wavefronts have to pass around such obstacles, which are immune to excitation themselves, and this can make a steadily propagating wave begin to rotate. Just this kind of behaviour has been seen in the BZ reaction when some obstacle like a physical barrier is placed in the way of a target wave (Fig. 3.25). In the heart, small areas of damaged tissue don't obviously present much of a threat in terms of impeding the flow of blood or the regular contraction of the rest of the heart. But this research on travelling waves shows that if they trip up the heart's spatio-temporal patterns, the result can be a switch from health to a life-threatening condition.

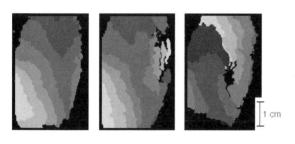

a

b

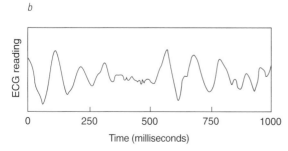

Fig. 3.23 A spiral wave developing in a rabbit heart, traced out by monitoring voltage-dependent dyes (a), and the associated electrocardiogram trace (b). (Image: Richard Gray, State University of New York Health Science Centre.)

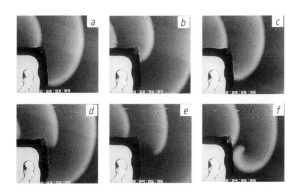

Fig. 3.25 The initiation of a spiral wave when a travelling wave in the BZ reaction encounters a barrier. (Photo: James Keener, University of Utah; from Agladze *et al.* (1994). *Science* **264**, 1746.)

Microbial crosstalk

The cells in any organ need to stay in touch and act in a coherent way, and in both the heart and the brain these interactions can result in regular oscillations and spatio-temporal patterns that are an important characteristic of the organ's healthy functioning. But of course some cells get by just fine on their own—bacteria, which are single-celled, are beyond doubt the most successful organisms on the planet. If we pride ourselves in surviving under conditions ranging from the burning Gobi desert to the frozen Arctic tundra, we should recapture some humility by observing that bacteria are to be found also in hot oil hundreds of feet below the ground, in superheated water around miniature submarine volcanoes, and amidst toxic radioactive waste.

But even for organisms as adaptable and resilient as bacteria, it sometimes pays to cooperate. For even bacteria need the basics—food, water, warmth. When these things become scarce, some bacteria adopt a remarkable survival strategy, in which we can see nature's own version of the BZ pattern-forming system. If times are tough—if food becomes scarce, say—evolution has taught the bacteria that they are then better off foraging in groups. So it becomes necessary for each bacterium to tell the others—who are blind, deaf and dumb—where it is.

Bacteria have evolved a clever way of doing this: they perfume themselves. The cells emit a chemical compound, called a chemoattractant, into the medium around them, much as animals emit pheromones to attract mates. Other bacteria can sense how much of this chemical signal is coming their way from different directions, and they start to move in the direction where

the concentration of the chemical rises most rapidly— in other words, they wriggle along the steepest gradient in chemoattractant concentration. This chemically stimulated movement is called chemotaxis.

Chemotaxis is not unique to bacteria, but is employed by other single-called organisms too, and is also utilized by our own body cells to form complex structures such as neural dendritic cells in the brain. It is a versatile mechanism for pattern formation, and nowhere is

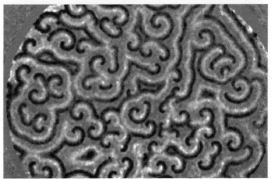

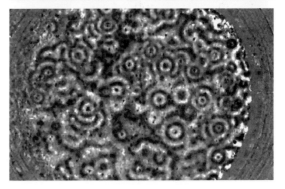

Fig. 3.26 Target and spiral patterns in colonies of the slime mold *Dictyostelium discoideum*. These patterns are generated when some cells emit periodic pulses of a chemical attractant, towards which other cells travel. (Photo: Cornelis Weijer, University of Dundee.)

this better illustrated than in colonies of the slime mold *Dictyostelium discoideum.* These single-celled organisms start up their chemical cross-talk when deprived of heat or moisture, and this communication allows them to aggregate into multicellular bodies that are more fit to survive hardship. Certain cells, called pioneer cells, release pulses of the compound cyclic adenosine monophosphate (cAMP), and nearby cells then follow this chemical trail to its source. Once a cell has emitted a burst of the chemoattractant, it falls silent for several minutes, as if recuperating from the exertion. This refractory period means that, when a slime mold colony starts to undergo chemotaxis, it behaves as an excitable medium. We saw earlier that the target and spiral waves of the BZ reaction turn out to be the *generic* patterns of excitable media, and so it may come as no surprise that the slime mold shows these patterns too (Fig. 3.26).

But the patterns are just a passing phase. Ultimately they fragment into distinct, branched islands of cells. The spirals and targets are simply a way of getting the self-organization and aggregation underway. Eventually the cells converge into isolated clumps, which thenceforth act cooperatively. Yet that is by no means the end of the story: the clumps pile up into mounds which can be regarded as individual multicellular organisms. Cells in different parts of the mound 'differentiate', which means that they are no longer equivalent but act out different roles. Each mound develops a bulbous body—the 'fruiting body'—perched on a long stalk. The fruiting body contains many spores, which can survive without food or water until such time as these become available again.

The initial stages of this aggregation process, involving travelling waves, can be modelled on the basis of precisely the same principles as the BZ reaction. The 'pacemaker' of the chemical waves is a biochemical reaction through which the signalling molecule cAMP is manufactured by the pioneer cells. cAMP is made from a molecule called adenosine triphosphate (ATP) in a reaction catalysed by an enzyme called adenylate cyclase, which is attached to the inside of the cell mem-

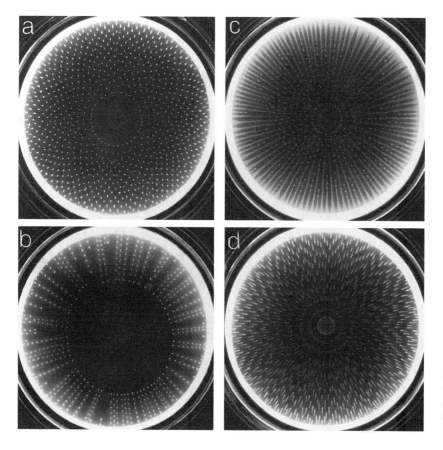

Fig. 3.27 Symmetrical patterns formed by the bacteria *Escherichia coli* in response to chemical signals. (Photos: Elena Budrene, Harvard University.)

brane. The cAMP diffuses out of the cell through the membrane. There it can continue to diffuse into the surrounding medium and excite nearby cells; but alternatively it can become involved in a feedback loop that regulates the formation of more cAMP. In the latter case, cAMP outside the cell interacts with another protein molecule in the cell membrane in such a way that the protein is stimulated into influencing the catalytic activity of the adenylate cyclase enzyme. In this way, cAMP produced by the enzyme can enhance the rate at which further cAMP is formed. This autocatalytic behaviour can give rise to bursts of cAMP production followed by quiescent periods.

Bacterial black holes

In 1991 Elena Budrene and Howard Berg of Harvard University expanded the repertoire of patterns that chemotaxis was known to generate. They reported astonishing patterns that developed in colonies of *Escherichia coli*, bacteria that live in the human gut, when the colonies grew in a semi-solid agar gel under life-threatening conditions—lack of food, too great an oxygen concentration, the presence of molecules that disrupt protein manufacture in the cells, or even just coldness (Fig. 3.27). Unlike the travelling waves of *Dictyostelium*, these patterns remain stable for long

periods. Budrene and Berg realized at once that the patterns were the result of chemotactic signalling between the bacteria, which emit a chemoattractant called aspartate under stress. But their sheer complexity and variety—much richer than the familiar targets and spirals of *Dictyostelium*—baffled everyone.

Nonetheless, it seemed likely that the patterns were again the result of a competition between a handful of basic processes: cell multiplication by division, cell migration (diffusion) in search of food, and cell clustering by chemotaxis once the local density of cells (and thus the local rate of chemoattractant formation) exceeds a certain threshold. Eshel Ben-Jacob and colleagues from Tel Aviv University in Israel attempted to reproduce the spot patterns in a cellular automaton model that captures these basic features of the bacteria's behaviour. I will describe this model in more detail in Chapter 5, where I show that it can mimic many of the branching patterns formed by growing bacterial colonies. In essence, the model postulates groups of bacterial cells that move *en masse*, consuming food, reproducing and emitting a chemoattractant if food becomes scarce. The model generates an expanding ring of cells, which cluster into spots behind the advancing front when they attract one another through chemotactic signalling (Fig. 3.28a). The spot patterns become

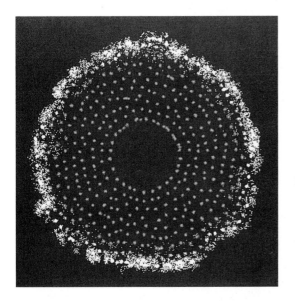

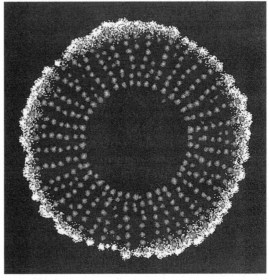

a

b

Fig. 3.28 Concentric spots (a) and radial patterns (b) of clustering bacteria can be reproduced in a cellular automaton model in which groups of cells migrate, reproduce, and respond to attractive and repulsive chemical signals from one another. (Images: Eshel Ben-Jacob, Tel Aviv University.)

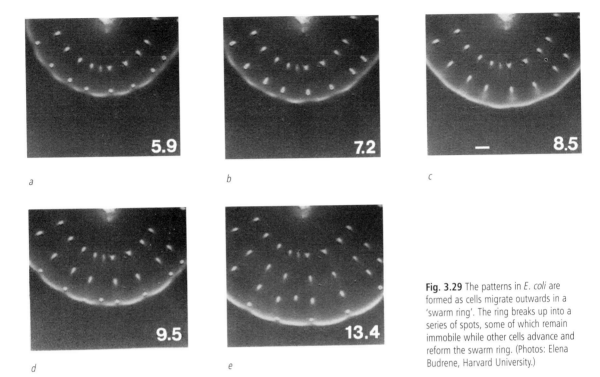

a   *b*   *c*

d   *e*

Fig. 3.29 The patterns in *E. coli* are formed as cells migrate outwards in a 'swarm ring'. The ring breaks up into a series of spots, some of which remain immobile while other cells advance and reform the swarm ring. (Photos: Elena Budrene, Harvard University.)

aligned along radial lines, as in the experiments, if a further component is added to the model: a repulsive interaction between the bacterial clusters, resulting from their emission of a chemical signal that warns other clusters to stay away. Radial spots then emerge from a delicate balance between attractive and repulsive interactions (Fig. 3.28*b*). Whether real *E. coli* bacteria exude a chemorepellent of this sort is, however, an open question. At the same time, Herbert Levine and Lev Tsimring at the University of California at San Diego proposed an alternative model that included the same kinds of interactions but described the bacteria's motions using reaction–diffusion equations akin to those first studied by Ronald Fisher in the 1930s, instead of invoking discrete cellular automata. This model also generates rings that break up into clusters behind the advancing front of the colony.

Budrene and Berg were meanwhile taking a closer look at what each clump of cells gets up to. In 1995 they reported that the symmetric patterns are the result of a very complex process in which the cells alternate between forging out into new territory and stopping to cluster into groups. The whole process, watched in time-lapse video, resembles a peculiar cycle of boldness and indecision.

First the bacteria disport themselves in a ring, which moves outwards from the central source (Fig. 3.29). Budrene and Berg call this a swarm ring, and it takes about a day to expand to the edges of a Petri dish. In order to produce the chemoattractant (aspartate), the cells need to be provided with the compound succinate in the culture medium—they carry out enzymatic chemical transformations which turn succinate into aspartate. If the amount of succinate is very low, there is little chemotaxis and the ring simply expands uniformly. But at higher concentrations of succinate, the swarm ring collapses suddenly (within a few minutes) into a series of more or less equally spaced clusters of cells around its perimeter, as if the advancing party of bacteria has paused and formed into little groups to discuss their next move. This instability of the swarm ring happens when the cells, which are excreting aspartate as they advance, have finally made enough of this attractant to trigger its spontaneous break-up into clusters, a process that breaks the circular symmetry.

Once one cluster forms, the parts of the ring to either side become unstable and clustering propagates in a kind of 'domino effect' all around the ring. But some cells are not drawn into these clusters, and they head out again in the swarm ring, multiplying as they go. As this cycle repeats, a trail of clusters in radial lines is created behind the advancing ring (Fig. 3.29e).

Although this sequence of events is not unlike that seen in the models of Ben-Jacob, Levine and their colleagues, Elena Budrene was not convinced that either represents a 'minimal model' whose ingredients are nothing more than the known biological properties of the bacteria. With colleagues Michael Brenner and Leonid Levitov from the Massachusetts Institute of Technology, she has developed a different model which suggests that patterning can result from the interplay of the following processes:

1. Diffusion and multiplication of the bacteria.

2. Production of the aspartate chemoattractant.

3. Diffusion of aspartate.

4. Consumption of succinate to make aspartate.

5. Chemotactic drift of cells towards regions of high aspartate concentration.

The researchers used Fisher-type reaction–diffusion equations to describe each of these processes. They showed that the expansion of the swarm ring is driven by the consumption of succinate: the cells move progressively outwards as they deplete all the succinate in the medium behind the ring. This drive to find fresh succinate even outweighs the lure of the chemoattractant, which serves simply to ensure that the ring remains a well defined circular band instead of getting smeared out by diffusion of the cells.

The instability leading to break-up of the ring into clusters, meanwhile, emerges from the model as a consequence of 'singular' solutions of the equations. These arise from a positive feedback process: above a certain threshold density of cells, they collectively produce so much attractant that any slight irregularity in density is rapidly amplified until the cell population becomes focused into a sharply defined, very dense point. That this instability might occur in chemotactic populations was first suggested in 1973, and it was later dubbed 'chemotactic collapse'. It can be loosely compared to the formation of an astrophysical black hole: as the density increases, so it encourages the accretion of even more mass, until the density at the central point blows up and, in the equations, becomes infinite. Of course, the cell density in these bacterial black holes cannot *really* become infinite; ultimately it is halted by the depletion of some essential substance such as oxygen in the 'singularity'.

So Budrene, Brenner and Levitov proposed that chemotactic collapse leads to the break-up of the swarm rings into a series of spots. They suggested several reasons why their model might provide a more realistic description of the process than the earlier ones. For one thing, it accounts for the marked difference in observed timescales for swarm-ring expansion and break-up. And it doesn't require any biochemical processes other than those already known. Furthermore, the formation of the patterns does not require fine-tuning of the model parameters—the clustering instability is a general and robust property of the equations, because the singular solutions are catastrophic and so pretty insensitive to the fine details.

One of the difficulties of modelling these bacterial tapestries is that they are so diverse. The bacterium *Salmonella typhimurium* has its own palette of patterns, including spots, concentric rings and spotty rings, while *E. coli* also displays amazing chevron structures (Plate 8). We will see still more, very different, patterns in Chapter 5. James Murray and Rebecca Tyson of the University of Washington, along with S.R. Lubkin of North Carolina State University, have proposed a model that purports to capture all of the pattern-forming mechanisms, and many of the patterns, seen in *E. coli* and *S. typhimurium*, while remaining faithful to the known biology. Their model is again based on reaction–diffusion, incorporates equations describing the diffusion and uptake of the nutrient (succinate) and chemoattractant (aspartate) and the migration and proliferation of cells. The researchers suggest that their model does more than just generate patterns similar to those observed in the experiments (you can make a reaction–diffusion model do just about anything, as we'll see in the next chapter)—it reproduces some of the features of the *way* the patterns appear, a hopeful indication that the visual coincidences are down to more than fortuity. But while the details will continue to be debated, it seems highly likely that a reaction–diffusion mechanism of some kind is capable of accounting for these bacterial kaleidoscopes.

In the beginning

Chemical waves may be with us from birth. Certainly that is the case if one is a frog: following fertilization, pulses of calcium ions are released over the surface of the egg. The reason for these calcium waves isn't entirely clear, but they are probably a kind of chemical signal that, like a radio wave, carries information through space encoded in its amplitude and frequency. In 1991 David Clapham from the Mayo Foundation in Rochester, New York, and colleagues observed *spiral*

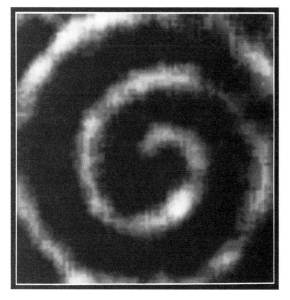

Fig. 3.30 Spiral waves of calcium travel across the surface of frog eggs when they are fertilized. The purpose of these waves in embryo development is as yet unclear. (Photo: David Clapham, Mayo Foundation, Rochester.)

waves propagating and annihilating on the surface of fertilized frog eggs (Fig. 3.30). These eggs behave as an excitable medium, like the heart, or like the plain old Belousov–Zhabotinsky brew.

Fig. 3.31 The spiral galaxy NGC 5236 in the southern sky. The structure of some spiral galaxies may result from a star-formation process with the characteristics of a reaction–diffusion system. (Photo: European Southern Observatory.)

At the other end of the scale, astronomer Lee Smolin at Pennsylvania State University has suggested that the formation of some spiral galaxies (Fig. 3.31) can be regarded as a kind of reaction–diffusion process. These cosmic spirals are so familiar to us now that it is worth remembering that not all galaxies are spirals: some are smooth elliptical blobs, some are lens-like disks (lenticular) and others have irregular shapes that defy easy categorization. But it is clear that galaxies have a strong tendency to form rotating spirals (our own Milky Way galaxy is of this type), and Edwin Hubble proposed in the 1920s that this is the pattern towards which more featureless elliptical galaxies evolve, via lenticular forms.

Galaxy structure has been studied for a long time, but there is still much to debate. In the late 1920s Bertil Lindblad suggested that the spiral structure was a natural consequence of the interplay between rotation and the gravitational interactions of the stars, but it wasn't until the 1960s that C.C. Lin and Frank Shu showed how a spiral galaxy formed in this way could avoid being pulled apart by centrifugal forces as it rotates. They proposed that the spiral arms are not rigid (stars are not either in them or out of them) but are waves of enhanced stellar density that sweep through the rotating disk.

Smolin points out that there are some spiral galaxies in which this idea of a density wave doesn't hold up: the spirals are instead tracing out regions of enhanced star formation. Why should stars form in one place in the galactic disk but not in another? Smolin proposes that the rate of star formation is influenced by positive and negative feedbacks. The positive feedback (autocatalysis) comes from dust produced in the atmospheres of stars that have formed already, which helps the clouds of interstellar particles to condense into new stars. Negative feedback comes largely from the ultraviolet light emitted by existing stars, which heats up the interstellar medium and makes it less likely to condense. In addition, Smolin argues that certain aspects of the star-formation process give it characteristics akin to a diffusion process. The galaxy then becomes a reaction–diffusion system, with all the pattern-forming potential that entails—making it permissible to view these cosmic pinwheels as gargantuan relatives of the whorls on a frog's egg.

Yes, say hello to life's universal patterns. We'll be seeing more of them.

BODIES

… and after another long time, what with standing half in the shade and half out of it, and what with the slippery-slidy shadows of the trees falling on them, the Giraffe grew blotchy, and the Zebra grew stripy, and the Eland and the Koodoo grew darker, with little wavy grey lines on their backs like bark on a tree trunk; and so, though you could hear them and smell them, you could very seldom see them, and then only when you knew precisely where to look.

Rudyard Kipling
The Just So Stories

When Rudyard Kipling explained how the animals of Africa acquired their markings, he was tapping into a universal mythology. The Native Americans, for instance, have their own tales of how the skunk became endowed with its two-tone tail. When people see patterns in nature as striking as these, they want some means of explaining them.

Darwin's theory of evolution gave us an answer for the modern age, and it was not so different in essence from Kipling's: the patterns enhance the animal's chances of survival. In the tree-dappled light of a tropical forest, a spotted leopard can merge with the surroundings, giving it a better chance of sneaking up on its prey. A striped zebra is better hidden amongst the vertical striations of the long grass and bushes on the veldt (Fig. 4.1), and an insect patterned to resemble a flower is at less risk from predators. In addition, patterns help animals to recognize other members of their species, an obvious requirement if the species is to propagate.

But as I explained in the introduction, this kind of explanation, while correct, is incomplete and ultimately rather unsatisfying. I shouldn't be surprised if some of you prefer the 'just so' explanation, which at least has something to say about how the stripes and spots *got there*. (The leopard's spots, you may recall, are the fingerprints of a solicitous Ethiopian.) A Darwinian explanation says nothing about this—it merely suggests that, once there, these patterns will stay because the animal is better off with them. Can it be that evolution

Fig. 4.1 Zebras in Kenya. (Photo: Michael and Sandra Ball.)

found its tortuous way to the striped zebra via a slow, gradual and random accumulation of mutations towards stripe-giving genes? What, then, were the intermediate stages? Were there proto-zebras that exhibited some other kinds of pattern? Darwinism in itself provides little guidance for answering such question—it does not allow us to deduce which patterns are possible and which are not, but simply provides a mechanism that explains why some persist in certain species but not others (differences in habitat, for example). Natural selection does not tell us what is on the palette; it is a tool for retrospective rationalization, and rarely if ever for prediction. Does nature really have an infinite choice of skin patterns, or must it select from just a few? And how do each of those arise?

These are some of the questions that I will look at in this chapter. We shall see that the puzzle of surface markings on organisms is ultimately bound up with the much broader matter of how bodies themselves are shaped, and why they take on the forms that they do. In a sense, similar mechanisms may be at play in both cases.

Yet while surface markings, like those on pelts and shells, are immediately recognizable as patterns, we might imagine that the *shapes* of bodies are functional rather than representational. Specific features would seem to be dictated by specific functions, they are tools rather than flags or camouflage. Surely we have eyes and hands not because these are elements of a body's 'pattern' as such but because we need them to see and to manipulate our environment? To put this another way: a particular stripe of a zebra's skin marking is merely a consistent part of the pattern, whereas (setting aside the bilateral symmetry of the body) a particular limb, like our left arm, doesn't obviously 'follow' from the form of the rest of our bodies—it seems to be a specialized element, not a generic one. We will see, however, that it can make a kind of sense to regard our body plans as biological patterns, albeit very complex and refined ones that derive from the sequential and hierarchical sub-patterning of simpler patterns. That this is so perhaps becomes more apparent by considering the forms of plants, which tend to have a higher degree of symmetry than mammals. A plant too has limbs, but these are much more obviously arranged in a regular, somewhat symmetrical manner. And it isn't so hard to identify animals to which the same applies—a starfish, for instance. There is a fertile tension inherent in the question of whether the form of living organisms should be regarded either as a haphazard assembly of components that together make up an evolutionarily viable being, or instead as a highly complex, spontaneously patterned form.

Frozen waves

If I am to be chronological, we must begin this exploration of biological pattern formation with just about the hardest question one could ask of it: how a body plan emerges in a fertilized egg. But the British mathematician Alan Turing, who considered this problem in the early 1950s, was one of the brightest scientific minds this century has seen, and didn't have much fear of hard questions.

Turing's work takes a decidedly tangential angle to much of the mainstream research on biological development, although in recent years the two points of view have shown signs of converging. We know that different tissues and organs in a multicellular body are characterized by differences in the genes that are 'active' in their constituent cells—basically, the cells of (say) the liver make use of some different genes (to manufacture different proteins) from the cells of bone marrow, even though the genetic content of both is the same. The cell types in these organs are said to be *differentiated*, and to show different regimes of gene *expression* (gene-to-protein conversion). Geneticists interested in development commonly seek to identify these differences in gene expression, and to investigate the protein products to see what role the proteins play in the function of the tissues.

This is a reductionist approach that helps us to understand the consequences of cell differentiation. It also shows us where to look for explanations of the origin of that differentiation: in the switching on or off of certain genes. But it is an approach that runs out of steam when we pare the problem of development back to its starting point. For in the beginning, an embryo is just a ball of identical cells, each with the same genetic constitution. How do the initial differences in gene expression arise from this uniform ball?

What we are faced with here is a question of symmetry breaking: somehow the (roughly) spherical symmetry of the multicellular embryo gets broken such that different parts of it follow different developmental pathways to become a head, a heart, a toe. What breaks the symmetry of the embryo? This is the fundamental question of morphogenesis, the study of the development of biological form.

At the beginning of the 1950s, symmetry breaking was not a new idea in physics, but no one had given much thought to how it might be relevant in chemistry or biology. At that time, Alan Turing was working on mathematical problems associated with computer theory at the University of Manchester. His work in this field was to become seminal, and underpins much of the present-day research into artificial intelligence. During the Second World War Turing was set to work as a code-breaker, and some of the techniques that he developed for unravelling German naval messages are still classified today. For his contributions in this area Turing was held in high esteem by the British intelligence organization, but his knowledge was also regarded as a national secret.

Turing's ultimate dream was to make a thinking machine, an artificial brain. His interest in brain structure and development led him to ponder on broader questions of biological development, and ultimately to the issue of morphogenesis. In 1952 he published a paper describing a hypothetical chemical reaction that could generate spontaneous symmetry breaking, leading to stable spatial patterns, in an initially uniform mixture of chemical compounds. This, he suggested, might provide a model for how patterning takes place in an initially spherical fertilized egg. Entitled 'The chemical basis of morphogenesis', this paper is undoubtedly one of the most influential in the whole of theoretical biology.

One of the remarkable things about Turing's idea was that he proposed that diffusion of the chemical species (called morphogens) through the medium in which they were dispersed could be the driving force for symmetry breaking. This goes against intuition: normally diffusion is seen as a mechanism for producing uniformity, for smoothing out inhomogeneities in a system. It was almost as if he was suggesting that diffusion could cause thoroughly dispersed ink in water to condense into concentrated ink droplets, rather than the reverse.

But in Turing's chemical system, diffusion is acting in competition with another process, namely an auto-catalytic chemical reaction. It is, in other words, an example of a reaction–diffusion system. We saw in the last chapter how these systems can generate non-stationary patterns, such as the travelling target and spiral waves of the Belousov–Zhabotinsky (BZ) reaction, when the system possesses the property of excitability. Turing showed that under certain conditions, *stationary* patterns can also arise in excitable reaction–diffusion systems. These generally take the form of spots or stripes of differing chemical concentrations.

Turing considered a process in which some chemical compound, say A, undergoes an autocatalytic reaction to generate more of itself: the rate at which A is generated depends on the amount of A already present. But within his scheme, A also activates in some way the formation of a compound B that inhibits the formation of more A. The key element for obtaining spatial patterns is that A and B diffuse through the reaction medium at different rates, so the effective ranges of their respective influences are different. This means that the A's and B's can dominate in distinct regions.

When Turing formulated this scheme, he had to rely on mathematics that could be performed with a pencil and paper, and so he chose to represent the chemical processes using the simplest mathematical equations possible. This forced upon him some compromises to get around the rather artificial results that these simple equations could sometimes generate. In particular, Turing had to make his equations *linear*, which implies that effects are proportional to their causes. Linear equations are easier to solve than non-linear ones, but in Turing's case it meant that the solutions—the distributions of his hypothetical chemical species—were unstable against perturbations. Although he was able to show that the scheme was capable of generating spatial patterns, Turing clearly felt hindered by the intractability of a more sophisticated analysis. He suggested that, while the difficulties were probably too great to allow for any all-embracing theory of pattern formation in these schemes, perhaps a 'digital computer' would enable one to investigate a few particular cases more accurately.

But it was only after his landmark paper was published that Turing seems to have begun to perceive the real key to his patterning mechanism, which is that it represents a competition between *activation* by compound A and *inhibition* by compound B. Moreover, the inhibitor B must diffuse more rapidly than A for patterning to occur. Thus, while activation and auto-catalytic production of A is a localized process, inhibition of A by B is *long-ranged*, because once formed in the vicinity of A, B can rapidly diffuse away to inhibit the formation of A elsewhere. At the same time, this rapid diffusion of B ensures that it does not inhibit the local formation of A—it is removed from the vicinity too quickly (Fig. 4.2). The whole scheme represents a subset of reaction–diffusion processes called activator–inhibitor systems.

What happens, then, is that once random fluctuations in the initial concentration of A trigger spots of enhanced A production through autocatalysis, the inhibitor B is produced and rapidly diffuses away to

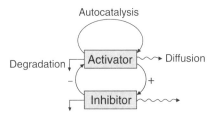

Fig. 4.2 How an activator–inhibitor scheme works. The activator generates more of itself by autocatalysis, and also activates the inhibitor. The inhibitor disrupts the autocatalytic formation of the activator. Meanwhile, the two substances diffuse through the system at different rates, with the inhibitor migrating faster.

suppress formation of the activator in the immediate surroundings. So an array of isolated spots of A is created, surrounded by regions rich in B in which formation of A is suppressed (Fig. 4.3a). If there are processes that remove the end products of the reaction at a steady rate and supply fresh sources of the reactants needed to make A and B, this pattern of spots can remain stable indefinitely. Alternatively, the regions of A production can merge into stripes that trace out a maze-like network (Fig. 4.3b).

Turing himself never used the terms 'activator' and 'inhibitor' however; instead, he regarded compound B as a rapidly diffusing 'poison'. It was not until 1972 that

Hans Meinhardt, then at the Max Planck Institute for Virus Research in Tübingen, Germany, and his colleague Alfred Gierer had the insight that short-ranged activation and long-ranged inhibition are the principal elements of Turing's patterns. With the benefit of computers to do the number-crunching, Meinhardt and Gierer were able to formulate Turing's mechanism using more complicated—and more physically motivated—non-linear equations, and to show how they could be plausibly related to the kinds of processes known to take place during real biological patterning and development.

Because of their stationary nature, it is tempting to regard Turing patterns as a kind of end product of the reaction, as if they were spatial differences in composition that have become 'frozen in' to the reaction mixture just as bubbles become frozen into ice. This is not so, however. The patterns are non-equilibrium patterns, and are *dynamic* in the sense that they are sustained by constant motion and reaction of the chemical compounds in the mixture. Similarly, the oscillation inside an organ pipe excites a fixed pattern of varying air density even though the molecules in the air continue to move around. The point is that Turing's patterns are maintained—the symmetry is broken—only so long as the system is driven *away* form equilibrium. They arise

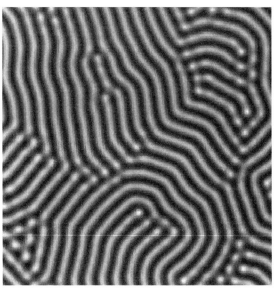

a   *b*

Fig. 4.3 The activator–inhibitor scheme can generate disordered patterns of spots and stripes. The composition of the system is different in the light and dark regions. (Images: J. Boissonade, University of Bordeaux.)

spontaneously from a homogeneous medium in a symmetry-breaking instability as the driving force away from equilibrium is increased.

By curious coincidence, Turing formulated his ideas on pattern formation in chemical systems at just the same time that Boris Belousov in the Soviet Union was discovering the peculiar oscillatory behaviour that some chemical reactions can exhibit. It was in 1951 that Belousov first observed the colour-changing properties of a chemical cocktail that was later to be refined into the BZ mixture. As we saw in the previous chapter, Belousov received scant reward for this discovery during his lifetime; but for his own contribution to chemical pattern formation, Turing was to see even less recompense. In the same year that he published his paper on pattern formation, he was arrested for sexual offences when an investigation into a burglary at his home led to the disclosure that he was homosexual. This supposed crime was enough for Turing to be compelled to take 'corrective' hormone treatment, and to be regarded as a security risk, placing restrictions on his freedom to travel. In 1954 the disgrace and constraints deriving from these charges led the 42-year-old Turing to commit suicide by ingesting cyanide.

Making striped paint

For almost twenty years after Turing published his paper, nothing happened. No new field of chemical pattern formation was born. Biologists studying morphogenesis took no heed. Yet today Turing's ideas are hailed as seminal. Why the delay?

For one thing, Alan Turing was years ahead of his time. The buzzwords that encapsulate the behaviour in his theoretical reaction–diffusion system—complexity, pattern formation, symmetry breaking, non-equilibrium systems, non-linearity—were either unheard of or regarded as rather specialized and obscure branches of science in the 1950s. It is only when the climate is right, when a sea change has taken place, that a genuinely new scientific idea can find wide acceptance. In the 1950s, almost all of chemistry was concerned with equilibrium processes, and most of mathematical science looked at non-linear problems with horror.

But there was also the matter of whether Turing's theoretical ideas had any relevance to the real world. He had to make a number of approximations to make his equations manageable, leaving open the question of whether real chemical systems would indeed show this kind of behaviour. Creating Turing patterns in a real reaction proved to be a tremendous challenge, and for a

time it began to look as if they might be merely mathematical phantoms. As for the relevance to morphogenesis—that is still an open question, although as we shall see, there are compelling reasons to believe that some of the exquisite surface markings of animals are Turing patterns or something very much like them.

When in the late 1960s and early 1970s chemists began to learn about the BZ reaction, the connection between this and Turing's reaction–diffusion scheme began to emerge. The connection was made more concrete in 1971 when Zhabotinsky (in Puschino, USSR) and Arthur Winfree (then at the University of Chicago) independently observed spiral waves in the BZ reaction. Winfree subsequently showed that the spiral is the result of an activator–inhibitor pattern-formation process in the poorly mixed BZ mixture. But although the BZ patterns are comparable to Turing structures insofar as they are both the result of non-linearities, autocatalysis and feedbacks in the chemical reactions that produce them, the mechanism that gives rise to BZ chemical waves is *not* the same as the instability that leads to Turing structures. For one thing, the BZ patterns are *travelling* waves, whereas Turing's patterns are stationary. In other words, the combination of localized activation and long-ranged inhibition is not by itself sufficient to guarantee stationary Turing patterns. They will be produced only if the response of the inhibitor to changes in the activator concentration is rapid (which means, in effect, that the processes that remove the inhibitor must be fast relative to those that remove the activator). If, on the other hand, the inhibitor sticks around for a long time, the system has a tendency to undergo BZ-like oscillations—and such oscillations can occur even if there is no difference between the rates of diffusion of the various chemical species, whereas such a difference is essential to Turing's mechanism. Travelling waves are generated, meanwhile, if the inhibitor diffuses less rapidly than the activator. And these waves need to be initiated by some local disturbance to the medium, rather than arising from a spontaneous, 'global' symmetry-breaking instability. From the perspective of morphogenesis these differences are critical, because Turing was looking to explain how stable, fixed structures can arise spontaneously in embryos.

In 1968 Ilya Prigogine and René Lefever from the University of Brussels, stimulated by the Prague conference at which the BZ reaction was given its first public airing to Western scientists, proposed a hypothetical scheme for an oscillatory reaction. This model, later

Box 4.1: The Brusselator

There are four steps in the Brusselator, which involve inter-conversions of molecules A, B, C, D, X and Y. A and B are the reactants, C and D the products, and X and Y are intermediates in this transformation:

$$A \rightarrow X \tag{4.1}$$

$$B + X \rightarrow C + Y \tag{4.2}$$

$$2X + Y \rightarrow 3X \tag{4.3}$$

$$X \rightarrow D. \tag{4.4}$$

Notice that equation 4.3 is autocatalytic, since two molecules of X give rise to three. Less obviously, it is also susceptible to inhibition, since the compound Y is needed for the reaction to occur but is consumed in the process. In other words, the fact that the reaction generates more X does not necessarily result in exponential growth because this X cannot itself undergo equation 4.3 if there is no Y left in the vicinity. Thus the rate at which Y diffuses to 'feed' equation 4.3 is crucial to the whole process, and this has the same significance as the diffusion of the inhibitor species in Turing's scheme.

called the Brusselator (see Box 4.1), possesses strong similarities to the Oregonator developed in 1974 by chemists at Oregon State University to account for the BZ reaction (page 55). Because it includes autocatalytic feedback, the Brusselator shows oscillations and bifurcations like those of the BZ reaction. But when it proceeds in an incompletely mixed system in which the various chemical species diffuse through the medium at markedly different rates, instabilities can occur that give rise to spatial patterns—variations in composition from place to place—which take the form of stationary Turing-type stripes and spots.

The hypothetical Brusselator scheme made Turing's ideas a little more concrete, and showed that oscillatory reactions similar to the BZ reaction (though not the BZ reaction itself!) might be able to produce Turing patterns under the right conditions. But it was not until 1990 that this was demonstrated experimentally. Patrick De Kepper and co-workers at the University of Bordeaux carried out an oscillatory chemical reaction involving chlorite and iodide ions and malonic acid in a thin layer of gel that was continuously fed from opposite directions with fresh reagents. This reaction, called the CIMA reaction, was developed by De Kepper and colleagues in the early 1980s as an alternative to the BZ reaction. Its oscillatory and pattern-forming behaviour can be made apparent by adding a colour-change indicator, starch—this changes from yellow to blue when it captures and binds the tri-iodide ions (I_3^-) involved in the reaction. Although in many ways similar to the BZ reaction, the CIMA reaction is closer to Turing's scheme because it has an explicit activator and inhibitor—the iodide and chlorite ions, respectively.

To turn an activator–inhibitor system into one capable of forming Turing structures, the two species must be made to diffuse through the reaction medium at very

different rates. This is hard to arrange, and explains why it took so long to find a suitable experimental system: in water, just about all small molecules and ions diffuse at more or less the same rate. But the Bordeaux researchers were able to introduce very different rates of diffusion in the CIMA reaction by conducting it in a polymer gel. The molecules of the starch indicator are themselves large polymers, and so they get entangled in the network of the gel with their captive tri-iodide ions. The chlorite ions (the inhibitor species) pass through the gel network unheeded; but the iodide ions (the activator species) are slowed down considerably, because they can keep getting stuck to the immobile starch/tri-iodide groups.

The researchers saw a colour change from blue to yellow along a strip where the various reagents meet and react. Under the right conditions this band broke up into rows of dots (Fig. 4.4). That these dots represented

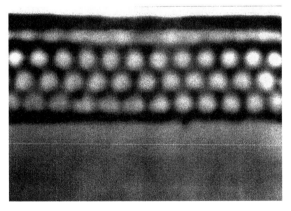

Fig. 4.4 Turing patterns in the CIMA reaction. The dot pattern is restricted to a strip where diffusing reagents meet. (Photo: J. Boissonade, University of Bordeaux.)

a genuine Turing pattern was confirmed by Irving Epstein of Brandeis University and coworkers in 1991, who performed theoretical calculations, based on the known mechanism of the CIMA reaction and the measured differences in diffusion rates of the activator and inhibitor. They showed that they could reproduce the patterns seen experimentally by invoking the Turing mechanism.

The next challenge was to grow Turing patterns over large areas. This was achieved in 1991 by Qi Ouyang and Harry Swinney from the University of Texas at Austin—their two-dimensional lattices of Turing structures contained thousands of the yellow dots (Fig. 4.5). The researchers showed that the pattern disappeared if the gel was warmed above 18°C, and that it reappeared when the gel was cooled. This abrupt and spontaneous patterning in response to a gradual change in conditions is what is expected of a Turing structure. Ouyang and Swinney were also able to demonstrate another of the enticing predictions of the Turing instability: the possibility of forming new stable patterns by changing the reaction conditions. By increasing the iodide concentration or lowering the malonic acid concentration, they broke the symmetry in a new way, forming stripes instead of spots (Plate 4). Their chemical leopard was transformed into a chemical tiger.

You will no doubt have noticed that the patterns in Fig. 4.4 and Plate 4 are ordered, whereas the spots and stripes in the theoretical calculations of a Turing system depicted earlier (Fig. 4.3) are less regular. Both periodic and disordered patterns are possible, depending on the precise parameters of the chemical reactants (such as their diffusion rates) and the nature of the way in which the patterns grow. I shall return to this point later. Notice, however, that in both cases a more or less uniform separation is maintained, on average, between the features (spots or stripes) of the pattern. This separation is set largely by the rate at which the inhibitor diffuses away from the centres of activation.

Coming alive

Swinney and colleagues, working with John Pearson from the Los Alamos National Laboratory, have found a curious kind of chemical pattern that might be considered a hybrid of the travelling waves of the BZ reaction and the stationary spots of Turing structures. In another type of oscillatory reaction–diffusion system called the ferrocyanide–iodate–sulphite reaction they have observed spots that grow and divide like replicating cells. Pearson first saw these life-like spots in numerical simulations of a reaction–diffusion system on the computer, in which they blossomed when the diffusion rates of the various chemical components were ascribed certain values. Swinney's group then discovered conditions under which the replicating spots would manifest themselves experimentally (Plate 9). The spots do not appear spontaneously; their formation has to be triggered by perturbing the mixture locally, for example, by shining ultraviolet light onto a part of it. The spots grow from a roughly circular shape, elongating into a dumbbell shape as they get bigger until finally they split into two circular spots, which then repeat the sequence (Fig. 4.6). But just as in life, these systems cannot support too abundant a birth rate. If the spots get too overcrowded, they annihilate each other—they 'die'.

Fig. 4.5 Extended Turing patterns in the CIMA reaction. (Photo: Harry Swinney, University of Texas at Austin.)

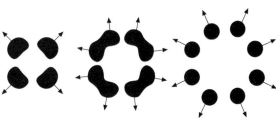

Fig. 4.6 Growth and division of replicating chemical spots. (After: Pearson *et al.* 1993.)

Skin deep

There is now no doubt that Turing's ideas about chemical pattern formation were visionary ones. But do they have anything at all to do with morphogenesis, which was what motivated him to propose his reaction–diffusion model in the first place? At present, this question remains open. Although some biologists are convinced that spontaneous patterning processes akin to, if not identical to, the Turing instability lie at the heart of the most fundamental aspects of embryo development, many regard the whole idea as an as-yet untested hypothesis that at best amounts to a curious sideshow. But one area of developmental biology in which Turing's ideas have made an undeniable impression is in the formation of the surface patterns of the living world: the leopard's spots (Plate 10), the zebra's stripes, the giraffe's blotches. Turing has become biology's answer to Kipling.

The beauty of all this is that the diverse range of pelt patterns and markings can be explained with the *same* basic mechanism. To my mind, this alone is a very good reason to believe that Turing was on the right track. William of Ockham would have reminded us that if we can account not only for the leopard's spots but also for the zebra's stripes and the giraffe's dapples with the same theoretical model, that is surely more satisfactory than having to construct a different model for each. The idea arguably makes sound evolutionary sense too: it economizes on the amount of (genetic) information needed to produce the pattern. The location and size of each of a zebra's stripes does not have to be specified by a personalized, paint-by-numbers genetic plan; all that the genes have to record is the blueprint for making the activator and inhibitor substances at the right stage in development.

The pelt patterns of mammals are mosaics of just a few colours—they are defined by hair colours that are either white, black, brown, or yellow/orange. The origin of these individual colours is well understood. The colour of the hairs that grow from a particular region of the skin is determined by pigment-producing cells called melanocytes that sit in the innermost layer of the skin's epidermis. The pigment, called melanin, is a light-absorbing protein that passes from the melanocyte into the hair. It comes in two forms: eumelanin, which turns hair black or brown, and phaeomelanin, which turns it yellow to orange.

Whether or not melanocytes produce melanin seems to be determined by the presence or absence of certain chemicals in or just below the epidermis. It is not yet known, however, what these chemicals are. During the late 1970s James Murray, a mathematician then working at the University of Oxford, proposed that the distribution of these chemical 'triggers' takes on a characteristic pattern owing to a Turing-like interaction of activator and inhibitor species during the first few weeks of embryogenesis. Thus at a very early stage the embryo acquires a 'pre-pattern' of chemical morphogens, which is later read out by the melanocytes when they respond to the presence or absence of these morphogens by making or failing to make pigments. It is rather like the trick in which a pattern of invisible ink, like lemon juice, is made visible by the heat of a candle flame.

To verify this model, one would at least have to identify the morphogens and their distribution in a pre-pattern in the growing embryo. This has not yet been achieved. But Murray took a different approach: he asked whether, if the basic mechanism were correct, it could produce the kinds of pattern features seen in nature.

The precise spatial pattern produced by a reaction–diffusion system that undergoes a Turing instability depends on a number of factors, such as the relative diffusion rates of the activator and inhibitor species. We saw above that different patterns can be produced by changing the reaction conditions (such as the temperature). Another strong influence on pattern selection is the size and the shape of the region in which the chemical process is occurring. This may seem a little odd—the outcome of most chemical reactions does not depend on whether the reaction is conducted in a narrow test tube or a round-bottomed flask. But the point about Turing patterns is that they are expressions of a wave-like modulation of the concentration of reacting species throughout the system, and like sound waves in an organ pipe, the kinds of wave (the *modes*) that can be supported are dependent on the dimensions of the container in which they are set up.

There is, in fact, a minimum size for which a reaction–diffusion system can generate a spatial pattern at all. The characteristic size of a feature of the pattern—the diameter of a spot, for instance—is determined by the diffusion rate (the 'range', if you like) of the activator. So if the system as a whole is about the same size as this, no pattern is evident—the concentrations are uniform throughout. As the system grows, increasingly complex patterns can be formed as the number of modes that can be supported increases (Fig. 4.7). (Although this analogy with acoustic standing waves is visually appealing, it should not be taken

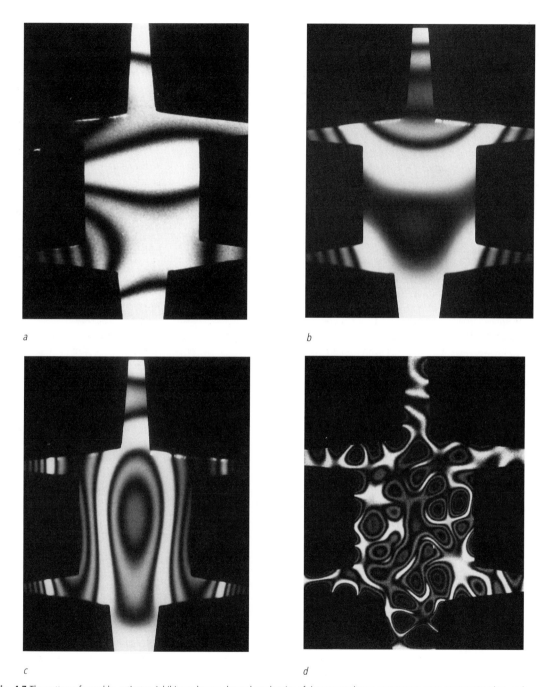

a

b

c

d

Fig. 4.7 The patterns formed by activator–inhibitor schemes depend on the size of the system: larger systems can support more 'modes', and so exhibit more complex patterns. This is analogous to the complexity of the vibrations excited by sound waves in surfaces of different sizes. Shown here are the acoustic vibrations excited in plates shaped to represent the body surfaces of mammals. The excitation increases in frequency from (*a*) to (*d*), which is equivalent to increasing the size of the plate. (Photos: James Murray, University of Washington, Seattle.)

Fig. 4.8 The patterns on animal tails may be either spots or bands, but bands always appear as the tail tapers towards the end—as seen here for a Geoffroy's cat (left) and an ocelot (right).

too literally. Standing waves such as those shown in Fig. 4.7 do *not* arise in the same way as Turing patterns—there is no local activation and long-ranged inhibition involved.)

Murray investigated whether this dependence on size and shape might plausibly account for the differences in pattern seen amongst animal tails. The tail is a good feature to study, since it can be modelled mathematically to a good approximation as a tapering cylinder, a fairly simple shape. Tail patterns come in just two basic varieties: bands running around the circumference, or spots. But just about all patterned tails end in a series of bands (Fig. 4.8).

When Murray performed calculations to see what patterns a reaction–diffusion system would generate on tapered cylinders, he found that both bands and spots could be produced. If the model tail is small, only bands are formed—these are essentially a one-dimensional pattern, since the variation in colour (that is, in pigment-stimulating activator chemicals) occurs only in one direction, along the tail's axis. If the tail is larger, however, more complex modes can be supported, and the patterns become two-dimensional (spots), varying around the circumference of the cylinder as well as along the axis (Fig. 4.9). So a transition from bands to spots may take place along the tail as it widens from the tip, just as is seen in the cheetah and leopard.

Murray found that inter-species differences between tail patterns can also be rationalized in terms of the known embryonic forms of the animals. The tail of the genet, for instance, is always banded along its entire length, whereas that of the leopard is mainly spotted, with bands just at its tip. To judge from the similar shape of the adult tails, there is no obvious reason why this

Fig. 4.9 The patterns produced on tapering cylindrical 'model tails' by an activator–inhibitor scheme depend on their size and shape. Small cylinders support only bands (stripes) (*a*), whereas spots appear on larger cylinders (*b*) as they widen. On a more slowly tapering tail (*c*), the transition from bands to spots is more clear. (After: Murray 1990.)

should be so; but in the respective embryos, the tail of the genet is thin and almost uniform in diameter and so supports only bands, whereas the embryonic leopard tail is fairly short and sharply tapered, and so will allow spots.

If indeed the markings of the adult animal are laid down by a chemical pre-pattern in the very young embryo, the timing of this pre-patterning stage can be crucial, since the size and shape of the embryo changes fast. This fact may be reflected in the differing stripe markings of the zebras *Equus burchelli* and *Equus grevyi*:

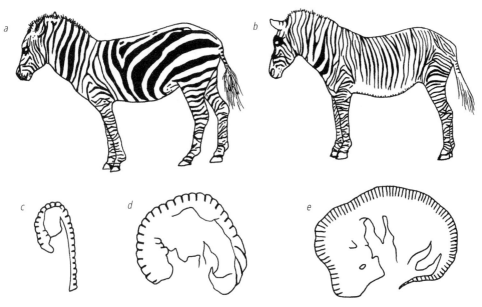

Fig. 4.10 The adult zebra *Equus grevyi* (*b*) has more and narrower stripes than the adult *Equus burchelli* (*a*). This is thought to be because the striped 'pre-pattern' is laid down on the embryo of the latter at an earlier stage: after 21 days for *Equus burchelli* (*c*), but after 5 weeks for *Equus grevyi* (*e*). The smaller embryo supports fewer stripes, and so by the time it is of comparable size (*d*), its stripes are wider. (Drawings by the author, after Murray 1989.)

the stripes of the former are broader and less numerous than those of the latter (Fig. 4.10). There is evidence to suggest that *Equus burchelli* acquires its pattern several weeks earlier in the gestation period than the latter. The same chemical mechanism, producing stripes of the same width, would then give the smaller *Equus burchelli* embryo (Fig. 4.10*c*) fewer stripes than the larger *Equus grevyi* (Fig. 4.10*e*). So when both have grown to a comparable size, the former has broader stripes than the latter.

I should add a cautionary note here: stripes are in fact not all that easy to make in Turing-type models, since they have a tendency to break up into spots. Murray assumed that stripes could survive in his model, but in

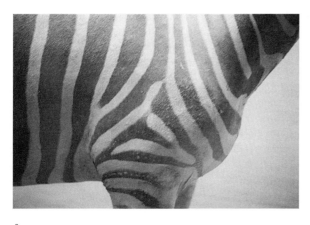

Fig. 4.11 The scapular stripes of a zebra, where the leg meets the body, form a kind of chevron pattern (*a*), which is reproduced in an activator–inhibitor model with this idealized geometry (*b*).

practice extra ingredients are commonly needed to ensure this. For example, stripes may be stabilized if there is an upper limit to the rate of autocatalytic production of the activator, so that this reaction can become 'saturated'.

Murray ventured to look at the patterns that would be generated by reaction–diffusion systems in more complicated geometries, such as the junction of the leg and body of a zebra. Here the same kind of modification of the stripe pattern is seen in all zebras—a kind of chevron pattern in which the bands of the leg blend with the stripes of the body. These markings are called scapular stripes (Fig. 4.11*a*). Murray considered a simplified two-dimensional approximation to the shape of the leg-body junction, and found that a system that generated stripes in the body and bands in the leg would also produce the chevron pattern at their junction (Fig. 4.11*b*).

So within Murray's model, if the chemical parameters in the reaction–diffusion system are much the same for all species (an assumption that is not unreasonable but not firmly supported either), then the size and shape of the embryo at the time of pre-patterning exert a dominant influence on the eventual pattern. One implication of this is that small animals with short gestation periods should have less complex pelt patterns than larger animals, because their smaller embryos support fewer modes. On the whole this seems to be borne out, perhaps most dramatically by the honey badger and the Valais goat, which exemplify the simplest kind of non-uniform colouration of all: an abrupt division into a white and a black half.

But it turns out that the apparent complexity of a pattern diminishes at the other end of the size range too, when the animal becomes very large. This is because, as more and more modes become possible on the patterned embryo, the features start to merge as the dividing lines between them become squeezed out. Thus, for instance, giraffes have very closely spaced spots with narrow light boundaries (Fig. 4.12); and elephants and hippopotami have no markings at all. More, in terms of skin markings, is less.

The giraffe patterns that Murray's model generates are blobs with rounded edges (Fig. 4.13*a*)—simply bloated versions of the leopard spots. But this is arguably not an accurate depiction of the spots on real giraffes, which—as you can see from Fig. 4.12—are more like irregular polygons separated by roughly straight lines of unpigmented hair. Hans Meinhardt, now at the Max Planck Institute for Developmental Biology in Tübingen, Germany, and colleague André Koch have developed a more sophisticated reaction–

Fig. 4.12 On large animals like the giraffe, the pelt pattern consists of very large features that almost merge, with narrow boundaries between them. (Photo: Michael and Sandra Ball.)

diffusion model that eliminates this deficiency. Their model incorporates an activator–inhibitor system in which the diffusion constants of the activator and inhibitor do not differ too greatly. Then, as I mentioned

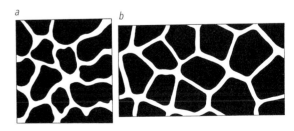

Fig. 4.13 A simple activator–inhibitor model for the giraffe's patterns produces large blotchy features that are only a crude approximation to the real pattern (*a*). A more sophisticated model in which travelling waves of activator and inhibitor throw a biochemical 'switch' to trigger pigment production generates more realistic polygonal shapes (*b*). (Images: (*a*) after Murray 1990; (*b*) after Koch and Meinhardt 1994.)

on page 81, the system does not generate stationary Turing patterns but travelling waves, rather like those of the BZ reaction. These waves become translated into a fixed spatial pattern by the interaction of the reaction–diffusion system with a biochemical switch: when the concentration of activator exceeds a certain threshold at any point in space, a chemical is generated there that stimulates melanocytes into producing melanin. Once this switch is thrown, it stays that way—melanin is produced even if the production of the activator subsequently ceases.

The production of activator is assumed to be initiated at several random points throughout the system. Chemical waves of activator then spread outward from these initial points, triggering melanin production as they go. But where the wavefronts meet, they annihilate each other, just as we see in the BZ reaction (Fig. 3.3). These annihilation fronts define linear boundaries between each domain of activator production, and so the system breaks up into melanin-producing polygonal domains separated by unpigmented boundaries (Fig. 4.13b)—a much closer approximation to the pattern seen on real giraffe pelts.

Meinhardt and Koch found that with a little fine tuning of model parameters they could also obtain a better approximation to the leopard's pattern too—these are commonly not mere blobs of pigmented hairs but rings or crescents (Plate 10); their model could generate structures like this (Fig. 4.14). Models of this sort, which involve two interacting chemical systems instead of the single reaction–diffusion system considered by Murray, are clearly able to produce much more complex patterns.

Hard stuff

Anyone who is happy to accept with complacency the view that animal markings are simply determined by Darwinian selective pressures has a surprise in store when they come to consider mollusc shells. The patterns to be seen on these calcified dwellings are of

Fig. 4.14 The leopard's spots are in fact mainly crescent-shaped features. An activator–inhibitor scheme that involves two interacting chemical patterning mechanisms can reproduce these shapes. (After: Koch and Meinhardt 1994.)

exquisite diversity and beauty, and yet frequently they serve no apparent purpose whatsoever. Many molluscs live buried in mud, where their elaborate exterior decoration will be totally obscured. Others cover their shell markings with an opaque coat, as if embarrassed by their virtuosity. And individual members of a single species can be found exhibiting such personalized interpretations of a common theme that you would think they would hardly recognize each other (Fig. 4.15).

Ultimately these patterns are still surely under some degree of genetic control, but they must represent one of the most striking examples of biological pattern for which there are often next to no selective pressures.[*] While this means that their function remains a mystery, it also means that nature is given free reign: she is, in Hans Meinhardt's words, 'allowed to play'.

It is tempting to regard shell patterns as analogous to the spots and stripes of mammal pelts, and some are indeed apparently laid down similarly in a global, two-

---

[*] There is nothing anti-Darwinian in this, however, since Darwin's theory does not insist that all features be adaptive.

Fig. 4.15 Shell patterns in molluscs can exhibit wide variations even amongst members of the same species. The shells of the garden snails shown here bear stripes of many different widths. (Photo: Hans Meinhardt, Max Planck Institute for Developmental Biology, Tübingen.)

Fig. 4.17 Oblique stripes are the result of travelling waves at the growth edge, periodic in both space and time. (Photo: Hans Meinhardt.)

Fig. 4.16 Stripes that run parallel to and perpendicular to the axis of the shell reflect profoundly different patterning mechanisms: in the former case (*top*), the stripes reflect a patterning process that is uniform in space but periodic in time; while the latter case (*bottom*) represents the converse. (Photo: Hans Meinhardt.)

dimensional surface-patterning process. But most are intriguingly different, in that they represent a historical *record* of a process that takes place continually as the shell grows. For the shell gets bigger by continual accretion of calcified material onto the outer edge, and so the pattern that we see across the surface of the shell is a trace of the pigment distribution along a one-dimensional line at the shell's edge. Thus stripes that run along or around the growth axis (Fig. 4.16), while superficially similar, are in fact frozen time-histories of qualitatively different patterning processes: one in which a spatially periodic pattern along the growing edge remains in place as the shell grows, the other in which bursts of pig-

mentation occur uniformly along the entire growth edge followed by periods of growth without pigmentation. Stripes that run at an oblique angle to the growth direction, meanwhile, are manifestations of a *travelling* wave of pigmentation that progresses along the edge as the shell grows (Fig. 4.17).

Thus we can see that shell patterns can be the product both of stationary patterns, analogous to Turing patterns, and of travelling waves, analogous to those in the BZ reaction—arising in an essentially one-dimensional system.

Hans Meinhardt has shown that both types of pattern can be reproduced by a model in which an activator–inhibitor process controls the deposition of pigment in the calcifying cells at the shell's growing edge. The stripe patterns in the lower shell of Fig. 4.16, for instance, are a manifestation of a simple, periodic stationary pattern in one dimension (Fig. 4.18*a*), an analogue of the two-

a

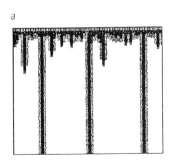

b

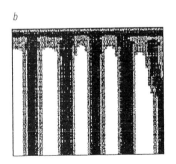

c

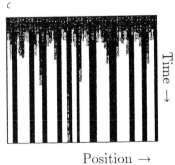

Time →

Position →

Fig. 4.18 Stripes perpendicular to the growth edge of the shell are the result of one-dimensional spatial patterning at the edge. The pattern gets 'pulled' into stripes as the shell edge advances (*a*). If the activator diffuses more rapidly, the stripes broaden (*b*). When the concentration of the activator rises until it 'saturates' (becomes limited by factors other than long-ranged inhibition), the spacing of the stripes becomes irregular (*c*). (Images: Hans Meinhardt.)

dimensional spot pattern of Fig. 4.2. The width of the stripes and the gaps between them can be acutely sensitive to the model parameters, particularly the relative diffusion rates of activator and inhibitor (Fig. 4.18*b*, *c*). So differences between members of the same species, like those seen in Fig. 4.15, might be the result of differing growth conditions, such as temperature, which alter the diffusion rates. Alternatively, Meinhardt has shown that such intra-species irregularities can arise if the pattern at the shell's growing edge becomes frozen in at an early stage of growth, for example if the communication between cells via diffusing chemical substances ceases.

As the pattern on a shell is a time-trace of the pattern on a growing edge, the full two-dimensional pattern depends on how the edge evolves. For example, the bands in Fig. 4.16 and the spoke-shaped patterns in Fig. 4.19 may be the result of just the same kind of periodic spatial pattern on the growing edge, except that in one case the edge curls around in a spiral and in the other it expands into a cone. When, however, the perimeter length of the edge increases as in Fig. 4.19, the change in dimension may introduce new features into the pattern, just as we saw earlier for the change in scale of patterned mammals. That is to say, as the expansion of the edge separates two adjacent pigmented regions, a new domain may be supportable between them (recall that the average distance between pattern features in an activator–inhibitor system tends to remain the same as

the system grows). That would account for the later appearance of new stripes in the conical shell shown on the right in Fig. 4.19.

When Meinhardt's activator–inhibitor systems give rise to travelling waves, the resulting trace on the shell is a series of oblique stripes, as an activation wave for pigmentation moves across the growing edge. We saw how such waves can be initiated in the two-dimensional BZ reaction from spots that act as pacemakers, sending out circular wavefronts. In one dimension these pacemaker regions emanate wavefronts in opposite directions along a line. So the resulting time-traces are inverted V shapes whose apexes point away from the growth edge. When two wavefronts meet on an edge, they annihilate one another just like the target patterns of the BZ reaction, and we then see two oblique stripes converge in a V with its apex *towards* the growth edge (Fig. 4.20*a*). Both features can be seen on real shells (Fig. 4.20*b*). This shows that even highly complex shell patterns can be produced by well-understood properties of reaction–diffusion systems—the complexity comes from the fact that we are seeing the time-history of the process traced out across the surface of the shell.

Occasionally one finds shells that seemed to have had a change of heart—that is to say, they display a beautiful pattern that suddenly changes to something else entirely (Fig. 4.21). An activator–inhibitor model can account for the patterns before and after the change, but to

Fig. 4.19 When the shell's growth edge traces out a cone instead of a spiral, a one-dimensional periodic pattern at the edge becomes a radial 'spoke' pattern. As the edge grows in length, new pattern features may appear in the spaces between existing spokes (*right*). (Photo: Hans Meinhardt.)

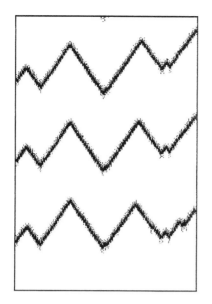

a

b

Fig. 4.20 Annihilation between travelling waves in an activator–inhibitor model leads to V-shaped patterns (*a*), as seen on the shell of *Lioconcha lorenziana* (*b*). (Photo: Hans Meinhardt.)

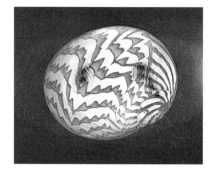

Fig. 4.21 Sudden changes in environmental conditions can restart the patterning process on shells, creating abrupt discontinuities in the pattern. (Photo: John Campbell, University of California at Los Angeles.)

account for the change itself we need to invoke some external agency. It seems likely that shells like this have experienced some severe environmental disturbance— perhaps the region became dry or food became scarce— and as a result the biochemical reactions at the shell's growing edge were knocked off balance by the tribulations of the soft creature within (remember that it is this creature, not the shell itself, that is ultimately supplying

the materials and energy for shell construction!). This sort of perturbation can 'restart the clock' in shell-building, and the pattern that is set up in the new environment may bear little relation to the old one. Like all good artists, molluscs need to be left alone in comfort to do the job well.

But is it real?

Biologists are hard to please. However striking might be the similarity between the patterns produced by these reaction–diffusion models and the real thing, they may say that it could be just coincidence. How can we be sure that the Turing mechanism is really at work in these creatures?

Ultimately the proof will require identification of the morphogens responsible, and that still has not been done. But in 1995, Japanese biologists Shigeru Kondo and Rihito Asai from Kyoto University staked a claim for a Turing mechanism in animal markings that was hard to deny. They looked at the stripe markings of the marine angelfish, a beautiful creature whose scaly skin bears bright yellow horizontal bands on a blue background. It is common knowledge that a reaction–diffusion system can produce parallel stripes; but what is different about the angelfish is that its stripes do not seem to be fixed into the skin at an early stage of development—they continue to evolve as the fish grows. More precisely, the pattern *stays more or less the same* as the fish gets bigger— smaller fish simply have fewer stripes. For example, when the young angelfish of the species *Pomacanthus semicirculatus* are less than 2 cm long, they each have three stripes. As they grow, the stripes get wider, but when the body reaches 4 cm there is an abrupt change: a new stripe emerges in the middle of the original ones, and the spac-

This must mean that the angelfish's stripes are being actively sustained during the growth process—the reaction–diffusion process is *still going on*. One would expect that, if the fish were able to grow large enough (to the size of a football, say), the effect of scale evident in Jim Murray's work would kick in and the pattern would change *qualitatively*. But the fish stop growing much short of this point.

Kondo and Asai were able to reproduce this behaviour in a theoretical model of an activator–inhibitor process taking place in a growing array of cells. This is more compelling evidence for the Turing mechanism than simply showing that a process of the same sort can reproduce a stationary pattern on an animal pelt—the mechanism is able to reproduce the growth-induced expansion of the pattern too.

But the researchers went further still. They looked also at the angelfish *Pomacanthus imperator*, which has rather different body markings. The young fish have concentric stripes that increase in number as the fish grows, in much the same way as the stripes of *P. semicirculatus*. But when the fish become adult, the stripes reorganize themselves so that they run parallel to the head-to-tail axis of the fish. These stripes then multiply steadily in number as the fish continues to grow, so that their number is always proportional to body size, and the spacing between them is uniform. New stripes grow from branching points which are present in some of the stripes—the stripe 'unzips' along these branching points, splitting into two (Fig. 4.23*a*). The calculations of Kondo and Asai, using the same reaction–diffusion model as for *P. semicirculatus*, generated this behaviour exactly (Fig. 4.23*b*). Their model also mimicked the more complex behaviour of branching points located at the dorsal or ventral regions (near the top and bottom

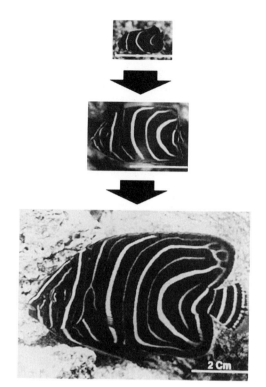

Fig. 4.22 As the angelfish grows, its stripes maintain the same width— so the body acquires more of them. This contrasts with the patterns on mammals such as the zebra or cheetah, where the patterns are laid down once for all and then expand like markings on a balloon. (Photo: Shigeru Kondo, Kyoto University.)

ing between stripes then reverts to that seen in the younger (2-cm) fish (Fig. 4.22). This process repeats again when the body grows to about 8 or 9 cm. In contrast, the pattern features on, say, a giraffe just get bigger, like a design on an inflating balloon.

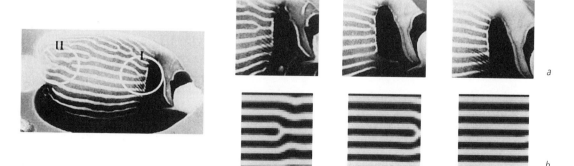

Fig. 4.23 The 'unzipping' of new stripes in *Pomacanthus imperator* (*a*; region I on the left) can be mimicked in a Turing-type model (*b*). (Photos: Shigeru Kondo.)

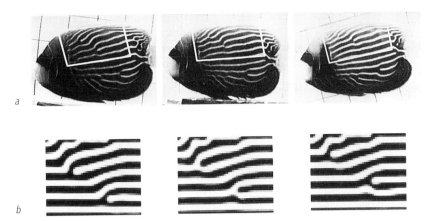

Fig. 4.24 Complex pattern reorganization in the dorsal and ventral regions of *Pomacanthus imperator* (*a*), is also captured by the model (*b*). (Photos: Shigeru Kondo.)

of the body) (Fig. 4.24). What is more, there was a rough correspondence between the relative times taken for these different transformations in the real fish and in the calculations (where 'time' means number of steps in the computer simulation).

It is hard to imagine that, given this ability of the reaction–diffusion model to generate the very complex rearrangements of the fish stripes, the model is anything but a true description of the natural process. Kondo and

Asai pointed out that since the reaction–diffusion process is apparently still going on in the adult fish (whereas it is assumed to take place only during the embryonic pre-patterning stage in patterned mammals), it might be a lot easier to identify the chemical species—the activator and inhibitor molecules—responsible in this case. That would provide incontrovertible proof that Alan Turing truly guessed how nature makes her patterns.

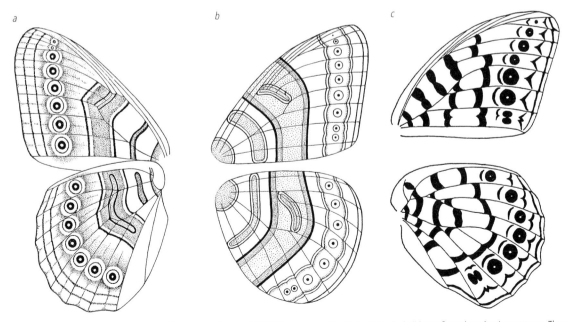

Fig. 4.25 The nymphalid ground plans of (*a*) Schwanwitsch and (*b*) Süffert represent the Platonic ideal of all butterfly and moth wing patterns. They both contain features from which almost all observed patterns can be derived. An updated version of the ground plan (*c*) takes more explicit account of the effect of wing veins. (Images: H. Frederik Nijhout, Duke University.)

On the wing

The animal-marking patterns considered so far are two-tone affairs: they involve the production of a single pigment by differentiated cells. But the natural world is replete with far more fanciful displays that are enough to make a theorist despair. Consider, for instance, the butterfly (Plate 11), whose wings are a kaleidoscope of colour. Not only is the range of hues fantastically rich, but the patterns seem to have a precision that goes beyond the zebra's stripes: they are highly symmetrical between the two wings, as though each spot and stripe has been carefully placed with a paint brush. Can we hope to understand how these designs have been painted?

That question was squarely faced in the 1920s by B.N. Schwanwitsch and F. Süffert, who synthesized a tremendous variety of wing patterns in butterflies and moths into a unified scheme known as the nymphalid ground plan. This depicts the most common basic elements observed in wing patterns in a single universal blueprint, from which a huge number of real patterns can be derived by selecting, omitting or distorting the individual elements. Although Schwanwitsch and Süffert developed their schemes independently, they show a remarkable degree of consistency (Fig. 4.25). The basic pattern elements are series of spots, arcs and bands that cross the wings from the top (anterior) to the bottom (posterior) edges. These top-to-bottom features are called symmetry systems, because they can be regarded as bands or sequences of discrete elements that are approximate mirror images around a symmetry axis that runs through their centre (Fig. 4.26). Even the most complicated of wing patterns can generally be broken down into some combination of these three or four symmetry systems lying side by side—although sometimes they are so elaborated by finer details that the relation to the ground plan is by no means obvious.

No butterfly is known that exhibits all of these elements, however; rather, the nymphalid ground plan represents the maximum possible degree of wing patterning that nature seems able to offer. The full range of wing patterns can be obtained by juggling with the size, shape and colour of selected elements of the plan.

The building blocks that make up these patterns are tiny scales on the wing surface that overlap like roofing tiles. Each scale has a single colour, so that looked at close up, every pattern has the 'pixellated' character of a television image (Fig. 4.27). Some of the colours are produced by chemical pigments—the melanins that feature in animal pelt markings, and other pigment molecules that give rise to whites, reds, yellows and occasionally blues (the latter are derived from plant pigments). But some scales acquire their colours by means of physics, not chemistry. They have a microscopic ribbed texture which scatters light so as to favour some wavelengths over others, depending on the match between the wavelength of the light and the spacing of the ribs. Most green and blue scales generate their colours this way, and it can result in the iridescent or silky appearance of some wing surfaces.

The wing pattern is laid down during pupation, when the surface cells of the developing wing become programmed to produce wing scales of a certain colour (whether it be by the production of pigments or of a particular surface texture). The challenge is to understand how this programming is carried out so as to express the characteristic distributions of spots and bands that each species selects from the nymphalid ground plan.

One important consideration is that the overall pattern appears to be strongly modified by the system of veins that laces the wing. Süffert's initial scheme did not

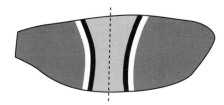

Fig. 4.26 The central symmetry system, a series of bands that runs from the top to the bottom of the wing. The mirror-symmetry axis is denoted by a dashed line.

Fig. 4.27 The wing patterns of butterflies and moths are made up from overlapping pigmented scales, each of a single colour. (Photo: H. Frederik Nijhout, Duke University.)

take this into account, but Schwanwitsch appreciated the importance of the veins. In some species, in fact, the wing pattern simply outlines the vein pattern with a coloured border. In general the stripes that cross the wing from top to bottom (particularly the broad band down the centre, called the central symmetry system: Fig. 4.26) are offset where they cross a vein. Schwanwitsch called these offsets dislocations, by analogy with the dislocations of sedimentary strata where they are cut by a geological fault. H. Frederik Nijhout of Duke University has proposed an updated version of the nymphalid ground plan which features these dislocations at veins much more prominently (Fig. 4.25c).

This classification of pattern elements helps immeasurably when we come to attack the question of how the patterns arise, because it means that we can focus on the handful of basic symmetry systems, and only afterwards need we worry about how these have become elaborated into the distorted forms that they might take in particular species. Take the central symmetry system, for example. In 1933 A. Kühn and A. von Engelhardt performed experiments to try to understand how this pattern element on the wings of the moth *Ephestia kuhniella* (Fig. 4.28a) came into being. The organization of this pattern—the fact that the bands run unbroken (albeit dislocated by the vein structure) from the anterior to the posterior wing edge—implies that the signal triggering it must be non-local: it must pass from cell to cell. So what happens if cell-to-cell communication is disrupted? To find out, Kühn and von Engelhardt cauterized small holes in the wings of the moths during the

first day after pupation to present an obstacle to between-cell signalling. They found that the coloured bands became deformed around the holes (Fig. 4.28b). After studying the effect of many such cauteries on different parts of the wing, they proposed that the bands of the central symmetry system represent the front of a propagating patterning signal—a 'determination stream'—which issues from two points, one on the anterior and one on the posterior edge (Fig. 4.28c).

This was a remarkably prescient idea, anticipating the idea of a diffusing chemical morphogen that triggers pattern formation. But Kühn and von Engelhardt didn't get it all right. For a start, a closer look at their cautery studies suggests that there are *three* sources of morphogen, not two, all of which lie on the mirror-symmetry axis of the central symmetry system. But more importantly, whereas they saw the bands as wavefronts, recent experiments suggest instead that the patterning is triggered when a smoothly varying concentration of the diffusing morphogen (not a sharp wavefront) exceeds a certain threshold and throws some kind of biochemical switch that induces a particular colouration.

Jim Murray has devised a reaction–diffusion system to model these experiments in which a morphogen, which switches on a particular gene in the wing cells, is released from two sources on the anterior and posterior wing edges. He found that the boundary of the gene-activated region of the wing mimicked the shapes of the deformed stripes quite well (Fig. 4.28d). Frederik Nijhout proposes that the cauterized holes don't just

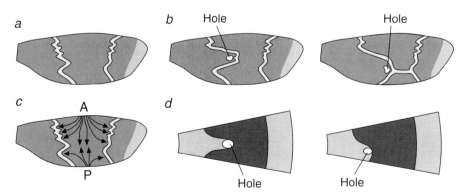

Fig. 4.28 (a) The moth *Ephestia kuhniella* has a central symmetry system defined by two light bands. (b) Kühn and von Engelhardt investigated the formation mechanism of these bands by cauterizing holes in pupal wings and observing the effect on the pattern. (c) They hypothesized that the disruptions of the pattern can be explained by invoking 'determination streams' of some chemical morphogen issuing from centres located on the anterior (A) and posterior (P) edges of the wing. (d) There is some correspondence between the pattern boundaries in these experiments and those generated in an idealized model in which a reaction–diffusion system switches on genes that fix the pattern. (After: Murray 1990.)

present obstacles to morphogen diffusion—they actually soak it up (that is, they are a morphogen *sink*). A model based on this assumption can explain all of the experimental results.

The idea that patterning is orchestrated by morphogen sources and sinks underpins all work on butterfly wing patterns today. Moreover, it appears that these sources and sinks are restricted to just a few locations: at the wing veins, along the edges of the wing, and at points or lines along the midpoint of the 'wing cells', the compartments defined by the vein network. Moreover, whereas Kühn and von Engelhardt assumed that their 'determination streams' issued across the whole wing, it is now clear that each wing cell has its own autonomous set of morphogen sources and sinks. So explaining the wing pattern as a whole can be reduced to the rather simpler problem of explaining the pattern in each wing cell, which is copied more or less faithfully from wing cell to wing cell.

The ingredients of a model for wing patterns can therefore be specified by a kind of hierarchical dismemberment of the full pattern. First, the nymphalid ground plan provides a kind of template onto which all actual patterns can be mapped, so that the underlying nature of pattern elements can be discerned. Then this pattern is regarded as an assembly of autonomous wing cells, each of which is itself a collection of pattern elements such as stripes and eyespots (ocelli) which are induced by 'organizing centres', sources and sinks of morphogens. The morphogens are assumed to diffuse through the wing cell, throwing biochemical switches where they surpass some critical threshold. And these organizing centres can lie only at the wing cell midpoints or at their edges (at veins or wing tips).

A general model for patterning that takes these principles as its starting point has been developed by Nijhout. It attempts to solve two mysteries: how do various combinations of sources and sinks create the vast array of pattern elements that we see, and how do these sources and sinks arise in the first place from a uniform sheet of cells?

The first question is the easier one, because Nijhout found that simply by selecting various combinations of sources and sinks located at the specified places he could obtain an endless variety of pattern features. He developed a 'toolbox' of sources and sinks that determine the concentration contours of a diffusing morphogen throughout the wing cell (Fig. 4.29a). As any of

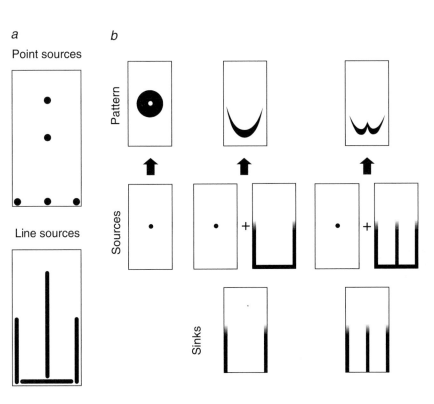

Fig. 4.29 A set of sources and sinks of morphogen (a) in an idealized wing cell (here shown as a rectangular unit with veins at the edges and the wing edge along the bottom) can be combined to generate many of the pattern features observed in nature (b). (After: Nijhout 1991.)

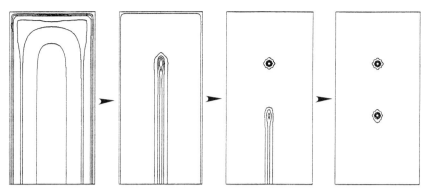

Fig. 4.30 The elements of the toolbox in Fig. 4.29*a* can be produced from an activator–inhibitor model in which an activator is released from the wing veins. The pattern of activator production (shown as contours) changes over time to a central line that retracts to leave isolated spots. (Images: H. Fredrick Nijhout.)

these contours can in principle represent the threshold above which the patterning switch is thrown, a single combination of 'tools' can generate a wide range of pattern features (Fig. 4.29*b*). Amongst these are most of those that appear in nature—and some that do not! What are we to deduce from the latter—that the model is flawed, or that butterflies don't make use of the full 'morphospace' of patterns available to them? The second possibility is quite feasible, because there may be certain types of pattern that simply don't help the evolutionary success of the creature.

So how are the sources and sinks put in place? This is a question that involves spontaneous symmetry breaking in the wing cell, and to answer it Nijhout invokes the activator–inhibitor scheme. To begin with, the only 'special' places in the wing cell are the edges, at the veins and at the wing tips. But of the tools in Fig. 4.29*a*, only one (the line source along the wing edge) tracks one of these special locations fully. Nijhout has shown that all of the other tools can be produced by an activator–inhibitor scheme in which an activator diffuses from the vein edges into an initially uniform mixture of activator and inhibitor. At first, this leads to *inhibition* of activator production adjacent to the veins (Fig. 4.30). Then a region of enhanced activator production appears down the wing cell midpoint. This retracts towards the wing cell edge, leaving one or more point sources of activator as it goes. The number and location of sources depends on the model parameters—the rates of diffusion and reaction. This model suggests that the location and shape of morphogen sources is therefore determined by the time during development when the pattern of the activating substance gets 'fixed' into a source region.

To really verify this model, we'd need to identify and to track the development and behaviour of putative morphogens. Ultimately this is a question of genetics—

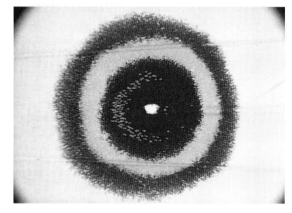

Fig. 4.31 The eyespot pattern is found on many butterfly and moth wings. It probably serves to alarm potential predators. (Photo: H. Frederik Nijhout.)

both the production of the morphogen and its influence on wing scale colour are under genetic control. Many genes have been identified that control certain pattern features in particular species, for example, by changing colours, adding or removing elements or changing their size. But how the genes exert this effect via diffusing morphogens is in general still poorly understood. One of the best studied pattern features is the eyespot or ocellus, a roughly circular target pattern (Fig. 4.31). These markings appear to serve as a defence mechanism, startling would-be predators with their resemblance to the eyes of some larger and possibly dangerous creature. The centre of the eyespot is an organizing centre that releases a morphogen, which diffuses outwards and programmes surrounding cells. Experiments by Sean Carroll of the Howard Hughes Medical Institute in Wisconsin and colleagues have elucidated the genetic basis of the patterning process. They found in 1996 that a gene called *Distal-less* determines the location of the eyespots. The gene is turned on (in other words, the Distal-less protein encoded by the *Distal-less* gene* begins to appear) in the late stages of larval growth, while the butterfly is still in its cocoon. That the *Distal-less* gene is involved in this process is something of a surprise, since in arthropods like beetles it is known to have a completely different role, determining where the legs grow.

Expression of the Distal-less protein occurs initially in a broad region around the tip of the wing, and the protein spreads by diffusion. Gradually, the production of the Distal-less protein becomes focused into spots, which define the centres of the future eyespots. This focusing is similar to that seen in Nijhout's model for the formation of morphogen sources (Fig. 4.30). Once the focal points have been defined, they serve as organizing centres for the formation of the concentric rings—and it seems that the Distal-less protein now does the organizing. It becomes expressed in an expanding circular field centred on the focal point, and this signal somehow controls the developmental pathways of surrounding cells, fixing within them a tendency to produce scales of a different colour to the background (Fig. 4.32). This process of differentiation of scale-producing cells around the eyespot focus is still imperfectly understood. But it seems clear that the diffusing morphogenetic signal (whether this be the Distal-less protein itself or some other gene product activated by it) controls the pattern but not the colour of the marking, since eyespot foci transplanted to different parts of the wing produce eyespots of different colours.

To me, one of the most astonishing things about the whole wing-patterning scheme is the way that evolution employs it as a paint-box to create highly specialized pictures. Some butterfly species have evolved patterns that mimic those of other species, because the latter are unpalatable to the former's predators. This kind of so-called Batesian mimicry is good for the mimic but bad for the species it imitates, once predators begin to wise up to the possibility of deception. So the two patterns become involved in a kind of evolutionary race as the mimic attempts to keep pace with its model's tendency to evolve a new set of colours. And the dead-leaf butterfly displays a particularly inventive use of the nymphalid ground plan, which it has gradually distorted and dislocated until the wing pattern and colouration acquire the appearance of a dead leaf—an example of a universal pattern corrupted into camouflage.

Written on the body

What, at last, of the patterns of body plans, which stimulated Turing in the first place? Can the complicated blueprint for our human shape really be imprinted on an embryo by chemicals that are blindly diffusing and reacting, activating and inhibiting?

This topic shows how a little knowledge can simply make life harder. In the eighteenth century no one was troubled by the question of how babies grow from embryos, because it was assumed that, naturally enough, all creatures start life as miniature but fully formed versions of their adult selves, and just grow bigger. People, it was thought, grow from microscopic homunculi in the womb, which possess arms, legs, eyes and fingers perfect in every detail. The problem with this idea, which was rather swept under the carpet, is that it entails an infinite regression: unless you are prepared to accept the formation of pattern from a shapeless egg at *some* stage, you have to assume that the female homunculi contain even smaller homunculi in their tiny ovaries, and so on for all future generations.

During the eighteenth century this idea was gradually dispensed with, but only in favour of an alternative that was really no more attractive. It assumed that egg cells need not be fully formed homunculi but were instead imbued with an invisible pattern that would find gradual expression as a mature organism. This was not

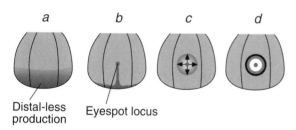

a   b   c   d

Distal-less production   Eyespot locus

Fig. 4.32 The formation and positioning of eyespot patterns is initiated by a gene product called Distal-less. This protein is at first produced over a broad region around the edge of the developing wing (*a*). It then becomes focused into narrow bands down the midpoint of one or more wing subdivisions (defined by the pattern of veins), ending in a spot which will form the centre of the eyespot (*b*). From this central locus issues a signal comprised of one or more other morphogens, which diffuse outwards (*c*) and eventually induce differentiation of the wing's scale cells into differently pigmented rings (*d*).

* The names of genes are conventionally spelt in italics, while the protein products derived from them have the same name but in normal typeface.

much of an advance because it still begs the question of where that patterning might come from.

To go from a spherical fertilized egg to a newborn baby, you have to break a lot of symmetry. Turing's mechanism provides a way to do that, but there is no reason to suppose that it is unique. Today's understanding of morphogenesis suggests that here, at least, nature may use tricks that are at the same time less complex and elegant but more complicated than Turing's reaction–diffusion instability. It seems that eggs are patterned and compartmentalized not by a single, global mechanism but by a sequence of rather cruder processes that achieve their goal only by virtue of their multiplicity.

The reference grid of a fertilized egg, which tells cells whether they lie in the region that will become the head, a leg, a vertebra or whatever, *is* apparently painted by diffusing chemicals. But there is no global emergence of a Turing-style pattern to differentiate one region form another; rather, the chemicals merely trace out monotonous gradients: high near their source and decreasing with increasing distance. A gradient of this sort differentiates space, providing a directional arrow that points down the slope of the gradient. Each of the chemical morphogens has a limited potential by itself to structure the egg, but several of them, launched from different sources, are enough to get the growth process underway by providing a criss-crossing of diffusional gradients that establish top from bottom, right from left. In other words, they suffice to break the symmetry of the egg and to sketch out the fundamentals of the body plan.

The idea of gradient fields as organizers of initial morphogenesis can be traced back to the beginning of this century: in 1901 Theodor Boveri advanced the idea that changes in concentration of some chemical species from one end of the egg to the other might control development. Experiments involving the transplantation of cells in early embryos led the eminent biologist Julian Huxley to propose in 1934 that small groups of cells, called organizing centres or organizers, set up 'developmental fields' in the fertilized egg that are responsible for the early stages of patterning over much larger regions. Transplanting these organizers to different parts of the fertilized egg was found to lead to new patterns of subsequent development, suggesting that the organizers exercise an influence on the cells around it while growth is occurring—the egg need not be pre-patterned before fertilization.

In 1969 the British biologist Lewis Wolpert moulded these ideas into a form that underpins most research on morphogenesis today. Wolpert asserted that the diffusional gradients of morphogens emanating from organizing centres provide *positional information*, letting cells know where they are situated in the body plan. Above a concentration threshold the morphogens switch on genes that set in train a series of biochemical interactions, leading to ever more patterning of the local environment and differentiation of cells into different tissue types.

One problem with the idea of a simple diffusional gradient as the patterning mechanism, however, is that once the single-celled egg has begun to divide into a multicelled body, the diffusing morphogens face the barrier of cell membranes. How can a gradient progress smoothly from cell to cell?

In the most extensively studied of developmental systems, the fruit fly *Drosophila melanogaster*, this problem does not arise. The fruit fly egg is unusual in that it does not become compartmentalized into many cells separated by membranes until a relatively late stage in the growth process, by which time much of the essential body plan is laid down. Like all developing eggs, the fruit fly egg makes copies of its central nucleus, where the genetic storehouse of DNA resides; but whereas in most organisms these replicated nuclei then become segregated into separate cells, the fruit fly egg just accumulates them around its periphery. Only when there are about 6000 nuclei in the egg do they start to acquire their own membranes.

For this reason, morphogens in the fruit fly embryo are free to diffuse throughout the egg in the first few hours after it is laid. After a short time, the egg develops

Fig. 4.33 The embryos of the fruit fly develop stripes soon after fertilization which eventually define the different body compartments. (Photo: Peter Lawrence, Laboratory for Molecular Biology, Cambridge; from Lawrence 1992.)

stripes (Fig. 4.33). These evolve into finer stripes, and as the egg begins to become divided into separate cells, these stripes mark out regions that will subsequently become different body segments: the head, the thorax, the abdomen and so forth.

As this striped pattern suggests, the first breaking of symmetry takes place along the long axis of the ellipsoidal egg. This is called the anterior–posterior axis, the anterior being the head region and the posterior the tail. The initial segmentation process seems to be controlled by three genetically encoded signals: one defines the head and thorax area, another the abdomen, and a third controls the development of structures at the tips of the head and tail. When the respective genes are activated, they generate a morphogen that then diffuses from the signalling site throughout the rest of the egg.

The head/thorax morphogen is a protein called bicoid, which is produced when the gene that encodes this protein is switched on. Production of the bicoid protein takes place at the extreme anterior end of the egg, and the protein diffuses through the cell to establish a smoothly declining concentration gradient (Fig. 4.34a). To transform this smooth gradient into a sharp compartmental boundary (which will subsequently define the extent of the head and thorax regions), nature exploits the kind of threshold switch that I have described earlier. Below a certain threshold concentration, bicoid has no effect on the egg, but above this threshold the protein binds to DNA and triggers the translation of another gene into its protein product, called hunchback. (More accurately, the bicoid protein promotes the formation of the intermediary hunchback RNA molecule from the *hunchback* gene on the chromosome—it is the RNA that is ultimately translated into a corresponding protein.) In this way, a smooth gradient in one molecule (bicoid) is converted into an abruptly stepped variation in another (the hunchback RNA) (Fig. 4.34b, c).

You may have noticed that this patterning mechanism seems to have cheated on the question posed at the outset: how does an initially uniform cell break its symmetry? OK, so the cell in this case is not quite so uniform—it already has a long axis and a short axis. But why should bicoid suddenly be produced at one end and not the other, or indeed in any one region of the cell and not others? The answer seems to be that the egg is acted on from outside in an asymmetric manner. Although the egg itself is initially a single cell, it begins its development as a part of a multicellular body. The single 'germ cell' that will grow into the egg becomes

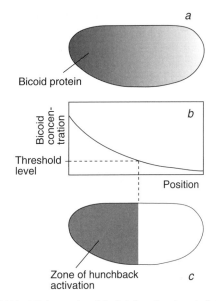

Fig. 4.34 The initial patterning of the fruit fly embryo is controlled by a protein called bicoid, which diffuses along the cell from the anterior end to set up a concentration gradient (*a*). Where the concentration surpasses a certain threshold, the bicoid protein triggers the formation of the so-called hunchback protein (*b, c*). Thus the smooth gradient in bicoid gives rise to an abrupt boundary of hunchback expression.

attached to follicle cells before fertilization, and within this assembly the follicle cells and other specialized entities called nurse cells provide nutrients for the egg cell's growth. The nurse cells deposit RNA encoding the bicoid protein at the anterior tip of the egg while they are still attached to one another, and the bicoid RNA starts to generate bicoid protein as soon as the cell is fertilized. So you see, I'm afraid that there is no wondrous spontaneous symmetry-breaking here as there is in Turing's mechanism—instead, a broken symmetry is passed from generation to generation.

The patterning of the posterior region of the fruit fly egg is controlled by a morphogen called the nanos protein. (*Nanos* is Greek for dwarf, and what with hunchbacks too, you can imagine that there are unfortunate deformities associated with the malfunctioning of these genes.) At some stage after longitudinal segmentation has taken place by the action of these morphogens, the egg has to break another symmetry, between top (where the wings will go) and bottom (where the legs and belly are). This is called the dorsoventral axis, and its direction is defined by a protein called dorsal. The mechanism by which dorsal does its job is rather more complicated than bicoid or nanos, however. The top–bottom gradient is not one in concentration of the dorsal protein—which is actually more or less uniform throughout the egg—but in the protein's location. Towards the bottom, it segregates more strongly into nuclei than into the cell's watery cytoplasm, while the reverse is true towards the top. There appears to be an underlying signal of still uncertain nature that determines whether or not the dorsal protein can find its way into the many nuclei in the egg; this signal is activated from the bottom (ventral) edge of the embryo. Again, the initial impulse for this symmetry-breaking signal seems to come from outside the cell—from a concentration gradient in some protein diffusing through the extracellular medium, which transmits its presence to the egg's interior by interactions at the cell membrane. The way the dorsal morphogen does its job is more complicated too. It is a double switch: above a certain threshold it inhibits the formation of RNA from a pair of developmental genes, whereas above a still higher threshold it promotes RNA formation from a second pair of genes. These gene products are themselves then involved in switching on other developmental processes. Moreover, other molecules called cofactors appear to be able to modify a gene's response to a morphogen, and the cofactors can establish their own concentration gradients. Already we are starting to see why molecular biology seldom lends itself to simple conceptual models: before too long, just about any biological process reveals itself as a sequence of many highly specific steps, in which proteins interact through convoluted pathways to regulate each other's formation.

Do these same initial processes of morphogenetic patterning by chemical gradients apply to other organisms, including us? It seems highly probable that they do, although as I say, most other organisms face the obstacle of cell-to-cell communication in early embryonic development. While there are probably chemical signalling molecules that act as morphogens by switching genes on or off according to their local concentration, they are presumably transmitted from their source region in a stepwise manner—one cell parcelling them out to another—rather than by smooth diffusion. Lewis Wolpert has proposed that morphogens make their way from clusters of cells called zones of polarizing activity (ZPAs) to convey positional information to surrounding cells. It was thought for some time that the small molecule retinoic acid might be a morphogen for limb development in vertebrates, as it appeared to be released from a ZPA at the posterior edge of the developing wing bud of chicks to define the front and back ends (anteroposterior axis) of the wing. But whether retinoic acid indeed has this role is still an open question.

Leg pulling

Not everyone, however, believes that development has to bow entirely to this kind of rigid genetic control. Jim Murray, working with George Oster from the University of California at Berkeley, has postulated a model for structuring and patterning of the body plan at much later stages of an organism's development that involves spontaneous instabilities much like those that give rise to chemical Turing patterns. Murray proposes that structures such as the characteristic hierarchical branching of limb bones or the regular positioning of feathers and scales are a consequence of the interplay of chemical signalling between cells and the mechanical forces that arise in response to these. There are two types of tissue cell: epithelial cells, which aggregate into sheets that constitute the fabric of skin and tissue, and mesenchymal cells, which can pull themselves around using finger-like protrusions called filopodia. Mesenchymal cells will move in response to a variety of stimuli, including gradients in chemical concentrations, in electric fields and in adhesive interactions with a substrate.

Murray and Oster's 'mechanochemical' model of morphogenesis proposes that these signalling mechanisms, particularly those involving chemical gradients set up by diffusion, cause mesenchymal cells to clump together. The traction forces caused by this aggregation, as the cells pull on the surrounding medium, can then establish instabilities that lead to further patterning. For instance, Murray and Oster propose that during limb development a spontaneous instability creates an aggregation of cells along the central axis of an initially uniform cylindrical limb (Fig. 4.35a), which will thicken into cartilage and eventually be mineralized into bone. This process is akin to the formation of a single Turing stripe. But the slightest ellipticity in the cross-section of the central cylindrical aggregate makes it unstable: the traction forces act to accentuate this ellipticity, making the limb flatten out (Fig. 4.35b). At a certain point, the flattening induces a symmetry-breaking bifurcation of the central condensation, causing it to branch (Fig. 4.35c). A subsequent cascade of bifurcations creates the segmentation of the aggregate into the characteristic bone patterns seen in limbs (Fig. 4.35d, e). Moreover, as a central aggregate gets longer and thinner, mechanical instabilities arise in the longitudinal

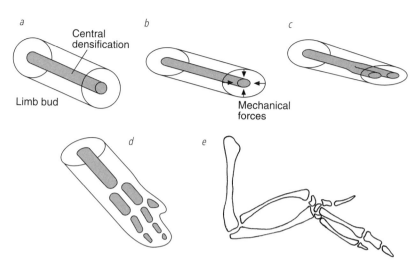

Fig. 4.35 A spontaneous instability in a developing limb bud, due to mechanical forces exerted by cells on their neighbours, creates an increase in cell density along the central axis of the cylindrical bud (a). Any deviation from a perfectly circular cross-section (ellipticity) is accentuated by the mechanical forces, causing the limb bud to flatten (b). When this flattening exceeds some threshold, a bifurcation takes place to produce two axes of densification (c). Subsequent bifurcations and segmentations (d) produce the structures that become cartilage and then bone, as seen in the limb of a 10-day-old chick (e).

direction (along the axis) which create segmentation of the digits.

Within this picture, the characteristic pattern of limb bones seen in many diverse large animals (Fig. 4.36)—

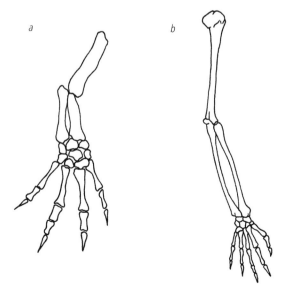

Fig. 4.36 The same sequence of bifurcations and segmentations as in Fig. 4.35e is seen in the bone structure of the limbs of many animals, including the salamander (a) and humans (b).

the division of a single radius into a bifurcated ulna and then into a series of segmented digits—is posited as an inevitable outcome of the physical forces that are acting, not a structure determined arbitrarily by genetics. The model of Murray and Oster—which, incidentally, has been advanced on a far more rigorous mathematical basis than the qualitative description given here—can also account for the polygonal patterning of feathers in birds and scales in fish and reptiles. Feathers are initiated from 'primordia', areas of thickening of the embryonic bird's epidermis caused by an aggregation of underlying dermal cells in the skin. The primordia are arrayed in roughly hexagonal patterns (familiar from the skin of the Christmas turkey), and in Murray and Oster's model these patterns are the mechanochemical equivalent of hexagonal Turing patterns (Fig. 4.3a), arising through spontaneous symmetry breaking.

If Murray and Oster are even partly right, these processes suggest that there are certain 'fundamental' structures of organisms that are not at all determined by the arbitrary experimentation and weeding out that evolution is thought to involve. Instead, these structures have an inevitability about them, being driven by the basic physics and chemistry of growth. If life were started from scratch a thousand times over, it would every time alight on these fundamental structures eventually. Within the parlance of modern physics, they are attractors—stable forms or patterns to which a system is drawn regardless of where it starts from. Within this picture are echoes of the ideas of the eighteenth-century zoologist Etienne Geoffroy de St Hilaire, who believed that there might be certain ideal, Platonic forms in living organisms, from which all other forms are derived by modifications of greater or lesser extent.

This is an extremely contentious idea, since at face value it challenges one of the central tenets of Darwin's theory: that evolution advances by selection from a pool of random mutants. The concept of morphogenetic attractors introduces an element of determinism to this randomness. But even if the protagonists of this concept turn out to be validated, that would not by any means bring Darwin tumbling from his pedestal. There is absolutely no question that natural selection operates in the real world and that it has produced the tremendous variety of organisms with which we share the planet. The idea that this process of mutation and selection might be modulated by other factors is not by any means new in itself, and is hard to doubt. Geological forces have undoubtedly shaped the evolution of the living world: continental drift has isolated sub-

populations of species and caused them to diverge, for example, and ice ages and at least one huge meteorite impact have profoundly altered survival prospects in the prehistoric world. No one argues, meanwhile, that nature's palette is not constrained by the rules of physics and chemistry. If the formation of patterns by symmetry-breaking proves to pose limitations on evolutionary choices, that will add just one more nuance to Darwin's towering achievement.

Patterns in bloom

Probably the best candidate system for the identification of Platonic forms in development is the arrangement of leaves on a plant stem. It isn't hard to imagine all sorts of ways in which leaves could be placed up the stem; but if you go out into the garden or park you will soon discover that there are just three basic patterns. Something seems to be placing rather severe constraints on the options.

Most commonly (in 80% of plant species), leaves execute a spiral up the stem, with each leaf displaced above the one below by a more or less constant angle

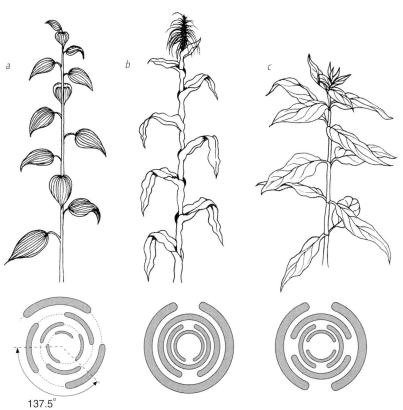

Fig. 4.37 Three distinct patterns can be identified in the arrangement of leaves around plant stems (phyllotaxis): (a) spiral, (b) distichous and (c) whorled. Below each drawing I have shown a schematic representation of the leaf pattern seen from above, with successive leaves depicted as smaller the farther they are down the stem.

(Fig. 4.37a). The potato plant, for instance, has this arrangement. The angle of offset is close to 137.5° in many different species, an observation that begs for an explanation. There is, we shall see, something a little spooky about this angle. A second arrangement, called distichous, places successive leaves on opposite sides of the stem, usually with the leaf wrapped almost fully around the stem (Fig. 4.37b). We could regard this as a form of spiral in which the offset angle is 180°. The third pattern, called whorled, has little clusters (whorls) of leaves—two or more—at regular intervals up the stem, with each whorl offset so that it sits over the gaps of the whorl below. A common whorled pattern juxtaposes two leaves 180° apart offset at an angle of 90° from the two below (Fig. 4.37c). Mint has this arrangement, and so does the stinging nettle. The formation of these patterns is called phyllotaxis ('leaf ordering'), and it turns out to have some remarkable mathematical properties.

When I first observed these arrangements for myself, I assumed that they were clever adaptations selected because they give the leaves maximum exposure to sunlight. You can be sure that arrangements that failed to do this would be selected *against*, but a closer investigation of phyllotactic patterns reveals that there must be more here than Darwinian selection from a random pool of possibilities. They have a mathematical structure in which we can surely see the fingerprint of some *physical* mechanism at work.

The arrangement of leaves along a stem provides us with a somewhat distorted version of the true growth pattern, which becomes extended along the stem axis. Plants grow from the tip of the stem, where one finds a

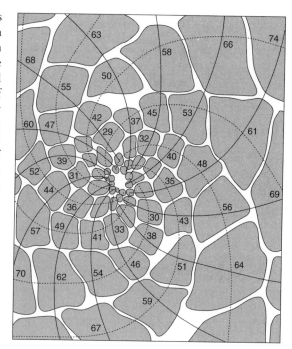

Fig. 4.39 The pattern of spiral phyllotaxis in the monkey puzzle tree. Here I show the projection of the pattern onto a two-dimensional plane, looking down the axis of the branch. Leaves are numbered consecutively from the youngest, and the two systems of spirals (solid and dashed lines) indicate leaves that are in contact with one another. (After: Goodwin 1994.)

bud of multicellular tissue called the meristem. Here cells are multiplying rapidly, and just behind the advancing tip (the apex), side buds called primordia begin to protrude one by one. These will subsequently develop into leaves (Fig. 4.38). There is a roughly constant time interval, called the plastochrone, between the formation of successive primordia, with a typical duration of one day. The leaf pattern is determined by *where* around the boundary of the apex the primordia appear. As the stem grows upwards, the positions of successive primordia trace out a spiral when seen from above. One can see this more clearly by projecting the leaf positions onto a plane perpendicular to the stem (Fig. 4.39). Here some of the leaves are numbered according to the sequence in which they developed, and lines are drawn through leaves that are in contact with one another. These trace out *two* systems of spirals, which twist in opposite directions. The double-spiral pattern is more immediately evident when the primordia develop not into leaves but into florets in a flower head (Fig. 4.40), since in that case they remain all in the same plane.

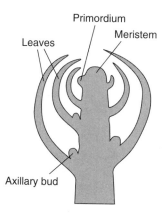

Fig. 4.38 The pattern of phyllotaxis is determined at the tip of the growing stem (the meristem), where the leaf buds (primordia) are initiated. (After: Koch and Meinhardt 1994.)

a

b

Fig. 4.40 The double spiral pattern of phyllotaxis is particularly clear in the arrangement of florets in a flower head (*a*) and of leaflets in a pine cone (*b*). (Photos: Scott Camazine, Pennsylvania State University.)

Golden wonder

The regularity of these spiral patterns has long been seen as the expression of mechanical laws that govern phyllotaxis. W. Hofmeister proposed in 1868 that each new primordium appears periodically on the apex boundary at an interval equal to the plastochrone, and in a position corresponding to the largest gap left by the preceding primordia. In other words, the primordia are simply trying to pack efficiently, just like atoms in a crystal. In 1904, A.H. Church took this idea further in a book called *On the Relation of Phyllotaxis to Mechanical Laws*, from which Fig. 4.39 is derived. And in 1979 H. Vogel performed computer calculations which showed that the preferred angle of 137.5° allows for the

optimal packing of primordia placed sequentially along a spiral. Yet there is a richness to the spiral patterns for which these simple packing considerations cannot fully account.

Travelling out along any one of the lines in Fig. 4.39, you will find that the leaf numbers differ from one another by eight along the dashed lines and by 13 along the solid lines. This construction permits a classification of the phyllotaxis pattern—it is denoted (8, 13). Examples from other monkey puzzle branches show other phyllotactic relationships—(5, 8), for instance, and (3, 5). To a mathematician, these pairs of numbers have a familiar ring. They are all adjacent pairs in a well-known mathematical sequence called the Fibonacci sequence, first defined in 1202 by the Italian mathematician Leonardo of Pisa, nicknamed *Filius Bonacci* or Fibonacci. Each term in the sequence is constructed by adding together the previous two, starting with 0 and 1. Thus, $0 + 1 = 1$, and the first three terms are 0, 1, 1. The next is $1 + 1 = 2$, then $1 + 2 = 3$, then $2 + 3 = 5$ and so on. The series runs 0, 1, 1, 2, 3, 5, 8, 13, 21, 34 ... .

Straight away we can see the adjacent pairs (3, 5), (5, 8) and (8, 13). But it turns out that the phyllotaxis classifications of leaves, petals or floret patterns in *any* plant species correspond to pairs in this series. A corollary of this is that the number of petals on most flowers corresponds to a Fibonacci number: buttercups have five, marigolds have 13, asters 21.

More mathematical spookiness follows. The ratio of successive terms in the Fibonacci series gets closer and closer to a constant value the further along the series one progresses: $8/13 = 0.615$, for example, and $13/21 = 0.619$. This ratio approaches a value of 0.618034 to the first six decimal places. This number was well known to the ancient Greeks, who knew it as the Golden Section. It can also be expressed as $(\sqrt{5}-1)/2$, where $\sqrt{5}$ is the square root of 5. To the Greeks, this was a harmonious, almost mystical constant of nature. If you want to draw a rectangle that can be subdivided into a square and a smaller rectangle with the same proportions as the original one (but reduced in scale), the ratio of the two sides must be equal to the Golden Section (Fig. 4.41*a*). These proportions were considered by the Greeks to be pleasing to the eye, and they based the dimensions of many temples, vases and other artefacts on this ratio. There is a long-standing idea that for a perfectly proportioned human body the ratio of the height of the navel to the total height (and also some other bodily proportions) is equal to the Golden Section. It is also related to the log-

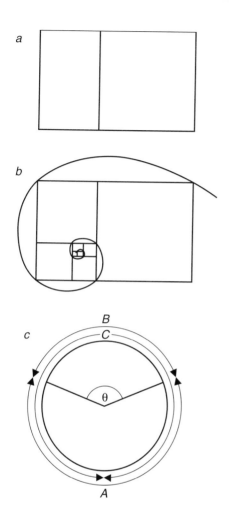

a

b

c

B
C
θ
A

Fig. 4.41 There is a rectangle of unique proportions that can be divided up into a square and a smaller rectangle that has the same proportions as the larger one (*a*). The ratio of equivalent sides of the larger and smaller rectangles is equal to the Golden Section. If we continue to divide the smaller rectangles in the same way, their equivalent corners trace out a logarithmic spiral (*b*). The Golden Angle is the angle at the apex of a segment of a circle of circumference *C* that sweeps out an arc of length *B* such that *B*/*A* = *A*/*C* (*c*). This angle (θ) is about 137.5°.

meter does to the circumference of the whole circle (Fig. 4.41*c*). The Golden Angle is that at the apex of the segment. And it is equal to 137.5°—the angle at which successive leaves are commonly offset along a plant stem in spiral phyllotaxis! This correspondence between the most common phyllotactic divergence angle and the Golden Angle was first identified, to their surprise, by the mathematicians L. and A. Bravais in 1837.

If this all seems like number-juggling akin to the numerology of end-of-the-world prophets, rest assured that it is mostly an expression of the same basic fact. Once we have established that leaves spiral up a stem with offsets of the Golden Angle, then all the rest—the relationship to the Fibonacci series and to the Golden Section—follows. Ian Stewart explains why in his book *Nature's Numbers*.

Phyllotaxis, therefore, contains a hidden mathematical pattern for which we are unlikely to find an explanation by rooting around in the genetics of plant developmental biology. It seems likely that there is some more universal basis to these observations.

That this is so was impressively demonstrated by the French physicists Stéphane Douady and Yves Couder in 1992. They performed an experiment in which they dropped tiny droplets of a magnetic fluid onto a disk covered with a film of oil, on which the droplets floated. The apparatus sat in a vertical magnetic field, which polarized the magnetic particles and caused them to repel one another. The researchers also applied a horizontal magnetic field, which was stronger at the periphery of the disk than at its centre—this pulled the

arithmic spiral (Chapter 1), which is traced out by the extremities of a series of rectangles growing in the successive proportions of the Fibonacci sequence (Fig. 4.41*b*). The Golden Section is commonly held to be one of nature's 'special' numbers, like π or *e*—but one particularly intimate to the geometry of life.

Now, the Golden Section has a 'Golden Angle' associated with it. This is most easily visualized by dividing a circle into a segment whose perimeter stands in the same ratio to the rest of the circle as the latter's peri-

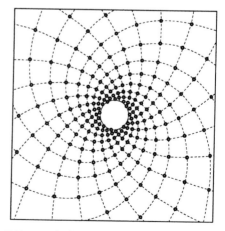

Fig. 4.42 Magnetic droplets moving from the centre to the edge of a round dish while repelling one another trace out spirals of the same kind as those observed in phyllotaxis. Yet here there are clearly only physical forces at play. (After: Douady and Couder 1992.)

droplets outwards towards the edge. Thus, as the droplets fell one by one, they were pushed out to the edges of the disk while repelling one another. When the droplets were added at a fast enough rate, they travelled outwards to form a spiral pattern just like those seen in phyllotaxis (Fig. 4.42), with successive droplets diverging at an angle of about 137.5°. Interestingly, when the rate of droplet addition was low enough, successive droplets diverged at 180° instead (since in this case each droplet was simply repelled by the previous one, the others being too far away)—the pattern then corresponds to distichous phyllotaxis (Fig. 4.37b). Under some conditions other divergence angles were seen, which correspond to other, more rare divergence angles seen between leaves that exhibit spiral phyllotaxis.

All very well—except that growing plants are not magnetic droplets! But what Douady and Couder were setting out to test was the idea that phyllotaxis at the Golden Angle is preferred because it allows the optimal packing together of primordia arranged around a spiral on the meristem. They suggested that their experiment, in which the droplets repel one another along spiral trajectories, reproduces these same packing effects. Their findings imply that a plant need not somehow 'know' from the outset that 137.5° spiral phyllotaxy is the best choice—on the contrary, the *dynamics* of the growth process automatically select this angle. If you like, each plant 'finds out' this solution as it grows.

This brings us back to attempts to capture the dynamics of patterned biological growth using reaction–diffusion models. Can such models reproduce the spiral phyllotaxis patterns?

There are at least two separate positioning mechanisms at work in this process. One must tell the primordia how far apart they should be along the stem's axis. This mechanism in effect specifies the interval between inception of primordia—the plastochrone. The other mechanism specifies where around the stem's circumference the primordia should develop—say, at a 137.5° angle from the primordium below for the case of a typical spiral phyllotaxis pattern. This is called the azimuthal position.

Experiments on plant growth dating back to the 1940s have shown that the axial position of a primordium is controlled by a chemical mechanism—specifically by plant hormones that are produced at the apex and transported towards the roots. Hans Meinhardt and André Koch have used this observation as the basis of a reaction–diffusion model in which the hormones act as inhibitors that repress primordia for-

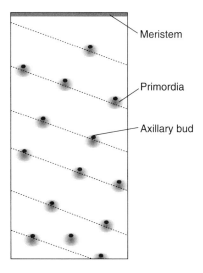

Fig. 4.43 Spiral phyllotaxis can be generated in a reaction–diffusion model of patterning on a cylindrical plant stem, here shown rolled out into a flat sheet. The spiral sequence of primordia is indicated by dashed lines. New primordia develop below the meristem at the top of the cylinder. (After: Koch and Meinhardt 1994.)

mation in a given region of the stem until the tip has grown far enough beyond this region for the hormone concentration to fall below a certain threshold value. Once this long-ranged inhibition becomes sufficiently weak, some local activator molecules switch on cell proliferation to induce the budding of a primordium.

In this model, a second activator–inhibitor mechanism controls the azimuthal position of the primordia. As this position is influenced by long-range inhibition, primordia cannot pack too closely together, just as spots in a Turing pattern cannot come too close. In a sense, this is an expression of the packing effects first suggested by Hofmeister. Meinhardt and Koch carried out calculations to find the primordia patterns that their model would produce on an idealized plant stem modelled as a narrow, hollow cylinder. They found that the primordia (and thus ultimately the leaves) became positioned along a spiral winding up the stem in a (2, 3) phyllotaxis pattern—one of the Fibonacci pairs observed in nature (Fig. 4.43). By making some simple and reasonable assumptions about how cells differentiated around the primordia, Meinhardt and Koch were even able to account for the formation of the little 'secondary' buds called axillary buds seen just above the developing leaf where it joins the stem in real plants. (I'm told to remove these from my tomato plants to ensure a good yield of tomatoes, and have often wondered why they were there in the first place.)

There is no direct evidence for this pattern-forming mechanism in phyllotaxis, although the role of plant hormones suggests that it is not unreasonable. But it shows that even quite complicated body shapes in living organisms can be plausibly explained by the chemical processes of self-organization and spontaneous pattern formation that Alan Turing dreamed up over four decades ago.

BRANCHES

*The ruddy clouds float in the four quarters of the caerulean sky
And the white snowflakes show forth their six-petalled flowers.*

Hsiao T'ung
Sixth century AD

In Mike Leigh's film *Nuts in May*, a pair of gauche campers named Keith and Candice Marie discover how a kind of modern-day vitalism colours our preconceptions about complex growth and form. They take a trip to a local quarry to search for fossils in the ancient Purbeck limestone of Dorset. At the prompting of a quarryman, the unsuspecting couple find a delicate, plant-like pattern traced out in the stone. 'Look at that—seaweed!' exclaims Keith. 'Yar, well, 'tis not seaweed, see', drawls the quarryman, adding that it is manganese oxide—a mineral—they are looking at. Keith is not convinced. 'It looks like a living organism to me.' 'Yar,' the quarryman rejoins, 'most people think that.'

You can see why Keith and Candice Marie jumped to the wrong conclusion. The structures they saw are called mineral dendrites, and they look for all the world like the kind of forms we associate with ferny plants (Fig. 5.1). Even the name derives from this source: *dendros* is Greek for tree. But the filigree structures contain no fossil material—they are made up of manganese or iron oxides, chemical deposits precipitated when a solution containing manganese or iron ions was squeezed through cracks in the rocks in the geological past.

It is scarcely surprising that we might think these deposits had a biological origin. The branched pattern finds countless echoes in the living world, from corals to leaf veins to the bronchial structure of the lung (Fig. 5.2). This ubiquity of branched formations in both the living and inorganic worlds begs the question of whether their formation can be ascribed some unifying features, in line with the idea of universality in pattern formation that I laid out at the start of this book. To turn that idea on its head, perhaps complex forms like these do not require the kind of complicated causes that only the living world can engender.

Ah, but not all branching forms are alike. Tree lovers know that much—they can identify a tree in winter simply from its silhouette on a hilltop, in which the pattern of branches has a characteristic form more or less

Fig. 5.1 A mineral dendrite of manganese oxide, found on the surface of limestone from Bavaria. (Photo: Tamás Vicsek, Eötvös University, Budapest.)

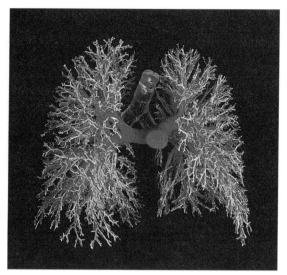

Fig. 5.2 The bronchial/arterial structure of the lungs has a highly branched, tree-like form. (Photo: Martin Dohrn/Royal College of Surgeons/Science Photo Library.)

Fig. 5.3 The branching patterns of trees are often characteristic of their species.

unique to each species (Fig. 5.3). The human mind is intriguingly adept at this sort of pattern recognition. I doubt if too many tree experts could give a precise explanation for how they distinguish one system of branches from another—they might be able to identify a few pronounced features such as the sharpness of the angle between diverging branches, but it wouldn't by any means amount to the kind of mathematical criterion that could be programmed into a computer to give it the same facility for telling apart an elm from a sycamore. We just seem able to 'sense' the pattern.

Yet in recent years, scientists have developed tools for assessing in a mathematically precise way the generic features of different branching patterns, and by doing so, have been able to provide clear and unambiguous criteria for distinguishing one such form from another. These tools have played a crucial role in allowing us to understand how branched forms grow, because only through them do we have a definite, quantifiable means of determining how close a given physical or biological model comes to reproducing the form observed in reality. As a result, the study of branching patterns has evolved into an exact science.

At the same time, a better appreciation has developed of the relationships between different types of branching pattern. Consider, for instance, the patterns shown in Fig. 5.4. All are generated in precisely the same apparatus, simply by varying the experimental conditions.

All of the forms are clearly branched, but it is equally apparent that they are qualitatively different from one another—as distinct as a naked oak from a poplar. So it appears that different branched forms can be the product of very similar formation processes.

And most intriguingly of all, we can recognize qualitatively similar branching patterns in a huge variety of different physical and organic systems. In Fig. 5.4 you might see echoes of mineral dendrites, of snowflakes, of ink blots. This leads us to the compelling conclusion that there is indeed something generic—something universal—about these forms, and by extension, about the rules for their formation. What are the rules?

Organic crystals

The mineral dendrites in Fig. 5.1 are crystals. But let's face it—this is not exactly what we are taught to believe

a   b   c

Fig. 5.4 Branching patterns in the Hele-Shaw cell, in which a bubble of a fluid such as air is injected under pressure into a more viscous fluid (see p. 118). The same apparatus can produce several different types of pattern, including symmetrical snowflake-like figures (*b*). (Photo: Eshel Ben-Jacob, Tel Aviv University.)

crystals look like. Think of mineral crystals and I wager that you will envisage the facetted, blocky shapes of gemstones (Fig. 5.5), in which the flat, bevelled faces simply reflect the regular 'greengrocer stall' stacking of the constituent atoms. Yet this regular stacking of atoms is present too in mineral dendrites, as one can verify by looking at the geometrical patterns produced by bounc-

Fig. 5.5 We normally think of crystals as possessing facetted, geometric shapes derived from the regular packing of their constituent atoms. How, then, can crystals acquire the 'organic' form of Fig. 5.1? (Photo: Steve Smale, City University of Hong Kong.)

ing X-rays off them. The mineral dendrites have apparently chosen to ignore this underlying geometric symmetry at the atomic scale and to grow instead into a ramified, 'organic' form.

One can grow crystals like this in the laboratory. One way is to use a process called electrodeposition, in which a crystalline deposit of a metal is grown at an electrode from a solution of metal ions by applying a voltage between this and a second electrode. The metal ions are positively charged—each of the metal atoms has lost one or more electrons, which are negatively charged particles. So these metal ions will be electrically attracted to a negatively charged electrode. If we allow a current to flow through the circuit, electrons can pass to metal ions drawn onto the electrode surface, and these are thereby converted back into neutral metal atoms. The atoms stack together in a crystal that grows outwards from the negative electrode.

This process is used to cover metals with a smooth veneer of fresh metal in the technique called electroplating. A steel component can be copper-plated by having it act as the negative electrode in an electrochemical cell in which it is immersed in a solution of copper ions. The voltage for electroplating is chosen so that the copper atoms are deposited at the negative electrode in a flat, slowly growing film that covers the entire surface evenly. But if a higher voltage is applied, the electrodeposit grows more quickly—that is to say, out of equilibrium. Then the smoothness is lost and instead we obtain a very irregular, branched deposit at the negative electrode (Fig. 5.6*a*). A closer, microscopic view of these branches reveals that they are after all

a

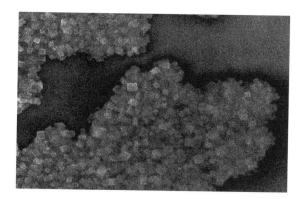

b

Fig. 5.6 (a) A branching metal formation produced by electrochemical deposition onto a central electrode. (b) Seen close up, the branches consist of conglomerates of tiny crystallites oriented at random. The image here is at a magnification of ×7580. (Photos: (a) Mitsugu Matsushita, Chuo University; (b) Vincent Fleury, Laboratory for Condensed Matter Physics, Palaiseau.)

contact. The clusters grow one particle at a time, and the particles simply diffuse through the air until they encounter a part of the cluster. In Chapter 4, I explained that particles undergoing diffusion execute a random walk as they stagger from collision to collision with the molecules of the medium in which they move. So there is no preferred direction from which new particles impinge on the cluster. Because the rate of growth is governed by the rate of diffusion of the particles, Witten and Sander called their model *diffusion-limited aggregation* (DLA).

It differs from the way in which regular, facetted crystals grow, in that there is no opportunity for the impinging particles to rearrange themselves so that they pack together most efficiently. This kind of rearrangement takes place at the surface of a crystal that is growing slowly, since the atoms at the surface can generally pass back into solution again, or move across the surface, until the most 'comfortable' arrangement is achieved (that is, the one with the lowest energy). In this way the crystal finds its way to a regular stacked arrangement of atoms. Because in DLA there is no chance of such reshuffling (particle attachment is irreversible), the surface of the growing cluster soon becomes very jagged and disorderly. The requirement that new particles stay where they first touch the cluster is a realistic one for clusters that grow very quickly, and it means that mistakes in packing get frozen in. This is why branched electrodeposits form when electrodeposition is conducted quickly (that is, at high voltages) while slower deposition gives smooth films.

composed of tiny facetted crystallites fused together in jumbled disarray (Fig. 5.6b).

In 1984 Robert Brady and Robin Ball from the University of Cambridge showed that a theoretical model developed three years earlier by American physicists Tom Witten and Len Sander could account for the shape of these branched electrodeposits. Witten and Sander had in fact set out to model the way particles of dust form aggregates in air. This they described as a growth process in which small particles form clusters by sticking together as soon as they come into

Fig. 5.7 A cluster grown by the diffusion-limited aggregation model. (Image: T. Rage and P. Meakin, University of Oslo.)

The branched clusters are another example of *non-equilibrium* structures.

The DLA process can be simulated on a computer by introducing particles one by one into a box from random points around its edges and allowing them to diffuse until they encounter and stick to a particle at the box's centre. This generates a cluster that grows steadily outwards from the central point, developing tenuous branches as it goes. Figure 5.7 shows the result when this process is conducted in two dimensions (in a flat plane). The cluster looks very similar to the structures created by non-equilibrium electrodeposition in a flat cell (Fig. 5.6a), something that was first recognized by Mitsugu Matsushita of Chuo University in Japan and co-workers in 1984. What Brady and Ball showed at much the same time was that this similarity persisted for three-dimensional growth too. They proposed that the mechanism of non-equilibrium electrochemical growth shares the same broad features as the DLA model—random diffusion of ions and irreversible attachment to the electrode deposit.

While there is no doubting the broad validity of this connection, the reality is much more complex. For one thing, in real electrodeposition the ions have to pick up electrons before they become part of the cluster, and the DLA model contains no such step. More fundamentally, unlike electrodeposition there is no electric field in the DLA model—as we'll see, the strength of the field can profoundly influence the shape of the branched electrodeposit. And while the diffusing particles in electrodeposition are ions, the particles that actually make up the deposit are, as we've seen, tiny crystallites containing many thousands of ions (Fig. 5.6b). A proper description of the process would therefore have to account for the formation of these crystallites (which seem to have the blocky appearance of normal crystals) followed by their irregular assembly into branched structures. Vincent Fleury of the Laboratory for Condensed Matter Physics in Palaiseau, France, has suggested that the formation of the crystallites takes place in an oscillatory manner: a crystallite is nucleated and sticks to the deposit, then a short interval later another does so, and so on. The branching pattern then arises from the interplay between this oscillatory crystallite growth and the randomizing thermal noise in the environment in which they appear.

From bumps to branches

These are complicated issues, so it is simplest to stay for now with the DLA model. It isn't hard to see why this model produces very imperfect, irregular clusters, since aggregation takes place following random diffusion. But why are the clusters branched? We could perhaps imagine instead the formation of a dense mass with a highly irregular edge, like a spreading ink blot. Why is this not what happens?

The answer is that the model possesses an instability that amplifies any small bumps or irregularities, causing them to extend into fingers rather than becoming smoothed out again. Look at the cluster in Fig. 5.7: it's not hard to imagine that a particle taking a tortuous, meandering path through the surrounding medium is likely to encounter one of the branch tips before being able to penetrate very far down the channels between them. So once they are formed, the branches tend to grow from their tips while the gaps in between them get ever less accessible to new particles.

Preferential growth at a tip ensures that any tiny bumps formed by chance at the cluster surface will have a tendency to grow faster than flat parts of the surface, because there is a better chance that a randomly diffusing particle will hit it (Fig. 5.8). And crucially, this growth advantage is self-enhancing—the more the bump develops, the greater the chance of new particles striking and sticking to it. The probability of this is always greatest at the very tip of the bump, since this is the most exposed part. So the slightest small bump soon grows into a sharp finger. Because irregularities are springing up by chance all over the surface all the time, the deposit becomes increasingly branched, with each new tip constantly sprouting extra appendages. Notice how essential to all of this is the random, diffusive motion of the particles: if they were instead all propelled towards the cluster along straight trajectories, the edge of the cluster would just grow uniformly along the direction of particle motion.

Fractal form

The DLA cluster in Fig. 5.7 certainly *looks* like the electrodeposit in Fig. 5.6a, but that doesn't constitute very good proof that the two are related in any fundamental way. Scientists don't consider it very good form to judge a model against reality by comparing qualitative aspects like appearances—they want to look for correspondences in hard, quantifiable terms. This is a requirement that dogs much of the work on pattern formation, because patterns are often, by their very nature, not very susceptible to precise numerical characterization.

Yet it turns out that even forms as apparently irregular as these branched aggregates have a measurable

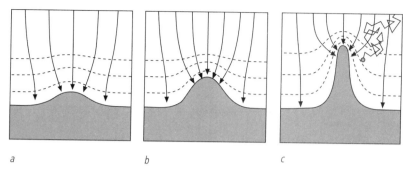

Fig. 5.8 In the DLA model, small protrusions on the surface of the growing cluster accumulate new particles faster than the surrounding flat surface, and so they become increasingly accentuated as growth progresses. Such protrusions will themselves acquire surface irregularities that will blossom into new fingers, and so the cluster quickly becomes highly branched. Here the *dashed lines* show contours of constant average density of incoming particles, and the *solid lines* show the average flow of the particles. Each individual particle takes a highly tortuous path, however—a random walk. One such is depicted in (*c*).

property that is almost as precise, reproducible and characteristic as the number of legs on an insect. It is called the fractal dimension, and is a measure of how densely packed the branches are. A cluster growing in two dimensions so densely that there are no gaps at all between the particles presents a solid black mass whose outer edge simply expands as the cluster gets larger. In that case, the number of particles in the cluster (N) increases in proportion to the area of the cluster, or equivalently, to the square of the cluster's size (its radius r, say):

$$N \propto r^2 \tag{5.1}$$

(where the $\propto$ symbol means 'is proportional to'). If, on the other hand, the branches are instead simply linear chains of particles, so that they form a many-pointed star like an asterisk, N increases in direct proportion to the size r:

$$N \propto r \tag{5.2}$$

For a DLA cluster, the rate at which N increases with size lies somewhere between these two extremes—N is proportional to r raised to some power d_f in between 1 and 2:

$$N \propto r^{d_f} \tag{5.3}$$

The value of d_f—called the *fractal dimension*—is 1.71 for two-dimensional DLA. This means that the cluster fills up the two-dimensional plane rather more completely than a star-like cluster but less fully than a dense, approximately circular cluster.

The fractal dimension is a robust property of the DLA growth process—it stays the same as the cluster grows

bigger, and two different DLA clusters, while differing in the precise positions and convolutions of their branches, will have exactly the same value of d_f. If, meanwhile, we change the rules that govern the growth of the clusters, for example by allowing new particles to make a few short hops around the surface before finally sticking irreversibly, we will obtain a branched cluster with a different value of d_f. Sometimes changes like this will produce very marked changes in the appearance of the clusters—they might develop very stout or very wispy branches, for instance—but the effect of other changes might be rather subtle, so that by visual inspection we will be unable to say whether the clusters are 'the same' or not. The fractal dimension provides a well-defined measure by which we can distinguish such differences.

In Fig. 5.9 I show another mineral dendrite, formed from manganese oxide in the plane of a crack that passes through a quartz crystal. Is this the same kind of cluster as that in Fig. 5.1? By eye, I wouldn't place bets. But by calculating its fractal dimension, we can pronounce confidently that the two are different—the earlier dendrite has a fractal dimension of 1.78, whereas for the latter it is about 1.51. You can perhaps see that the smaller the fractal dimension, the wispier the cluster.

Branched electrodeposits like that in Fig. 5.6*a* commonly have a fractal dimension of about 1.7, and this can give us confidence that their mechanism of formation shares something in common with the DLA process. (In three-dimensional growth, a DLA cluster has a fractal dimension of about 2.5, while Brady and Ball showed that electrodeposits grown in three dimensions have a fractal dimension of around 2.43.) What

Fig. 5.9 Another mineral dendrite, this time formed inside a quartz crystal. (Photo: Tamas Vicsek.)

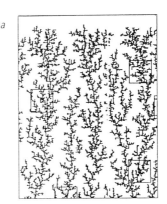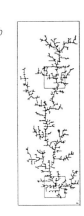

Fig. 5.10 Mineral dendrite patterns generated by a reaction–diffusion model in which particles diffuse by random walks, as in DLA, before encountering each other and reacting to form the dark deposit. The model is able to generate dendrite forms with fractal dimensions of 1.78 (*a*), similar to that in Fig. 5.1, and 1.58 (*b*), close to that in Fig. 5.9. (The square boxes denote regions selected for determining the fractal dimension.) (From: Chopard *et al*. 1991.)

about the mineral dendrites? You might have guessed from their shapes alone that DLA would be a good model for their formation process too, but we now discover that two mineral dendrites can have fractal dimensions not only different from that of a DLA cluster but from one another.

Swiss physicist Bastien Chopard and colleagues have shown that these observations can be rationalized in terms of a more sophisticated adaptation of the DLA model, in which the ions that form the mineral dendrite diffuse through cracks in the surrounding medium of the rock and then undergo a chemical reaction. In this model the process by which dissolved manganese ions permeate through the rock and react with oxide ions to form the dark deposit is emulated by two soluble chemical species A and B that diffuse through the medium and react to form a dissolved compound C when they encounter one another. If enough C accumulates in a particular region, the solution becomes over-saturated and C precipitates in the form of a dark deposit D, which then stays put. If, on the other hand, a single C particle encounters a cluster of D, it too will precipitate as D. Although couched in different terms, this is actually a reaction–diffusion model like those encountered in Chapters 3 and 4; but whereas those were constructed in terms of smoothly varying *fields* of chemical reagents, here the model contains discrete particles that diffuse and react.

Chopard and colleagues found that the model generates fractal clusters much like real mineral dendrites (Fig. 5.10). Although they look similar to DLA clusters, the fractal dimension of the model clusters varies depending on the concentration of species B: the researchers were able to generate simulated mineral dendrites with fractal dimensions of 1.75 and 1.58 (close to the values for the two natural samples shown here) by changing this concentration.

Fractals everywhere

Fractal objects have become quite the fashion. Probably the most famous are the mathematical fractals discovered in the 1970s by Benoit Mandelbrot at IBM's research centre in Yorktown Heights, New York. They are exemplified by the bulbous black kidney shape of the so-called Mandelbrot set (Fig. 5.11). This object, with its familiar accoutrements of wispy tendrils and furiously (and spuriously) coloured spirals, erupts out of an abstract mathematical plane when one plots on the plane points corresponding to the solutions of a mathematical equation (the images are typically decorated with colours that denote the number of iterative computational steps needed to converge on a particular solution to the equation for each point on the plane). These baroque patterns have been depicted and described at length in several popular books, and I shall make just two observations here. First, nothing much like them is seen in nature—if, as Mandelbrot has suggested, they are 'monsters', then they are of a par-

Fig. 5.11 The Mandelbrot set, a mathematical fractal that defines the boundary between the 'basins of attraction' for solutions to an 'iterated mapping' equation. That is, you start at a point in the plane and move to a new point calculated from the mapping equation. By applying this mapping again and again, you move towards one of two basins of attraction whose boundary forms the fractal perimeter of the kidney shape.

ticularly Platonic variety. The fractals that we see in the natural world do not generally have the 'symmetry' evident in the Mandelbrot set (which is the product of a rather esoteric and exact mathematical procedure); they are irregular, like a branching DLA deposit, because they are formed in a noisy, random environment. This noisy aspect of natural fractals is one of the central messages of Mandelbrot's seminal book *The Fractal Geometry of Nature*.

Second, I should explain why the Mandelbrot set and its ilk are fractals at all, since they don't look much like DLA clusters. What both the Mandelbrot set and DLA clusters have in common—along with all other breeds of fractal structure—is the property of scale invariance. That is to say, they look more or less the same on all size scales. Take a region of a DLA cluster and magnify one corner 10 times, and you'll see a convoluted, branched cluster that looks much the same as the original region. Do the same again to a corner of the magnified region, and you get the same result. What this means is that you can't judge the scale of a fractal structure from appearances, because there's no natural yardstick. In contrast,

you can quite easily assess the scale of an aerial photo of a town because there will be features, like cars, houses and roads, that provide a known measure of length. (Because many geological structures are fractals—as we shall see—geologists often leave a hammer evident in photographs of rock faces to provide a reference scale.) Like a DLA cluster, the Mandelbrot set exhibits the type of scale invariance known as self-similarity, which implies that as you magnify one region of it again and again, that ominous black bulb keeps reappearing like a malformed Russian doll.

Once you get the hang of what a fractal structure looks like, you see them everywhere. In trees, certainly, and in their roots too. In rivers, mountain ranges, clouds, coastlines. Small wonder that Mandelbrot was moved to proclaim that fractals are 'the geometry of nature'. And without a doubt the branched fractals typified by DLA clusters do represent one of nature's universal forms, and are splendid examples of how complex, 'life-like' forms can be the product of relatively simple and entirely non-biogenic processes. In the face of all of this, however, it is worth making two cautionary points:

1. Not all branched patterns are fractal (later we shall see some that are not).

2. Just saying that a structure is fractal doesn't bring you any closer to understanding how it forms. There is not a unique fractal-forming process, nor a uniquely fractal kind of pattern. The fractal dimension can be a useful measure for classifying self-similar structures, but does not necessarily represent a magic key to deeper understanding.

Squeeze patterns

Spanish physicist Juan Manuel Garcia-Ruiz recounts how he was assailed by fractal branches even as he sat down for a quiet cup of coffee in the Hotel Los Lebreros in Seville. Across the broad plate-glass windows of the coffee shop were creeping three of Mandelbrot's monsters, like virtual plants growing in the glass (Fig. 5.12). The window panes each consisted of three laminated glass sheets, separated from one another by thin films of a polyvinyl plastic. The laminates were imperfectly sealed at their edges, so that air could find its way between the sheets. In the heat of Seville, the plastic films had become soft and viscous, and the air had pushed its way through the film in a bubble whose advancing front had broken up into a delicate tracery of fingered branches, for all the world like the tendrils of DLA clusters. Recognizing the characteristic fractal

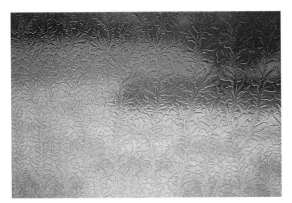

Fig. 5.12 Viscous fingering patterns in the layered window pane of the Hotel Los Lobreros in Seville. These patterns are formed as air penetrates into the plastic film between the glass panes. (Photo: Juan Manuel Garcia-Ruiz, University of Granada.)

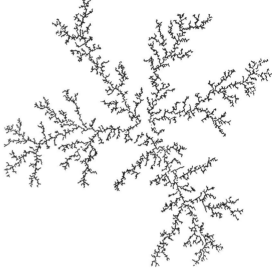

Fig. 5.13 Viscous fingering occurs as a fluid such as water or air advances into a more viscous fluid, such as oil, held within a porous rock. The advancing front is unstable and breaks up into fingers, which hampers oil extraction. Here I show the pattern produced when air displaces oil in a model porous medium. (Image: Roland Lenormand, Institut Français du Pétrole, Rueil-Malmaison.)

pattern, Garcia-Ruiz took photos and analysed them to deduce that the fractal dimension was about 1.65.[*]

This process in which an air bubble sprouts branching fingers as it forces its way under pressure into a surrounding, viscous medium is called viscous fingering. It is a much studied phenomenon, because it is relevant to some very practical problems in engineering. For instance, oil is often extracted from oil fields by injecting water into the oil-containing porous rock through a borehole. The idea is that the water, which does not mix with oil, advances in a front that pushes the oil to the wells at the edge of the field. But if viscous fingering occurs, the water front breaks up into narrow fingers and the efficiency with which oil is displaced and recovered is very low (Fig. 5.13).

Clearly, viscous fingering can produce branching patterns similar to those seen in DLA. At face value this coincidence might appear specious, because the way in which these patterns form seems to bear little resemblance to the DLA process. There is no gradual addition of solid particles to advancing tips; rather, the tips are comprised of a fluid, and they advance into a surrounding, more viscous medium by the pressure exerted at the interface of the two fluids. But a closer inspection of the problem reveals that the two processes share mathe-

matical features in common, and so can give rise to similar patterns.

The branching mechanism of viscous fingering involves an instability that renders any bulges at the interface unstable against elongation. We saw above that the same is true for DLA. The origin of the fingering instability was identified in 1958 by P.G. Saffman and Geoffrey Taylor, who studied viscous fingering using an apparatus devised by a nineteenth-century British naval engineer named Henry Hele-Shaw. The aim of its inventor was to study how water flows around a ship's hull, but the device, now called a Hele-Shaw cell, has provided some of the most important insights into branching patterns. It consists of two horizontal, parallel plates with a fixed narrow gap between them. The top plate is made of some transparent material and has a hole in the centre, through which a less viscous fluid such as air or water can be injected into a more viscous fluid such as glycerine or oil held between the plates. Anyone can make a Hele-Shaw cell, and you can do it yourself if you want a first-hand look at the patterns that viscous fingering can generate—details are given in Appendix 5.

The rate at which the edge of the air bubble (say) moves into the oil depends on the rate at which the pressure drops from its value at the interface (where it is

---

[*] After reporting these findings, Garcia-Ruiz received several messages from others who had observed similar patterns elsewhere. One of them asked why they are always observed in five-star hotels, to which Garcia-Ruiz's answer was: 'They are the ones with enough money to buy anti-crack windows'.

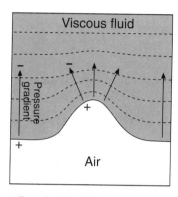

Fig. 5.14 The Saffman–Taylor instability. As a bulge develops at the advancing fluid front, the pressure gradient at the bulge tip is enhanced and so the tip advances more rapidly. (Contours of constant pressure—isobars, like those in weather maps—are shown as dashed lines.) This amplifies small bulges into sharp fingers. Compare this to the growth instability in DLA (Fig. 5.8).

highest) to the value in the bulk of the oil. If we think of a model analogy in which the pressure is equivalent to the height of a hill and the motion of the air bubble is equivalent to the motion of a ball, the ball accelerates more rapidly down the hill the steeper it is—in other words, it is the *gradient* that determines the rate of advance. Saffman and Taylor pointed out that the gradient in pressure around a bulge at the air/oil interface gets steeper as the bulge gets sharper. This sets up a self-amplifying process in which a small initial bulge begins to move faster than the interface to either side. The sharper and longer the finger gets, the steeper the pressure gradient at its tip and so the more rapidly it grows (Fig. 5.14).

This instability is called the Saffman–Taylor instability. In 1984, Australian physicist Lincoln Paterson pointed out that the equations that describe it are analogous to those that underlie the DLA instability described by Witten and Sander. So it is entirely to be expected that viscous fingering and DLA produce the same kind of fractal branching networks. Both are examples of so-called Laplacian growth, which can be described by a set of equations derived from the work of the eighteenth-century French scientist Pierre Laplace. Within these deceptively simple equations are the ingredients for growth instabilities that lead to branching.

But tenuous fractal patterns directly comparable to those of DLA occur in viscous fingering only under rather unusual conditions. More commonly one sees a subtly altered kind of branching structure: the basic

pattern or 'backbone' of the network has a comparable, disorderly form, but the branches themselves are fat fingers, not wispy tendrils (Fig. 5.15; compare 5.12). And under some conditions the bubbles cease to have the ragged DLA-like form at all, and instead advance in broad fingers that split at their tips (Fig. 5.4*a*). This sort of branching pattern is called the dense-branching morphology, and is more or less space-filling (two-dimensional) rather than fractal. Why then, if the same tip-growth instability operates in both viscous fingering and DLA, do different patterns result?

All viscous fingering patterns differ from that of DLA in at least one important respect—they have a characteristic length scale, defined by the average width of the fingers. This length scale is most clearly apparent at relatively low injection pressures, when the air bubble's boundary advances quite slowly. Then one sees just a few fat fingers that split as they grow (Fig. 5.16). There is a kind of regularity in this so-called tip-splitting pattern—the fingers seem to define a more or less periodic undulation around the perimeter of the bubble with a characteristic *wavelength*. But a length scale is apparent in the widths of the fingers even for more irregular patterns formed at higher growth rates (for example, Fig. 5.15). For the self-similar DLA cluster (Fig. 5.7), on the other hand, there is no characteristic size—it looks the same on all scales.

Eshel Ben-Jacob of Tel Aviv University explained the reason for these differences in the mid-1980s: between the air bubble and the surrounding viscous fluid there is

Fig. 5.15 Viscous fingering has a characteristic length scale, which determines the minimum width of the branches. So the fingers are fatter than the fine filaments of DLA clusters. (Image: Yves Couder, Ecole Normale Supérieure, Paris.)

Fig. 5.16 At low injection pressures, the length scale of viscous fingering is quite large, and the advancing bubble front then has a kind of undulating shape with a well-defined wavelength.

Fig. 5.17 When surface tension is included in the DLA model, it generates fat, tip-splitting branches like those in viscous fingering. Here the bands depict the cluster at different stages of its growth. (Image: Paul Meakin and Tamás Vicsek.)

an interface with a *surface tension*. As I explained in Chapter 2, the presence of a surface tension means that an interface has an energetic cost. Surface tension encourages surfaces to minimize their area. Clearly, a DLA cluster is highly profligate with surface area—the cluster is about as indiscriminate with the extent of its perimeter as you can imagine. This is because there is effectively *no* surface tension built into the theoretical DLA model—there is no penalty incurred if new surface is introduced by sprouting a thin branch. In viscous fingering, on the other hand, there will always be a surface tension (provided that the two fluids do not mix), and so there would be a crippling cost in energy in forming the kind of highly crenelated interface found in DLA. The fat fingers represent a compromise between the Saffman–Taylor instability, which favours the growth of branches on all length scales, and the smoothing effect of surface tension, which washes out bulges smaller than a certain limit. To a first approximation, you could say that the characteristic wavelength of viscous fingering is set by the point at which the advantage in growth rate of ever narrower branches is counterbalanced by their cost in surface energy.

The relation between DLA and viscous fingering is made very apparent when DLA growth is conducted in a system where a surface tension *is* built in. The surface tension has the effect of expanding the cluster's branches into fat fingers (Fig. 5.17). Ben-Jacob showed that the generic branching pattern in such cases is the dense-branching morphology. Conversely, a wispy DLA-like 'bubble' can be produced experimentally in the Hele-

Shaw cell by using fluids whose interface has a very low surface tension.

Physicists Johann Nittmann and Gene Stanley have shown that, somewhat surprisingly, the fat branches of viscous fingering can be generated instead of the tenuous DLA morphology even in a system with *no* surface tension. They formulated a DLA-type model in which they could vary the amount of 'noise' (that is, of randomizing influences) in the system. In their model the perimeter of the cluster can grow only after a particle has impinged on it a certain number of times (in pure DLA just one collision is enough). This reduces the tendency for new branches to sprout at the slightest fluctuation. Nittmann and Stanley found that, when the noise is very low, the model generates fat branching patterns (Fig. 5.18a), which mutate smoothly to the DLA-type structure as the noise is increased (Fig. 5.18b,c). This suggests that one way to impose a DLA-like pattern on viscous fingering in a Hele-Shaw cell is to introduce a randomizing influence (that is, to make the system more 'noisy'). A simple way of doing this is to score grooves at random into one of the cell plates until it is criss-crossed by a dense network of disorderly lines—this was how the pattern shown earlier in Fig. 5.4c was obtained. The lesson here is that noise or randomness can influence a growth pattern in pronounced ways.

a

b

c

Fig. 5.18 Dense-branching patterns appear in DLA growth even in the absence of surface tension, when the effect of noise in the system is reduced by reducing the sticking probability of the impinging particles (*a*). As the noise is increased (from *a* to *c*), the branches contract into the fine tendrils of the DLA-type pattern. Again, contours denote different stages of the growth process. Note that, despite their differing appearance, all of the patterns here have a fractal dimension of about 1.7. (Images: Gene Stanley, Boston University.)

The six-petalled flowers

Just as random noise can jumble up branching growth, so can an underlying symmetry have the opposite effect of introducing order. Take another look at Fig. 5.4*b*, which is a viscous-fingering pattern formed in a Hele-Shaw cell in which one plate has been scored with a regular hexagonal lattice of grooves. The sixfold symmetry of the underlying medium shows up clearly in the pattern, whose branching form is reminiscent of a snowflake.

The beautiful, symmetric complexity of snowflakes (which share such hexagonal symmetry) has captivated scientists for centuries. Their hexagonal character was apparently known to the Chinese almost two millennia

before Western natural philosophers became aware of it. Around 135 BC Han Ying wrote with astonishing perception that 'Flowers of plants and trees are generally five-pointed, but those of snow, which are called *ying*, are always six-pointed'. (About five-pointed flowers we have heard already in the previous chapter.) Yet as late as 1555, the Scandinavian bishop Olaus Magnus could be found claiming that snowflakes display a variety of shapes, including those of crescents, arrows, nails and bells. The Englishman Thomas Hariot seems to have been the first in the West to note the six-pointed shape, in 1591; but it was not until 1611 that this fact became common knowledge, when Johannes Kepler wrote a treatise entitled *De niva sexangula* ('*On the Six-cornered Snowflake*'). Herein Kepler pondered over the mysterious origin of this shape. Although lacking the theoretical

a

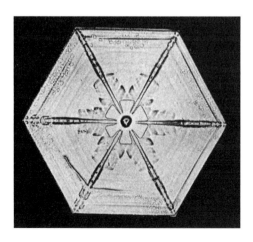

b

c

Fig. 5.19 Snowflakes are symmetrical branching patterns of infinite variety. (Photos: from Bentley and Humphreys 1962, kindly provided by Gene Stanley.)

tools and concepts needed to make much impact on the problem, Kepler did have the remarkable insight that the hexagonal symmetry must result from the packing together of constituent particles on a regular lattice. The symmetry, he said, was a consequence of their 'Patterns of contact: for instance, square in a plane, cubic in a solid'. At a time when atoms and molecules were barely conceived of, this was truly a leap of inspired imagination.

Modern techniques for analysing crystal structures have now shown us that water molecules do indeed pack together on a regular lattice that, looked at from certain directions, has sixfold symmetry (which is to say that it looks the same when rotated through a sixth of a full revolution). Astonishingly, we can see in this an echo of ancient Chinese wisdom about the cosmic schemes of nature: the number six was associated with water (then seen as one of the fundamental elements), and the scholar T'ang Chin wrote 'Since Six is the true number of Water, when water congeals into flowers they must be six-pointed'.

Everyone now believes that the hexagonal symmetry of snowflakes is a manifestation of this deep-seated symmetry in the crystal structure, just as the cubic shape of table-salt crystals reflects the cubic packing of its constituent ions. But that is only a small part of the problem—by analogy with other crystals, we might then expect ice crystals to be dense polyhedra with hexagonal facets, whereas instead we find these flat, highly branched and infinitely varied natural sculptures (Fig. 5.19).

Just how varied they are becomes evident from a glance through *Snow Crystals* by amateur photographer William Bentley and his colleague W.J. Humphreys. This astonishing book documents thousands of snapshots of snow crystals captured and photographed by the authors shortly after the turn of the century. A book of the same title published in 1954 by Japanese physicist Ukichiro Nakaya adds about 800 more snapshots to the family album, each one an individual. Nowhere in these two books will you find two identical snowflakes. From where does nature obtain this ability to turn out endless variations on a theme?

There is still no complete, universally accepted answer to that question. Indeed, in 1987 Johann Nittmann and Gene Stanley began a paper on snowflake patterns by confessing that 'There is no answer to even the simplest of questions that one can pose about snowflake growth, such as why the six arms are roughly identical in length and why the overall pattern of each arm resembles the five others'. Nor, they added, are we quite sure why snowflakes are (mostly) flat.

But although ice seems to be unique in forming these highly symmetrical flakes, regularly branched crystals analogous to a single snowflake arm may be seen in many other solidifying materials, including metals crystallizing from a melt (Fig. 5.20a), salts precipitating from supersaturated solution, and electrodeposits (Fig. 5.20b). These structures, known as dendrites, are generally formed when solidification is rapid—that is, far from equilibrium. For metals freezing from their melt, for instance, rapid solidification can be induced by cooling the molten metal suddenly to far below its freezing point. Slow growth of crystals close to equilibrium gives instead compact, facetted shapes. (I should point out that these dendrites are *not* the same as the mineral dendrites mentioned at the start of the chapter, which instead have a more random DLA-like structure—unfortunately researchers in different fields have been rather inconsistent with the 'tree' metaphor.)

Dendrites clearly represent another of nature's universal growth patterns. They typically have a rounded tip, like the prow of a boat, behind which side-arms sprout and grow in a Christmas-tree pattern. The Soviet mathematician G.P. Ivantsov developed in 1947 an explanation for the form of the tip, whose gently curved sides have a shape that mathematicians describe as parabolic (it's the same shape as the trajectory of a stone thrown through the air and falling under gravity). Ivantsov analysed the case of rapid solidification of a molten metal, an important problem in metallurgy. He showed that the interface between the solid and the melt can advance in a whole family of parabolic shapes: all possible parabolas are allowed, on the condition that the thinner they become, the more rapidly they advance (Fig. 5.21). So thin, needle-like tips should shoot rapidly through the melt, while fatter bulges make their way forward at a more ponderous pace.

But in the mid-1970s, Marshall Glicksman and co-workers at the Rensselaer Polytechnic Institute in New York performed careful experiments which showed

a

b

Fig. 5.20 Regularly branched dendrites are formed in crystals grown from the melt (a) and in electrodeposition of metals (b). (Photos: (a) Lynn Boatner, Oak Ridge National Laboratory, Tennessee; (b) Eshel Ben-Jacob.)

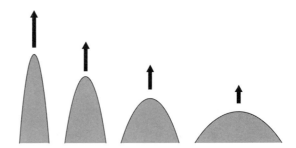

Fig. 5.21 A simple analysis of the solidification of a metal from its melt suggests that parabolic tips, like the end of a dendritic arm, should be stable. They will advance with a velocity that increases as the tips get narrower.

that, even if one ignores the problem of the side branches and focuses just on the shape of a dendrite tip, there must be something missing from Ivantsov's solution. They found that instead of a family of parabolic tips, only one single tip shape was seen during rapid solidification: for the same degree of undercooling (that is, cooling below the molten metal's freezing point), the same tip would be seen each time. For some reason, one of Ivantsov's family was 'special' for a given set of experimental conditions.

The problem was even worse than this, however, because in 1963 two Americans, W.W. Mullins and R.F. Sekerka from Carnegie Mellon University, presented theoretical arguments for why *none* of Ivantsov's parabolas should be formed. They showed that the slightest disturbance to a parabolic tip should cause it to break up into a mass of random branches. This so-called Mullins–Sekerka instability allows small bulges at the edge of the advancing solid to grow rapidly into thin fingers—it is yet another example of a Laplacian branching instability.

It works like this. When a liquid freezes, it gives out heat. This is called latent heat, and it is the key to the difference between a liquid and its frozen, solid form at the same temperature. Ice and water can both exist at 0°C, but the water can become ice only after it has become less 'excited' by giving up its latent heat.

So in order to freeze, an undercooled liquid has to throw away its latent heat. The rate of freezing depends on how quickly heat can be conducted away from the advancing edge of the solid. This in turn depends on how steeply the temperature drops from that of the liquid close to the solidification front to that of the liquid further away—the steeper the gradient in temperature, the faster heat flows down it. (It may seem odd that the liquid close to the freezing front is actually warmer than that further away, but this is simply because the front is where the latent heat is released. Remember that in these experiments *all* of the liquid has been rapidly cooled below its freezing point but has not yet had a chance to freeze.)

If a bulge develops by chance (that is, because of the random fluctuations—noise—in the system) on an otherwise flat solidification front, the temperature gradient becomes steeper around the bulge than elsewhere, because the contours of constant temperature get pressed closer together (Fig. 5.22). So the bulge grows more rapidly than the rest of the front—and the sharper it gets, the steeper the gradient and so the more rapidly it grows. The situation is mathematically equiv-

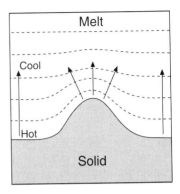

Fig. 5.22 The Mullins–Sekerka instability makes protrusions at a solidification front unstable. Because the temperature gradient (revealed here by dashed contours of equal temperature) at the tip of the protrusion is steeper, heat is conducted away faster and so solidification proceeds more rapidly. This self-enhancing instability is entirely analogous to those in Figs 5.8 and 5.14.

alent to the Saffman–Taylor instability in viscous fingering, with the pressure gradient in the latter case playing the same role as the temperature gradient here. If the Mullins–Sekerka instability alone acted on a rapidly advancing solidification front, an initially circular crystal might be expected to develop into a tenuous shape like a DLA cluster. But at the interface between a real solid and its melt there is again a surface tension, and this moderates the effect of the instability, just as it does for viscous fingering, by imposing a minimum size limit on the fingers.

A singular problem

In 1977 James Langer at the University of California and Hans Müller-Krumbhaar at Jülich in Germany threw all of these ideas together in an attempt to understand Glicksman's observation that a single dendrite tip is selected from all of Ivantsov's solutions. (They still ignored the question of how the symmetrical side branches form.) Perhaps, they suggested, while the Mullins–Sekerka instability renders the fattest parabolic tips unstable against splitting, and surface tension makes very thin tips too energetically costly, there is an optimal tip width at which the two effects balance to allow a 'marginally stable' Ivantsov-like parabolic tip to grow.

At first it looked as though this might be the answer. But problems soon became apparent. For one thing, their solution didn't take any account of the atomic structure of the solidifying substance, which was modelled just as a featureless solid. This meant that all solids were predicted to produce the same dendrites, whereas in practice the shape of a dendrite varies from one material to another. But even more disturbingly, it became clear that the researchers had underestimated the effect of surface tension, which was incorporated into their theory as merely a minor perturbing influence. In fact, a closer look at the problem by Eshel Ben-Jacob, Jim Langer and co-workers in the early 1980s showed that surface tension causes a 'singular' perturbation of the Ivantsov parabolas, which means that a small effect amplifies itself until it changes the whole game. What happens is that surface tension makes the tip of the dendrite cooler than the regions to either side—so the tip starts to slow down. Eventually, the tip forks into two new fingers, which then themselves split subsequently. This repeated tip-splitting doesn't give a dendritic growth shape at all, but instead the dense-branching morphology.

Back to square one. Researchers knew that dendrites *do* have roughly parabolic tips, decorated symmetrically with side branches, but the theories kept throwing up instabilities that led to randomly branched fingering patterns. The solution to this dilemma, it turned out, had been staring them in the face all along.

The whole reason why dendrites are so captivating is that they are so symmetrical. The arms of a snowflake do not shoot out any old how, but in a regular, sixfold pattern. The side branches do not sprout in any direction—all point the same way, at 60° to the main branch. (Dendrites of solidified metals often sprout side arms at right angles instead.) It had been long assumed that this regularity was an echo of the symmetry of the crystal structure—even Kepler felt that some underlying symmetry in the arrangement of constituent particles was responsible. But no one had guessed that it was to this symmetry that the dendrites owed their very existence.

Because of the symmetrical packing of atoms in the crystal structure, not all directions are the same for a growing crystal. That is why facetted crystals have the characteristic shapes that they do: some faces of the crystal grow faster than others. This non-equivalence of directions is called anisotropy (recall that an isotropic substance is one that looks the same, and behaves in the same way, in all directions).

The anisotropy of crystals means that properties like surface tension differ in different directions. In 1984, Ben-Jacob, Langer and their co-workers showed that, for Ivantsov parabolas growing in certain 'favoured' directions picked out by the anisotropy of the material's crystal structure, the effect of surface tension no longer renders the tip colder than the adjacent regions, and so

tip splitting does not occur—the parabolic tip remains stable as it grows. Thus dendritic branches will grow outwards from an initial crystal 'seed' only in these preferred directions, which are determined by the underlying symmetry of the crystal's atomic structure. The snowflake grows six arms. This special role of anisotropy in stabilizing the growth of a particular needle crystal was identified independently at the same time by David Kessler at Rutgers University together with Joel Koplik and Herbert Levine at the Schlumberger Doll Research Center in Ridgefield Connecticut. The idea gained support in 1985 when Ben-Jacob and colleagues showed that viscous fingering in the Hele-Shaw cell, which typically produces the dense-branching morphology (Fig. 5.4*a*), generates dendritic 'snowflake' patterns (Fig. 5.4*b*) when anisotropy is introduced by scoring a regular lattice of grooves into one plate.

Anisotropy also explains why a dendrite develops side branches. The roughly parabolic main branch is continually at risk of developing small bulges on its flanks through random fluctuations, and these then have the potential to grow through the Mullins–Sekerka instability. But again, only bulges that grow in certain directions will be stable. And there is only one kind of dendrite tip, for a given set of growth conditions, that grows fast enough to avoid being overwhelmed by these side branches. So a particular dendrite, with side branches sprouting in particular directions, is uniquely selected from amongst the possible growth shapes. Of course, because the side branches are initiated by random events, no two dendrites are identical; but all have recognizably the same general shape and features.

Arms control

While these ideas account for the features of most dendrites, the shapes of snowflakes remain the subject of some controversy. Snowflakes are just *so* symmetrical that some researchers believe we need something more to explain them. In particular, all six arms in any one snowflake appear to be almost identical, both in length and in the pattern of side branches (see Fig. 5.19*a*, for instance). How is this possible, if each arm is to be regarded as a dendrite whose side branches are determined by random events?

Early attempts in the mid-1980s to describe snowflake formation using the concepts developed for dendritic growth side-stepped this tricky question by simply building the sixfold symmetry into the models, which were set up so that they could *only* produce identical arms. No justification was given for why the arms should be identical. But in reality there appears to be something almost magical at play here—the tip of each arm seems somehow to 'know' what all the others are doing!

Implausible as that might sound, there *is* a way in which remote parts of crystals can communicate with one another. Every crystal has a characteristic set of vibrations that involve synchronized oscillations of all the atoms about their equilibrium positions in the lattice. You know how two people walking down a street will tend to fall in step with each other? An array of atoms can act rather like that, oscillating coherently like a whole battalion of soldiers walking in step. These coherent motions of the entire lattice are called phonons. They put distant parts of the lattice in touch with one another: a disturbance in one place may spread coherently by modifying a phonon vibration, just as a soldier who alters his pace in a marching battalion might gradually change the pace of all the other marchers. In 1957, Dan McLachlan suggested that phonon vibrations induced by the appearance of a side branch on one arm of a snowflake might bounce around the crystal and ultimately create disturbances at symmetrically equivalent positions on the other branches. The phonon is rather like a 'standing wave' of the sort established in organ pipes, which impress a periodic variation on the density of the air inside. McLachlan's idea is a promising one, but still lacks firm experimental support.

But Johann Nittmann and Gene Stanley propose that we should not get too caught up in trying to account for the apparent symmetry of snowflakes. They have pointed out that in fact no two branches of a snowflake are *exactly* alike, and suggest that almost perfectly regular snowflakes are the exception rather than the rule. Our eyes can be fooled into thinking that snowflakes are 'perfect' simply because each arm has side branches

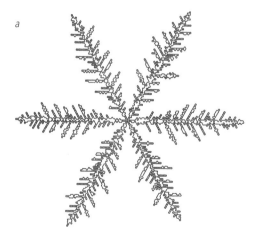

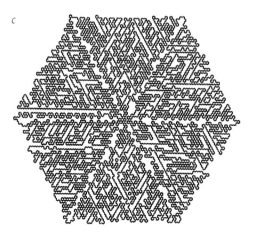

Fig. 5.23 How symmetrical are snowflakes? A DLA-type growth model that includes nothing more than sixfold anisotropy to produce the correct branching angles of 60° is able to generate snowflake-like clusters. There is nothing in this model to ensure that all branches are the same, and indeed they are *not* the same; but our eyes are fooled into seeing more symmetry than there really is by the uniformity of the branching angles. As the model is modified to make the depths of the 'fjords' more accessible, the snowflakes become denser (*b*, *c*—compare Fig. 5.19). (Images: Gene Stanley, Boston University.)

diverging at the same (60°) angle and because the *envelopes* traced out by the tips of each arm have the same shape. Nittmann and Stanley showed that a purely random DLA-type model of particle aggregation can give rise to snowflake-like shapes when sixfold anisotropy is included by requiring each particle to sit at a lattice point on a hexagonal (honeycomb) lattice. Even on this regular lattice a pure DLA process produces randomly branched patterns like that in Fig. 5.7, because the noise inherent in the DLA process over-whelms the effect of the underlying symmetry. But by reducing the noise in the same way as described on page 120, Nittmann and Stanley grew a snowflake-like cluster (Fig. 5.23*a*). What's more, they were able to generate denser clusters (Fig. 5.23*b, c*) analogous to some real snowflakes (Fig. 5.19*b, c*) by adding to the model a way of enhancing the probability of particles attaching with-in deep 'fjords' in the cluster (remember that this is usually unlikely in normal DLA because the particles tend to stick near a branch tip before they can get so far inside). This change to the model was admittedly a bit of a fix for which there was no clear justification—but it showed that even a random model can give growth patterns with a range comparable to that of real snow-flakes, provided that the randomness is not so great that it overwhelms an underlying symmetry. You'll see in Fig. 5.23 that none of the branches is identical, even though at a glance they do look similar. But the general Christmas-tree shape is preserved in all of them, and their lengths are more or less the same, simply because both the main branches and their respective side branches grow at roughly the same rates—the random-ness actually ensures this, because it gives no one branch any reason to grow faster than the others.

I hope you can now see that a wide variety of branch-ing structures can arise in non-equilibrium growth processes from the subtle interplay between relatively few physical phenomena: fingering instabilities, anisotropy, noise, surface tension. Changes in one of these factors can lead to qualitatively different growth patterns, either by a gradual transformation from one to the other (as in, for instance, Fig. 5.18) or by an abrupt transition (Fig. 5.24). What's more, we can iden-tify similar processes operating in apparently different systems, like electrodeposition and viscous fingering, and so can explain why similar growth patterns are seen. In the next chapter I show that these same ideas carry over when growth is turned on its head: there I shall consider how things break apart rather than how they grow. But to conclude here, I want to return to a ques-

Fig. 5.24 Branching patterns, like that shown here in electrodeposition, can undergo abrupt changes in shape as the growth conditions are varied. Here the change took place as the electric-field strength (given by the voltage drop between the edge of the cluster and the edge of the triangular cell, divided by the distance between them) exceeded a certain threshold during growth. (Photo: Eshel Ben-Jacob.)

tion that will recur throughout this book: to what extent can these ideas help us to understand biological form?

Tree and leaf

If physicists are going to draw so heavily on the tree (*den-dros*) metaphor in their descriptions of branching pat-terns in non-living systems, you might think that they should be able to tell us something about the shapes of real trees. But therein lies a problem of another order altogether. A tree is a form with a purpose. There are many problems that a tree must solve if it is to survive. How can it pump water from the roots to the leaves? How can it support its own tremendous weight? How to maximize its light-gathering efficiency? How to grow tall enough to compete for light with its neighbours, with-out becoming too massive for the roots to bear? In the face of these dilemmas, there is little chance that a simple physical model will tell all about the shape of a tree.

Besides, there are many ways to describe a tree. You could work from the cellular level, explaining how the cellulose fabric is synthesized from carbon dioxide and water and how it is woven into the composite matrix of the multi-layered cell walls, like glass fibres set in resin. You could choose an engineering perspective, explain-ing how the material properties of wood enable a branch to support its own weight or to flex in the wind. Or a hydrodynamic description, in which vertical and horizontal cellular channels carry water and sugar-rich fluids to and from the extremities, pumped by evaporation from the leaves. Or an ecological viewpoint, explaining how the tree harmonizes with the chemical and biological rhythms of its environment…

So even though the forms of trees have been a rich source of inspiration to physical scientists who think about fractal growth, I feel one must admit that the contribution of ideas about branching growth in physics to our understanding of trees, to *dendrology* itself, is not profound. In particular, rather little connection has been made between tree development and concepts such as growth instabilities, noise and so forth. The one respect in which the concepts developed in this chapter do have some value, however, is in the *description* of tree forms. Even if the various factors influencing tree growth are too numerous and too complicated to account for, we can attempt to develop mathematical models that, while ignoring the biology and mechanics, nevertheless aim to reproduce the essential shapes of trees. As I indicated in the first chapter, this approach sometimes allows one to make an informed guess at the primary factors determining form, for which one would hope to identify corresponding parameters in the mathematical model.

For branching patterns in particular, attempts to provide mathematical descriptions of shape and form unfold along rather different lines than we are used to in classical geometry. Such models are in fact more properly regarded as prescriptions rather than descriptions—they do not provide geometrical labels of shape like 'circle' or 'octahedron', but instead sets of rules, called *algorithms*, for generating characteristic but non-unique forms.

What does that mean? Well, you can describe the shape of a planet (spherical) or a salt crystal (cubic) easily enough, but you'd be hard pushed to assign a similar label to a cypress tree. 'Branched' is not specific enough, and 'tall and branched' does little better. To give an accurate geometrical description, you'd need to specify all of the branches and all of their angles and lengths—to paint in words a picture of the complete tree (and then only of *that* cypress tree!). You end up, in other words, like Sartre's Antoine Roquentin in *La Nausée*, horribly fixated on the particulars of the structure. But an algorithmic prescription provides an alternative—it tells how to generate a whole set of branched figures, all looking recognizably like a cypress. The word 'algorithm' comes from the name of the ninth-century Moorish mathematician Muhammad ibn Musa al-Khwarizmi, who incidentally also bequeathed to math-

ematics the word 'algebra' and the concept of zero. An algorithmic approach to generic form is what underpins much work on mathematical fractals.

Leonardo da Vinci suspected (although without formulating it in quite these terms) that there are algorithmic rules governing tree growth. For example, he suggested that at branching points the rule is that the central trunk is deflected by some angle when a side branch occurs on its own, but is not deflected if two side branches are positioned opposite one another. Is that true? To a degree, but it depends on the size of the side branch—single small ones cause next to no deflection. Wilhelm Roux attempted to specify these rules more precisely around the end of the nineteenth century, by identifying the following principles:

1. When the central stem forks into two branches with equal width, they both make the same angle with the original stem.

2. If one branch of the fork is of lesser width than the other (so that it can be regarded as a side branch, the wider one being a continuation of the main stem), then the thinner branch diverges at a larger angle than the thicker.

3. Side branches small enough that they do not deflect the main stem appreciably diverge at angles between 70° and 90°.

Roux in fact developed these rules while studying arterial networks, but in the 1920s Cecil Murray made them more quantitative and extended them to trees too. Murray proposed that, for arteries, they could be understood according to the principle of least work (which we'll encounter in the next chapter): the energy required to drive blood to the point reached by a side-branching artery is minimized if narrow branches diverge at large angles and wide ones diverge at shallow angles. And as trees are themselves a kind of vascular system too, through which water and sap are pumped—well, why shouldn't the same parsimonious principle of least work apply here too?

Murray's algorithmic rules generate somewhat realistic-looking 'trees' when used to create a randomly branched network. Another algorithm for making tree-like branching structures was proposed by H. Honda in 1971, and runs as follows (Fig. 5.25):

1. Every branch forks into two 'daughter' branches at a single branching point.

2. The two daughter branches are shorter than the 'mother' branch by constant ratios r_1 and r_2.

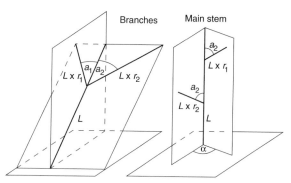

Fig. 5.25 Rules for creating tree shapes algorithmically, proposed by H. Honda. Branches are specified by the length ratios and angles shown on the left, except for those that diverge from the main trunk. In the latter case, the rules on the right apply. Notice that the latter specify a kind of spiral phyllotaxis with angle α. (After: Prusinkiewicz and Lindenmayer, 1990.)

3. The two daughter branches lie in the same plane as the mother branch (the branch plane), and diverge from it at constant angles a_1 and a_2.

4. The branch plane is always such that a line lying in this plane perpendicular to the mother branch is horizontal. (This is the trickiest of the rules to envisage, but is explained in the figure.)

5. An exception to (4) is made for branches diverging from the main trunk, which observe the length ratios specified in (2) but branch off individually at a constant angle a_2, with a divergence angle of α between consecutive branches.

With a few minor modifications this algorithm produces a whole range of branching patterns closely mimicking those of real trees (Fig. 5.26). Further modifications to account for the influences to which

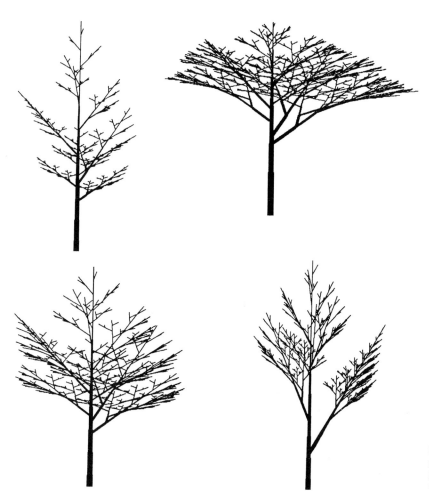

Fig. 5.26 Trees generated from the rules in Fig. 5.25. (Images: from Prusinkiewicz and Lindenmayer 1990.)

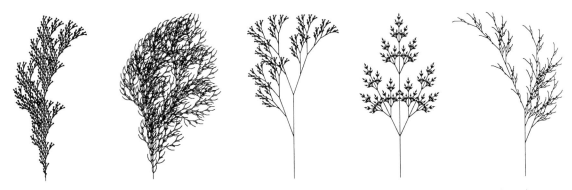

Fig. 5.27 Plants and ferns generated by deterministic branching algorithms. The same motifs recur again and again at different scales in these structures, but the regularity is evident to greater or lesser degrees. (Images: from Prusinkiewicz and Lindenmayer 1990).

real trees are subjected—wind, gravity, the need to arrange leaves for optimal light harvesting—give increased realism. Honda's algorithm is deterministic—it prescribes the branching pattern fully once the ratios and angles are fixed. Other algorithms used to generate life-like trees in computer art employ random elements to create more irregular forms. In nature, a certain randomness enters into the branching patterns as a consequence of such things as breakages, collisions between branches, growth stunting due to the shade of an overlying canopy, and the mechanical influences of the elements. Another class of deterministic algorithms, called L-systems by Przemyslaw Prusinkiewicz of the University of Regina, will generate plant- and fern-like structures (Fig. 5.27). These algorithms have spawned some stunning computer art; but they have not yet clearly extended the ideas of Roux and Murray in terms of explaining *how* it is that these branched patterns appear in such profusion in our hedgerows. Ultimately one might hope that appropriate rules for tree-growing algorithms will be derived from models of phyllotaxis mentioned in the previous chapter, augmented by other deterministic or random elements to account for the external, environmental influences to which a growing tree or shrub is subjected.

Networking

Branching structures in living organisms—in lungs, blood vessels, neurons and the vein systems of leaves—are so tantalizingly similar in many respects to those observed in the inorganic world that for many researchers it is hard to resist drawing some analogy, or even suggesting that there must be fundamental similarities

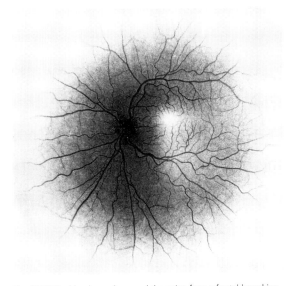

Fig. 5.28 The blood vessels around the retina form a fractal branching network with a fractal dimension of about 1.7. (Photo: Fereydoon Family, Emory University.)

between the growth mechanisms. Consider, for example, the system of blood vessels in the human retina (Fig. 5.28). Fereydoon Family and co-workers at Emory University in Georgia have shown that this branching structure has a fractal dimension of around 1.7—very similar to that of DLA clusters.

But this does not imply that blood-vessel formation (called angiogenesis) is at root entirely (or even slightly!) analogous to the DLA process. The biology of angiogenesis is complicated, and doesn't always generate a diverging, randomly branched structure—often

But even before we begin to worry about the finer points of shape and form, it would be folly to assume that one can simply map a mathematical procedure like the DLA algorithm onto biological growth. Life's structures have a purpose, and if they don't evolve to fulfil it with at least some modicum of efficiency, there will be a strong selective pressure towards modification. So in general complicated biochemical mechanisms have evolved to make sure that the architecture is up to the task.

Vascular systems, for instance, have to deliver fluids (such as blood) to their host tissues while those tissues are themselves growing in size. This means that the growing tissue and the existing vessels have to communicate with one another so that, once a region of new tissue develops too far from an existing vessel, it can broadcast its need for new vessels to supply it. This happens by a mechanism very much akin to the process of chemotaxis that bacteria use to 'talk' to one another (Chapter 3). In angiogenesis the remote tissue cells begin to produce proteins called angiogenic factors (AFs), which diffuse out into the surrounding tissue. Such cells are said to be ischemic. When these chemical messengers reach a nearby vessel, they trigger it into sprouting a new limb, which grows in the direction of increasing AF concentration—that is, towards the source. When two blood vessels, growing from different directions towards a region of AF production, meet at its source, they undergo anastomosis, fusing end to end to form a single vessel.

The similarity in fractal dimension of retinal vessels and DLA clusters led Fereydoon Family and colleagues to conclude tentatively that at the very least this might reflect the central importance of diffusion in both growth processes. Mark Gottlieb of Arizona State University has attempted to go further by concocting a simple model that takes into account some of the specific biological processes known to control vascular growth. He modelled the host tissue as a checkerboard lattice of cells, interlaced with a system of blood vessels. To mimic the growth of the host tissue, he allowed the size of the whole checkerboard array to increase. After each growth step, the distance of each cell from a blood vessel is determined, and if this distance is too great then the cell becomes ischemic and a new vessel is added, reaching from the nearest existing vessel to the centre of the ischemic cell. If two vessels are equally distant from an ischemic cell, they both sprout new vessels, which meet end to end in the ischemic cell. Finally, existing vessels grow wider as the host tissue expands, so that older vessels become broader than new vessels. This

a

b

Fig. 5.29 The branches in vascular systems are often interconnected, as seen here in the veins of a leaf (*a*) and of a Caribbean sea fan (*b*).

the vessels are interconnected in more complex ways. Blood vessels and the veins in leaves (so-called vascular networks) commonly form closed loops (Fig. 5.29), which means that there is more than one possible route for getting from one point to another. The reconnection between two branches in a vascular system is called anastomosis. In DLA-type branching, in contrast, loops are almost entirely absent and there is just one path that will take you from the 'root' to any particular branch tip. So a vascular system is more like the London underground system than like a tree: if you want to go from A to B, you often have a choice of several possible routes.

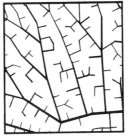

Fig. 5.30 A simple model of angiogenesis called the vascular tree model can generate vascular networks like those seen in living systems. (After: Gottlieb 1993.)

model (in which much of the biology is rather crudely added 'by hand') produces fractal networks (Fig. 5.30) that resemble those seen in real vascular systems of both animals and plants.

Hans Meinhardt, on the other hand, sees in the development of biological vascular networks a pat-terning process under the control of diffusing chemical signals that should be explicable by a reaction–diffusion model. He has developed an activator–inhibitor scheme in which short-ranged activation allows branches to grow and divide, while long-ranged inhibition makes the advancing tips of branches avoid each other. But in this model a growing tip is less strongly repelled by a filament that already exists, and so this repulsion can sometimes be overwhelmed by that between growing tips, allowing a tip to reconnect with an existing branch in an anastomatic event. Meinhardt's model represents a rather rare example of a convergence between work on reaction–diffusion systems, which are commonly invoked to explain periodic or pseudo-periodic pattern-ing, and work on branched growth patterns, which are more typically approached using DLA and related clustering models strongly influenced by noise.

Scaling up

Regardless of exactly how vascular systems are formed, there may be a deep connection between their fractal structure and their biological function. Ecologists James Brown and Brian Enquist from the University of New Mexico, in collaboration with physicist Geoffrey West of the Los Alamos National Laboratory, have proposed that the way in which metabolic rates of living organ-isms vary with their body size is a consequence of the fractal nature of their fluid distribution systems: the cardiovascular and respiratory systems of animals, for example, and the vascular systems of plants.

The relationship between metabolic rate and size is a long-standing puzzle. It is common knowledge that the rate of a creature's heartbeat decreases as its body size increases: babies' hearts beat faster than those of adults (they also breath faster), and the heartbeats of small creatures like birds are more rapid still. For a wide variety of organisms, the heartbeat rate turns out to be proportional to the inverse of the body mass raised to the power $\frac{1}{4}$. The metabolic rate of individual cells in an organism—the rate at which they consume energy—follows the same mathematical law. In other words, big organisms have a slower metabolism.

What's more, the total metabolic rate—the net rate of energy consumption of the whole organism—varies as the $\frac{3}{4}$ power of body mass. And the cross-sectional area of aortal arteries in mammals and of tree trunks varies in the same way with body mass. These relationships are examples of so-called allometric scaling laws, and they are obeyed by organisms ranging from microbes to whales. Now, you'd *expect* large creatures to use up more energy than small ones, but it isn't obvious that the same scaling law should be followed over such a huge range of sizes. Still more puzzling are the actual values of the powers in the scaling laws: they all seem to be multiples of $\frac{1}{4}$. If the biological parameters—heartbeat and so forth—were related to how quickly fluids could be distributed in the body, you'd expect the relationship to depend on the body's dimensions, which vary as the $\frac{1}{3}$ power of body mass. (This might be easier to see from the inverse relationship: the body mass is directly pro-portional to the body volume, which varies as the cube—the 3rd power—of the body's linear dimen-sions.) In the same way, the time taken to travel at con-stant speed across a cube-shaped box depends on the $\frac{1}{3}$ power of the box's volume. You'd therefore think that all these scaling laws should come with powers that are multiples of $\frac{1}{3}$, not $\frac{1}{4}$.

Enquist, Brown and West sought for an answer to this puzzle in the fractal networks of the distribution sys-tems. They modelled these as systems of tubes which become progressively thinner at each branching point. The model networks are constrained by two require-ments. First, all of them (regardless of size) have to end in tubes of the same size. These terminal branches can be considered to be the analogues of the smallest capil-laries in cardiovascular systems, whose size is geared to that of the organism's individual cells—which varies lit-tle regardless of the total body size. Second, the network is structured so that the amount of energy required to transport fluids through it is minimized. This echoes the rationale of Cecil Murray for his algorithms for tree structure.

For plant vascular systems, the passages of the network are in fact bundles of vessels of the same cross-section. At each branching point, the bundles split into thinner bundles with fewer vessels in each. For this situation, the researchers showed that the $\frac{3}{4}$ scaling law of metabolic rate with body mass (that is, with volume supplied by the vascular network) falls out quite naturally from an analysis of the geometric properties of the energy-minimizing network. For mammalian distribution networks, on the other hand, the situation is rather more complex, and a $\frac{3}{4}$ scaling law is obtained only when the model includes the facts that the fluid flow is pulsed (due to the pumping of the heart) and the tubes are elastic. Most importantly, these relationships apply only for fractal distribution networks—non-fractal systems show $\frac{1}{3}$-power scaling with size, not $\frac{1}{4}$-scaling.

This can't be the whole answer to allometric scaling laws—for one thing, they are obeyed by organisms that don't have branched distribution systems—but it posits an intriguing significance for fractal networks in the living world. James Brown suggests that it is in fact the ability of fractal networks to provide an optimal supply system to bodies of different sizes that enables living organisms to show such a huge range in body shapes and sizes. This range extends over 21 orders of magnitude—21 levels of magnification by 10. Perhaps we would not have this diversity, from bacteria to whales, without the special characteristics of fractal branching patterns.

Life in the colonies

There is at least one area of biology that has genuinely proved in recent years to be a rich playground into which ideas from non-living branching systems can be freely exported: the growth of bacterial colonies. Watching a bacterial colony grow is like watching a city expand into an urban sprawl, except that it happens in days rather than decades. The inhabitants of the colony multiply (although bacteria can achieve this simply by cell division rather than by the more complicated strategies we humans must employ), and what drives this multiplication and growth is a supply of food. As well as eating and generating offspring, bacteria share other tendencies with us. They can move around, thanks to long, wavy tentacles called flagella that propel them

through a fluid; and they can communicate with one another, in particular by sending out chemical signals as described in Chapter 3. All of this means that a growing bacterial colony must be regarded as a rather complex social structure, and it's not at all obvious that we should expect any similarities with growth behaviour in non-living systems.

And yet, when they set out to study bacterial growth in the late 1980s, Mitsugu Matsushita and H. Fujikawa of Chuo University in Japan found that colonies of the bacterium *Bacillus subtilis* evolved into patterns that looked very much like DLA clusters (Fig. 5.31a). Is life

a

b

Fig. 5.31 (a) Fractal, DLA-like growth of a colony of the bacteria *Bacillus subtilis*. (b) Two adjacent colonies suppress each other's growth in the region between them, just as would be expected for DLA growth. (Photos: Mitsugu Matsushita, Chuo University.)

for once simpler than expected, or is the apparent resemblance coincidental?

The Japanese researchers showed that the similarities were more than skin deep. For one thing, the branching colonies had the same fractal dimension as DLA clusters, about 1.7. And they showed some of the features that would be expected of a DLA-type process—for instance, two adjacent colonies seemed to repel one another, with suppression of growth in the region between them (Fig. 5.31*b*). But why should bacterial growth be like diffusion-limited aggregation?

Matsushita and Fujikawa grew these colonies in flat, circular Petri dishes containing a water-saturated gel made of a substance called agar. They injected a few bacteria into the centre of the dish, added some of the nutrients needed for growth, and let nature take its course. By varying the conditions under which growth occurred, they found that they could obtain colonies with very different shapes. They looked at the effect of changing just one of two variables—the concentration of nutrient and the hardness of the gel—while keeping everything else constant. Because the bacteria could not penetrate through the gel, the colony could grow only by pushing back the gel at its boundary. The more agar they added to the growth medium, the harder the gel was—it could vary in consistency from jelly-like to rubbery. And the harder the gel, the harder it became for the colony to expand.

The researchers observed fractal, DLA-like colonies under conditions where the gel was hard and nutrients were scarce—the most challenging situation that their bacteria faced. If the amount of nutrient is increased in these hard gels, the colonies become much denser, but still with an irregular perimeter (Fig. 5.32*a*). This is called an Eden-like growth mode, after the mathematician M. Eden who observed it in 1960 in one of the first ever computer models of biological growth. If the gel is made softer at low nutrient levels, the pattern changes from DLA-like to one that more closely resembles the dense-branching morphology (DBM) (Fig. 5.32*b*). But if conditions are rendered highly favourable—plenty of nutrient and a soft gel—the colony expands in a single dense mass, with no branching (Fig. 5.32*c*).

So here is another growth process in which distinct patterns are selected under different conditions. The DLA-like pattern can be accounted for in an arm-waving way by noting that, under conditions of low nutrient levels, the rate at which the bacteria multiply is limited by the rate at which nutrients can diffuse through the gel medium to reach them. This diffusion-limited process might then be susceptible to the same kind of branching instability that we encountered for simple aggregation. The Japanese researchers were also able to give an explanation for why the growth patterns changed rather abruptly as the gel became harder: there

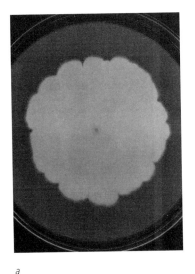

a   *b*   *c*

Fig. 5.32 The morphology of the bacterial colony depends on the conditions under which it grows: the amount of nutrient, and the hardness of the gel medium. (*a*) 'Eden'-like growth. (*b*) The dense-branching growth mode. (*c*) Compact, non-branching growth. (Photos: Mitsugu Matsushita.)

is a certain degree of hardness beyond which the bacteria simply cannot move. Under a microscope, they could see that bacteria in the DBM-like and dense colonies were swarming about, while those in the DLA-like and Eden-like colonies just sat there. In the latter case, the colony expands as the sheer mass of multiplying cells forces the gel back, whereas in the former case the growth is very much faster because of a constant battering of individual cells against the gel. Matsushita and Fujikawa confirmed the importance of cell movement in pattern selection by growing colonies of *Bacillus* mutants that lacked flagella and so could not swim around—in that case they observed just the DLA and Eden patterns no matter how soft the gel was.

Invasion of the mutants

Physicists who have worked on branching patterns in physical phenomena, such as viscous fingering and dendritic growth, have been attracted to the growth patterns seen by Matsushita and Fujikawa because they seem to offer a model system through which one might try to expand the concepts learned in the physical sciences to embrace the living world. Bacteria are undoubtedly more complex than, say, solidifying metals—there is more than enough going on in a single bacterium to keep microbiologists busy for years to come—but nonetheless it has proved possible to perform experiments on pattern formation in these systems under well-controlled conditions and obtain reproducible results. This means that, even without a detailed knowledge of microbiology, physicists can hope to make some headway in understanding the growth processes.

Eshel Ben-Jacob and co-workers in Tel Aviv have looked at the patterns formed by *Bacillus subtilis*. Although they saw the same kind of growth patterns as the Japanese researchers, they also noticed something new: occasionally, a colony that was happily advancing in one pattern would suddenly sprout new branches from one or more points on its perimeter that would show a different kind of growth (Fig. 5.33). If cells from the new growth pattern were extracted and used as the seeds of a new colony, that colony too would exhibit the new pattern—even though, under the same conditions, the initial colony had begun growing with a different pattern. It was as though the initial colony had suddenly mutated, and the offspring of the mutants had inherited the tendency to form a different pattern.

And that, suggested Ben-Jacob and colleagues, was probably just what was happening. Unlike aggregating

Fig. 5.33 A transition to a new growth mode at one point on the perimeter of a growing colony of *Bacillus subtilis*. The transition permits an episode of explosive expansion. (Photo: Eshel Ben-Jacob.)

metal atoms or smoke particles, bacteria can mutate. When a cell divides, the daughter cells can have a different genetic constitution to the parent cell, owing to mistakes made in duplicating the parental DNA for the progeny. These genetic mutations happen all the time—some may be fatal to the progeny, some may have no observable effect, but just occasionally a mutation will give the new cell a better chance of surviving than the parent cell. When this happens, the mutant has a reproductive advantage: it is better able to survive under adverse circumstances, and so reproduces more prolifically. This is exactly how Darwinian natural selection works.

Ben-Jacob's group suggested that what one was seeing in these bursts of new patterns growing from old was natural selection in a Petri dish. Some chance mutation of the bacteria in the new pattern gives them superior fitness, and a consequence of this is a change in the mode of branching growth. The new pattern might be an incidental outcome of some other fitness-enhancing characteristic; or it might be that *the pattern itself* gives the bacteria a competitive advantage.

Suddenly the possibilities for pattern-forming processes blossomed. Whenever a mutation of this sort took place, cells could be extracted from the mutant pattern to constitute an entirely new strain of *Bacillus* with new pattern-forming potential. Some of the mutant patterns were familiar—a dense-branching sub-colony might burst forth from a compact Eden-like colony, for example (Fig. 5.33). Mutants of this type

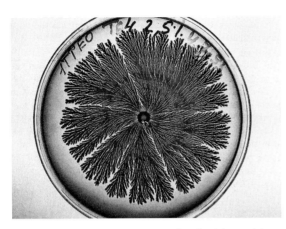

Fig. 5.34 The tip-splitting (*T*) morphotype of *Bacillus*. (Photo: Eshel Ben-Jacob.)

were quite common, and Ben-Jacob and colleagues called them the tip-splitting or *T* morphotype (Fig. 5.34). But other mutant patterns were unlike anything seen in non-living systems: one consisted of elegant hook-like twists that all curved in the same direction, creating a colony reminiscent of a Chinese dragon (Plate 12*a*). This was denoted the chiral or *C* morphotype ('chiral' derives from the Greek word for hand, as these hooks can twist either in a left- or right-handed sense). And yet another mode of growth was shown by the vortex or *V* morphotype: here tendrils develop in the wake of small, roughly circular droplets of cells (Plate 12*b*). Under the microscope, the researchers could see that the cells in the droplets were all rotating in a spiral vortex.

Faced with this richness, the challenge of developing a model to account for the growth patterns looked immense. But Ben-Jacob and colleagues started by focusing on the DLA-like and tip-splitting patterns, since in non-living systems these are relatively well understood. As Matsushita and Fujikawa had found, these patterns become sparser and increasingly DLA-like when the colonies are grown in harder gel or with less nutrient. Can a simple model capture this behaviour?

The Israeli scientists, in collaboration with Tamás Vicsek and Andras Czirok in Budapest, assumed that the most significant facts to include in a model were that the bacteria move, feed and reproduce. So they adopted the following rules:

1. The bacteria move at random.

2. While food is available, the bacteria feed at a steady rate.

3. If they eat enough, they reproduce (one cell splits into two); if the food runs out, they stop moving.

4. The dispersal of food (nutrient) throughout the system takes place by diffusion, and so is governed by well-known mathematical equations.

We encountered this kind of modelling in Chapter 3; but there I talked largely about *continuous* models, where the bacterial colony is described as if it were some kind of fluid of varying density obeying the equations of diffusion and 'reaction'. In such a description, branching instabilities may emerge in much the same way as those of viscous fingering in a fluid. In contrast, Ben-Jacob and colleagues adopted a particle-like model, akin to the DLA model of Witten and Sander, in which each particle is governed by the rules above. But a single colony might contain as many as ten billion individual 'particles' (cells)—far too many for a computer to cope with. So they grouped cells together into 'walkers', each containing many thousands of cells, and assumed that these walkers moved around according to the same rules. Thus the walkers, not individual cells, were the fundamental particles of the model. They moved about on a regular underlying lattice. To model the advance of the colony into the gel, the researchers allowed the boundary of the colony to advance onto a new lattice point only when that point had been struck a certain number of times by the moving walkers. By varying this number, they could simulate the effect of making the gel harder.

With nothing more than these elements, Ben-Jacob, Vicsek and colleagues found that their computer model of the growing colony produced the branched patterns seen in the experiments. The lower the concentration of food, the more tenuous the branches become (Fig. 5.35). So this part of the gallery of growth shapes, at least, seems to be open to explanation by a model that takes no account of the detailed biology of the bacteria.

But there was one complication. In the experiments a curious thing happens: when the amount of nutrient is very low, the colony suddenly becomes denser again. This does not happen in the model—the branches just go on getting thinner as food became scarcer. The researchers figured that it is at this point, when things look really desperate for the starved bacteria, that they start to do something only living 'particles' can do: 'talk' to each other. As we saw in Chapter 3, the language of bacteria is a chemical one: they communicate by emitting chemicals, which then guide the cells' direction of motion in the process known as chemotaxis. Like the

Fig. 5.35 The 'walkers' model of bacterial growth makes a few simple assumptions about the rules that determine the movement and multiplication of the bacteria. It generates DLA-like branching patterns that become increasingly sparse as the gel medium becomes harder (bottom to top) or as nutrients become scarcer (right to left). (Image: Eshel Ben-Jacob.)

slime mold *Dictyostelium discoideum*, the cooperative aggregation induced by chemotaxis of *B. subtilis* leads to the formation of differentiated cell types and to spores which are then released when conditions are more favourable.

So the researchers added to their model a simple description of chemotaxis. Any walkers that become immobile due to lack of nutrients emit a chemical signal at a fixed rate, which diffuses into the surrounding medium. Mobile walkers consume this substance at a fixed rate, and their random walks develop a bias so that they are more likely to move in the direction of *decreasing* levels of the chemical—away from the signalling walker. (Notice that this is a repulsive interaction, in contrast to the attractive chemotaxis employed by

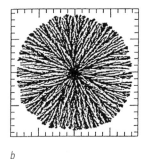

a   b

Fig. 5.36 When communication between bacteria—chemotaxis—is included in the walkers model, it generates a change of growth morphology from DLA-type (a) to a denser branching mode (b) at low nutrient levels. (Images: Eshel Ben-Jacob.)

Dictyostelium discoideum). These ingredients do indeed produce a switch to denser branching patterns at very low nutrient levels, as seen experimentally (Fig. 5.36).

In dendritic growth, anisotropy in the underlying growth process due to the symmetry of the crystal's atomic structure becomes manifest as symmetric branches in the growth pattern. Can anything like this be seen in bacterial growth? There is no obvious way to introduce anisotropy at the microscopic level, because while they are mobile the cells simply dash around at random. But Ben-Jacob's team found that they could make the gel medium itself anisotropic by stamping it with a symmetric pattern that leaves an imprint on the gel surface. A colony grown in a gel stamped with a six-fold lattice of grooves develops a dendritic pattern—a kind of bacterial snowflake (Fig. 5.37a). This can be mimicked in the 'walkers' model by allowing the colony perimeter to advance after fewer 'hits' against the gel in some directions than in others. This modification produces sixfold patterns at relatively high nutrient levels, but the sixfold symmetry fades away to give more random branches at low levels (Fig. 5.37b). In the experiments, on the other hand, the effect of the anisotropy still shines through even with very little nutrient around. The researchers concluded that again some kind of chemical signalling must be operating at very low nutrient levels, and they found that they could recapture the sixfold symmetry under these conditions by introducing the repulsive chemotactic interaction into the model. There is no evidence that real *Bacillus subtilis* employ a mechanism anything like this,

a

Fig. 5.37 Bacterial snowflakes. (*a*) Growth in a gel that has been stamped with a set of grooves with sixfold symmetry. This anisotropy of the growth medium guides growth in certain preferred directions. (*b*) The model is able to capture this behaviour; but the sixfold pattern starts to disappear at low nutrient concentrations (decreasing to right). (Images: Eshel Ben-Jacob.)

b

however, so one can't yet be sure that this apparent success of the model is not just a happy coincidence.

What about the more complex patterns—the chiral and vortex growth modes? One can reproduce these by adding to the walkers model the assumption that the *C* morphotype cells do not tumble around at random but rotate preferentially in one direction. Bacterial flagella are coiled filaments which are known to coil with a certain handedness, like the left- or right-handed thread of a screw. Under certain conditions this coiling might confer a preference to the direction of the cell's rotation, and this microscopic effect could provide a singular perturbation, like surface tension or crystal anisotropy, that becomes amplified until it is manifest in the large-scale growth pattern.

The merry-go-round motions of the *V* morphotype are a particularly striking phenomenon, bringing to mind the collective swooping of a flock of birds. But this sort of behaviour in bacteria is by no means unprecedented: in 1916 W.W. Ford reported vortex motion in colonies of *Bacillus circulans*, from which

their name obviously derives. Apparently the mutations of the *V* strain of *Bacillus* give them a similar tendency to execute pirouettes. Tamás Vicsek proposed that this behaviour might be modelled by treating the colony as a collection of 'gliders' rather than walkers: instead of meandering along a random path, each of the collective bacterial 'particles' is assumed to propel itself over a surface with a well-defined velocity at any instant. But this velocity can change, either because the glider gets entrained in the motion of its neighbours or because it alters its motion in response to a chemotactic signal. The model is quite complicated, but it does reproduce the vortex motions. How these vortices create the fingering patterns seen in Plate 10 is another matter, and remains just one of the many challenges to this sort of modelling.

Does all of this mean that a process as complex as the growth of a living colony can be understood in much the same way as the formation of a dendrite or the advance of an air bubble? At present, that is partly a matter of taste. Eshel Ben-Jacob notes with chagrin the comments of biologist Jim Cowan, who had this to say about those who attempt to develop simple models of complex systems:

> They say 'Look, isn't this reminiscent of a biological or physical phenomenon!' They jump in right away as if it's a decent model for the phenomenon, and usually of course it's just got some accidental features that make it look like something.

Whether the models that have been developed so far for bacterial growth share anything more than acci-

dental features with the patterns seen experimentally is still an open question. To my mind the correspondences are impressive, but the difficulty is in knowing whether one is adding by hand the most relevant elements of the process, while not overlooking others that are equally important. There is a lot going on in biological systems, for sure! Yet what we have seen so far is just a beginning. However crude the present models, they promise that a marriage of physics with biology will surely have much to tell us about the ramifications of growth and form.

BREAKDOWNS

But evil are
The paths, for crookedly
Like horses go the imprisoned
Elements and ancient laws
Of the earth.

Friedrich Holderlin
Mnemosyne III

Children are very good at making patterns. One at which they are particularly adept is shown in Fig. 6.1. It is the result of probably the easiest of the experiments that I propose in this book, and requires only readily available apparatus: a football and a window, a stone and a car windscreen, a dart and a mirror. You might be happier, though, just to take the results for granted.

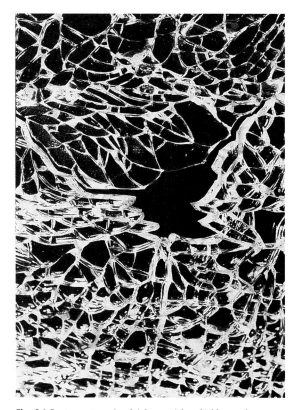

Fig. 6.1 Fracture patterns in a brittle material are highly complex. Shown here is a pattern formed in shattered windscreen glass. (Photo: Mehau Kulyk/Science Photo Library.)

When things break down, they tend to do so spectacularly. I don't just mean that bridges fall down, that ships sink, that cities are levelled by earthquakes. I mean that we may be left with stunningly rich patterns, sometimes on an awesome scale (Fig. 6.2). Cracks are amongst the most familiar of branching patterns. They are also amongst the least well understood.

It is only very recently, in fact, that scientists have begun to understand why it is that cracks form at all. For a long time, the science of fracture and failure of materials limped along with just a handful of basic concepts, most of which failed to provide any predictive power for what was seen in the real world. This was much more than an academic embarrassment: society wanted strong, tough materials, but scientists still had no clear concept of what it was that made a material tough. They would earnestly apply what seemed to be sensible criteria, only to end up with substances 'about as strong as stale hard cheese'. On the other hand, their experience with materials that were genuinely strong sometimes flew so much in the face of what seemed like common sense that they had some persuading to do. The materials scientist James Gordon, to whom I owe the acerbic quote above, recalls the response of a British Air Marshal in 1943 to the idea that Lancaster bombers were to have glass-fibre domes: '*Glass!—Glass!* I won't have you putting glass on any of my bloody aeroplanes, blast you!'.

Fig. 6.2 A section of the San Andreas fault, a fracture in the Earth's crust. The ridges that branch from the main fault line (*diagonal*) are the result of many fault movements. (Photo: François Gohier/Science Photo Library.)

From a modern perspective, we can now see why fracture is so hard to understand. It is a *non-linear* phenomenon: the effect does not follow in proportion to the cause. All the attendant difficulties follow from this. A tiny crack can prove catastrophic. The smallest of perturbations can grow into a major instability. Events occur over a very wide range of length scales. The precise behaviour in any one case depends on details that might be too small or too numerous to track.

I don't, however, want to paint too bleak a picture. We are certainly no longer scrabbling around in the dark as we were during Gordon's wartime exploits. There is still no complete theory of fracture that explains crack growth based on elementary physics, and there may never be one; but we have a pretty good idea of what it takes to make a material strong, and bridges do not collapse nor aircraft fall apart with anything like the regularity that they once did. But I am concerned here primarily with patterns, not engineering. Fracture

provides plenty of those—some that will be familiar from the last chapter, others that are new. And I think that they will show us that intuition about what cracks do is not always our best guide.

Breaking glass

We can understand the Air Marshall's feeling about glass aeroplanes, because we know how readily glass shatters. But with a little more knowledge, we can understand why a physicist might think that, on the contrary, few materials could be better for making aircraft than glass. It consists of disordered silicon dioxide, the same stuff as sand and quartz but melted to break up the regular crystal structure and then cooled quickly so that the atoms become all but immobile before they can pack together in an orderly manner. The chemical bonds between silicon and oxygen atoms are extremely strong, not far off the strength of those between carbon atoms in diamond. You need to expend a lot of energy to pull them apart. So glass should be nearly as strong a diamond, shouldn't it?

The Air Marshall was wrong about glass fibres—they are very tough. But the physicist is wrong too about glass—it breaks rather easily. What is going on?

A stiff, brittle material like glass is tough so long as cracks cannot be initiated. But they can start from the tiniest of beginnings, and from such little seeds grow flaws that shoot through the whole material. Window glass inevitably contains innumerable tiny scratches on its surface, any one of which can act as the initiation point of a crack that spreads with non-linear vigour. Gordon helped to show in the 1950s that only very minor scraping or scratching contact between glass and another surface is sufficient to create elaborate surface cracks. The reason glass fibres are so strong is simply that they have a much smaller surface area than a plate of window glass, and so have far fewer of these microscopic flaws. The thinner they get, the fewer flaws there are.

But why, even if there are tiny scratches to initiate a crack, does it then grow with such awesome speed, if the bonds are really so strong? A.A. Griffith had a critical insight into this problem in the 1920s, while he was laying the foundations of glass-fibre technology by drawing heated glass rods into thin threads. Knowing the energy contained in a single chemical bond in glass, it was a simple matter to calculate what the theoretical strength of glass ought to be, assuming that breaking the glass means breaking all those bonds. The puzzle was that the observed strength was typically about a hundred times smaller than that calculated from the amount of energy needed to break all the chemical bonds through a fracture. But Griffith found that the strength of very thin glass fibres starts to approach this theoretical limit. So the question was not so much why glass fibres are strong but why normal glass in bottles and windows is so weak. How can a minor scratch confined to the surface be responsible for such catastrophic failure?

Griffith built on the work of G.E. Inglis, who in 1913 helped to explain why British ships were falling apart by showing mathematically that the stresses around a hole in a material can be much greater than those through the rest of the material: there is 'stress concentration' at the flaw. Griffith realized that the same would apply on scales too fine to see—at microscopic scratches on the surface of a brittle material. Inglis had shown that the enhancement in stress at the narrow end of an elliptical hole, relative to the stress far from the hole, was related to the square root of the ratio of the length of the hole to its radius. For our purposes, all this means is that sharper, longer holes give more stress concentration than short, broad ones. Griffith showed that, if one considers a microscopic crack in a material one-thousandth of a millimetre long, whose tip cleaves through just one chemical bond at a time, the stress at the crack tip is about two hundred times that elsewhere. So a stress that is two hundred times smaller than that required to break the chemical bonds in a perfect material will be enough to set the bonds snapping at the tip of a crack of this sort. Notice too that if the crack were to grow without changing its width, the stress concentration at the tip gets even greater as the length increases.

This is why brittle materials crack so readily even if the bonds holding them together are strong. Mostly what I shall now be concerned with are the patterns that these cracks make as they grow. But I should just mention briefly that not all materials are consigned to catastrophic failure from their inevitably imperfect surfaces. Metals like copper and iron generally do not undergo brittle failure, but are ductile—they stretch and bend in response to stress. Crack tips in these materials still suffer the same stress concentration, but ductile materials

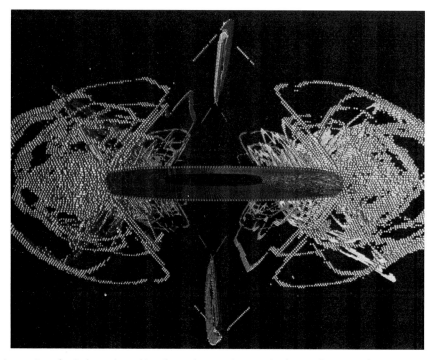

Fig. 6.3 Dislocations—mismatches in the regular stacking of atoms in a crystal—are emitted in complex patterns from a propagating crack in a ductile material. Here I show a computer simulation of a slot-like crack advancing horizontally through a thin sheet of copper. The sheet fills the plane, but only atoms at the surface of the crack, or those involved in dislocations, are shown. The dislocations veer off in unpredictable directions. (Image: Brad Holian, Los Alamos National Laboratory.)

are able to relieve this extra stress by releasing imperfections called dislocations from the crack tip. Dislocations are flaws in the otherwise regular stacking of atoms in the material, where rows of atoms are imperfectly aligned but remain nevertheless bonded to one another. They are rather like the zip heads of a double zipper— the flaw can travel along rows of atoms, but the rows are stuck together again behind them. Because dislocations carry off the energy of a crack without causing the material to fail, they slow down a crack's progress and cause it to become blunt. The pattern of dislocations at a crack tip in a ductile material can be complex and unexpected—the dislocations begin by heading off from the tip at a sharp angle but can then veer off into arcs that form halos around the tip (Fig. 6.3).

Jagged edge

Griffith's work suggests that a long, narrow crack initiated at a notch in a brittle material will cut like a knife straight through the material when it is stressed. This, however, is the exception rather than the rule. A glass cutter can make a clean, straight break through a sheet of glass by first scoring a shallow scratch along the path of the intended fracture, but without this guidance the result is more likely to be that shown in Fig. 6.1. At some stage the crack veers from the straight and narrow, and may begin to throw out a network of branches.

Careful experiments on brittle materials have revealed that there are typically three stages in the way that a crack grows. At 'birth' it accelerates from the initiating notch to reach, in less than a millionth of

a second, a speed of around 200 metres per second—a substantial fraction of the speed of sound in such a material. During 'childhood' the crack continues to pick up speed smoothly while staying straight, and the fractured surfaces that it leaves in its wake are smooth and mirror-like. But once the crack speed exceeds a certain threshold, the mid-life 'crisis' stage sets in—the velocity suddenly starts to fluctuate wildly and unpredictably and the crack tip veers to either side of its previous path. The fracture surfaces therefore become rough and sprout a forest of small side branches (Fig. 6.4).

Does this behaviour sound familiar? I showed in the previous chapter how the tip of a growing crystal finger can develop an instability that induces it to sprout the side branches characteristic of a dendrite. What we now see here is a side-branching instability that has a velocity threshold—it sets in only when the advancing tip gets faster than some critical speed. In 1951 Elizabeth Yoffe at Cambridge University performed a mathematical analysis of the way in which the stresses around a crack tip depend on its speed, and found that as the tip approaches the speed of sound, the stress field ahead of the tip starts to contract in the direction of motion and develops bulges pointing in different directions. These new stress concentrations might then force the tip to deviate from its straight path. John Willis at Cambridge showed in the 1960s that in fact the largest stresses at the tip of a fast crack point at *right angles* to its direction of motion, suggesting that it should be constantly veering off in a perpendicular direction. Computer simulations of fracture, in which a crystal lattice of atoms is pulled apart, support these ideas, revealing that a crack tip begins to develop tributaries when it moves fast enough (Fig. 6.5). Once the branching instability has set in, pulling harder doesn't actually increase the speed of the crack, because most of the energy in the crack is dissipated through the creation of the side branches rather than helping to speed up the tip. In fact, pulling harder can turn out to be counter-productive: so greatly does it increase branching that the tip itself starts to slow down.

Slow motion

Yoffe's analysis helps to explain why a rapidly growing crack should become unstable to steady growth above a certain velocity, close to the speed of sound. It implies that windows would not shatter into innumerable jagged shards, but would simply split cleanly, if only the cracks did not travel so fast. But the Japanese researchers A. Yuse and M. Sano of Tohoku University

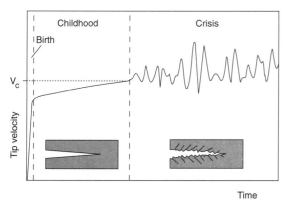

Fig. 6.4 Cracks in a brittle material accelerate rapidly at first ('birth'), and then level off to a steady velocity ('childhood'). If the velocity exceeds some critical threshold, however, this steady motion gives way to wildly fluctuating growth speeds ('crisis'), at which point the crack throws out side branches as it proceeds.

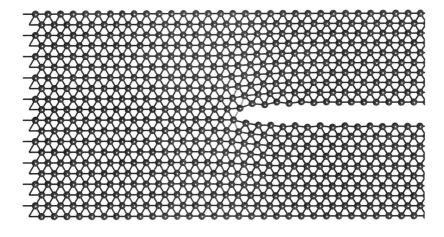

a

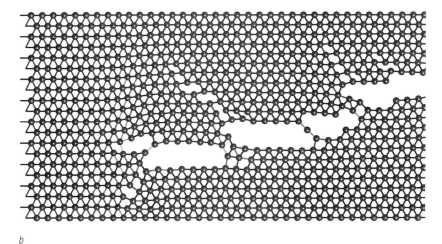

b

Fig. 6.5 Computer simulations of a crack advancing through a two-dimensional atomic lattice show that the crack front remains smooth below a critical velocity (*a*) but becomes irregular and branched above this threshold (*b*). (Images: Michael Marder, University of Texas at Austin.)

showed in 1993 that even cracks that grow very slowly may evolve into complex patterns. They developed a method for growing cracks at just a few centimetres per second—much slower than those passing through a brittle material as it shatters. What they found was an astonishing sequence of crack patterns.

Yuse and Sano sent cracks through flat strips of glass by using heat to induce the stress. They lowered the strips slowly through a heater into a bath of cold water, plunging the temperature by tens or hundreds of degrees centigrade. As anyone knows who has mistakenly put a hot glass dish from the oven into cold washing-up water, this abrupt cooling can shatter glass. When hot, the material expands so that the atoms are on average further apart; when cooled, it shrinks. This disparity between the separation of atoms in hot and

cool parts of the glass sets up large stresses, which can then propagate as cracks from the cool to the hot regions. The stresses are created only in the region of the material across which the temperature changes rapidly, and so the cracks propagate only in this region. In Yuse and Sano's experiments, the interface of the hot and cool regions moved along the glass strip at a speed controlled by the rate at which the strip was lowered through the heater and into the water bath. So by varying this lowering rate the researchers were able to control precisely the speed of a crack initiated from a notch in the lower end. And the stress in the plate could be varied by altering the temperature difference between the hot plate and the water bath.

They found that for very slow speeds (about a millimetre per second) the cracks were generally

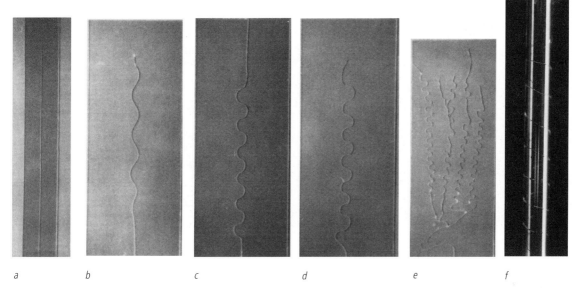

a   *b*   *c*   *d*   *e*   *f*

Fig. 6.6 Even cracks that propagate slowly may develop growth instabilities. A straight crack (initiated at a notch) passes through a hot glass plate lowered slowly into a water bath (*a*). At higher speeds (and for a temperature drop of between about 60 and 150°C), the crack becomes oscillatory, with a constant wavelength (*b*). At still higher growth speeds the amplitude of the wiggles 'saturates', so that they become distorted from a sine-wave shape (*c*, *d*). And for larger temperature drops, the cracks split into branches, which may have wiggles of their own (*e*). When a glass cylinder is used instead of a flat plate, cracks in the 'oscillatory' regime are not wavy, but instead thread around the cylinder in a helix (*f*). (Photos: M. Sano and A. Yuse, Tohoku University.)

perfectly straight (Fig. 6.6*a*). But above certain thresholds in speed and temperature drop, the linear crack became unstable and began to wiggle—not at random, but in a steady oscillation with a well-defined wavelength (Fig. 6.6*b*). Who would have thought that you could cut such a perfect wavy path through a glass sheet just by cracking it? There is clearly a pattern selection process going on here that makes one wavelength of the wavy instability preferred over the others, just as there is for viscous fingering (p. 119), dendritic growth (p. 123) and some of the fluid patterns seen in the next chapter (see Fig. 7.28, for example). The Japanese researchers found that the wavelength increased in direct proportion to the side-to-side width of the glass strip, suggesting that the disturbance of the stress field by the edges of the strip was important in selecting the pattern. For a hypothetical infinitely wide strip the wavelength would be infinite too, which would mean that, once the wavy instability set in, the crack would show the bizarre tendency to head off from the vertical direction at a fixed angle, never to return. The researchers couldn't lay their hands on an infinitely wide glass strip, but they could nevertheless find one without edges: a glass tube. And here they found this very behaviour: the crack set off at an angle from the axis of the tube and kept travelling

around it at this angle, thereby cutting out a perfect helix (Fig. 6.6*f*). Again, it seems amazing to me that you can cut a glass tube into a coil just by cracking it. Helical cracks are rumoured to be found in frozen natural-gas pipelines in Alaska, sometimes winding their way around the pipes for miles.

As wavy cracks in the descending glass plates get faster, the oscillations become more pronounced, until finally they start to distort from pure 'sine waves' and develop kinks (Fig. 6.6*c*, *d*). What happened at even faster speeds depends on the temperature difference between the heater and the bath (which controls the crack-inducing stress). For a small temperature drop (between about 60 and 100°C), straight cracks actually *reappear* at fast speeds. But for temperature drops above about 180°C, wavy cracks at low speeds rapidly give way to branching cracks, whose patterns can be very complex. A single initial crack first forks into two, and these might then each fork into a further two branches and so on. Sometimes the branches develop wavy instabilities before forking (Fig. 6.6*e*). Yuse and Sano were able to map out the boundaries of the straight, wavy and branching regimes as the temperature drop and crack speed (that is, the descent speed of the strip) were altered (Fig. 6.7).

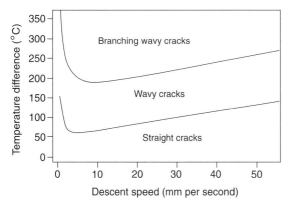

Fig. 6.7 The boundaries between different crack patterns depend on the speed of crack propagation (*horizontal axis*) and the temperature drop (*vertical axis*).

a

Although they weren't able to understand fully the reasons why waves and branches develop in a straight crack, they were able to show that these instabilities have some of the characteristics of the Hopf bifurcation described in Chapter 3 (p. 67). James Rice and B. Cotterell of Harvard University deduced in 1980 the conditions under which a straight crack becomes unstable, and Michael Marder of the University of Texas at Austin has shown how these conditions are met in the experiments of Yuse and Sano when the temperature difference between the hot and cool parts of the glass strip (which sets up the stresses that allow the crack to propagate) gets large enough.

A matter of chance

The crack pattern that forms as a material fractures is a complex product of many factors, including the microscopic structure of the material (whether it is crystalline, glassy, granular, porous, fibrous and so forth), the distribution of the stresses applied, and the speed with which crack tips propagate. One of the most common features of a great many crack patterns, however, is that they are fractal: the network of cracks defines a structure whose dimensionality is not a whole number. Take the two-dimensional crack pattern in Fig. 6.8*a*, for example: here the network of fractures has spread from a single central focus. The highly branched nature of this network might put you somewhat in mind of the fractal DLA cluster in Fig. 5.7, and indeed the crack network has the same property of self-similarity, so that it looks the same at different scales of magnification.

b

Fig. 6.8 (*a*) A crack spreading through a brittle material in two dimensions. (*b*) The network of fault lines that surrounds the San Andreas fault. This too has fractal characteristics.

But other fracture patterns can be less obviously fractal too. Take, for example, the network of fault lines that surrounds the San Andreas fault in California (Fig. 6.8b). This fault system is not a radically divergent network of cracks; indeed it is not even fully connected—many faults stand alone, isolated from the others. These isolated faults can arise when stresses are transferred through the Earth's crust until they reach a point of relative weakness. For all that this fault system looks different from the branching network of Fig. 6.8a, nevertheless a mathematical analysis of the pattern shows that it too is fractal.

Although the fractal nature of cracks is widespread, there is no unique theory that accounts for it. Yet there have now been so many models proposed for fractal crack networks that theorists are almost spoilt for choice. The essential feature of most of these models, however, is that the fracture process involves a strong dash of randomness. It is not hard to justify this: most real materials have microscopic structures that embrace a considerable degree of randomness. Rocks are typically haphazard compactions of grains of many different sizes and shapes, welded together at their boundaries. Metals too, while possessing crystalline orderliness at the atomic scale, are at larger scales agglomerates of many domains, each with their crystal planes pointing in different directions. Cement and porous rocks like sandstone are shot through with random networks of pores. Hard, brittle plastics contain a tangle of polymer chains that are partly aligned but partly entangled and disordered.

Models of fracture in disordered materials seldom try to capture any of these specific kinds of randomness, but instead typically seek to introduce disorder into the way that the bonds break between particles joined together in a regular, orderly lattice. One of the most popular of these models was introduced by Lutz Niemeyer, Hans Jurg Wiesmann and Luciano Pietronero at the Brown Boveri Research Centre in Baden, Switzerland in 1984 to describe the phenomenon of dielectric breakdown. In electrical devices such as capacitors, an electrical voltage is applied across a layer of insulating material called a dielectric. If this voltage exceeds a certain threshold, the dielectric can no longer hold back the flow of charge between the two charged terminals, and a spark discharge crackles through the material. This is dielectric breakdown, and it usually spells doom for an electronic device in which it occurs. The discharge has a branched, lightning-like appearance (Fig. 6.9a), and indeed atmospheric lightning (Fig. 6.9b) is itself closely related to dielectric breakdown—air acts as an electrical insulator between a charged

cloud and the ground. Dielectric breakdown can be regarded as a kind of 'electrical fracture' of a material. In some cases, it is accompanied by *real* fracture, as the material is shattered by the flow of charge. The break-

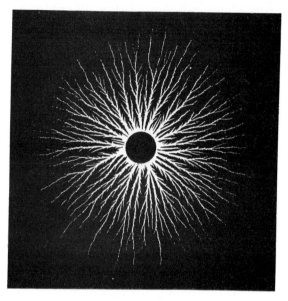

a

b

Fig. 6.9 Electrical discharges are branched formations that resemble crack patterns. (a) A discharge pattern due to dielectric breakdown on the surface of a glass plate. (b) A lightning discharge. (Images: (a) after Niemeyer et al. 1986; (b) UCAR, Boulder, Colorado.)

Fig. 6.10 Lichtenberg figures are produced by the passage of an electric discharge through a solid medium such as Plexiglass. The current cracks and vaporizes the medium, leaving behind a replica of its route. (Photo: Kenneth Brecher, Boston University.)

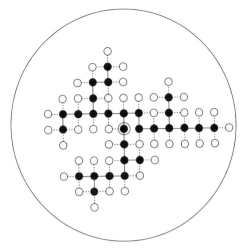

Fig. 6.11 In the dielectric breakdown model, the electrical discharge advances between adjacent points on a regular lattice. At each step there are several directions in which the discharge can flow from each point already reached—the former are shown here as white circles, and the latter as black, connected by a solid line showing the path of the discharge. The discharge was started at the centre of the circle, and is moving towards the lower electrical potential at the edge. The white points that experience the highest electric field have the greatest probability of being selected. (After: Niemeyer *et al*. 1984.)

down pattern is then left frozen into the material (Fig. 6.10). Discharge fracture patterns like these were studied in the eighteenth century by the German scientist Georg Christoph Lichtenberg, and are commonly called Lichtenberg figures. Lichtenberg, incidentally, invented electrostatic printing and was the first person to show conclusively that lightning is an electrical phenomenon, by carrying out a hazardous version of the kite-flying experiment popularly (and probably apocryphally) attributed to Benjamin Franklin.

Niemeyer and colleagues chose to model the dielectric breakdown process by considering a regular checkerboard lattice on which charge could flow from point to point in straight lines. In their model the discharge advances in a series of discrete time steps: at each step, it progresses one lattice site further from each of the lattice points through which it has already passed. From most of these points there are several possible new points to which the discharge can flow in the next step (Fig. 6.11). Which way does the discharge flow? Niemeyer and colleagues assumed that the discharge passes at random to any of the next accessible points at each time step, but with a bias that depends on the size of the electric field at that point. This is a reasonable thing to assume, since the larger the electric field between the discharge's boundaries and any next access-

ible point, the larger is the chance that the spark will flow that way.

So there is an element of chance in all of this. The next advance of the discharge at any point along its length will not necessarily be towards the lattice point that has the highest electric field—it is just more *likely* to go that way. Now, it turns out that the electric field around the tips of the branching discharge is higher than that in the valleys and clefts of the branches. So advance from the tips is more likely than advance from the interior of the 'spark'. Does this sound familiar? Remember that growth was also more likely at the tips of clusters formed by diffusion-limited aggregation than in their recesses, and for an entirely analogous reason—the probability of a new particle striking the tips was higher. It is no surprise, then, that the dielectric breakdown model can give ramified, tenuous discharge patterns that look very much like DLA clusters (Fig. 6.12). The simplest relationship between the probability of the discharge flowing to a white point and the electric field at that point assumes that these two quantities are directly proportional. Discharge patterns generated by this model have a fractal dimension of 1.75, which is almost the same as that of DLA clusters.

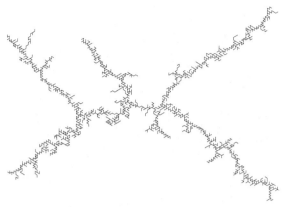

Fig. 6.13 Crack formation can be modelled by a modified form of the dielectric breakdown model that allows bonds to stretch and relax. This can generate more tenuous, almost one-dimensional branching patterns. (Image: Paul Meakin, University of Oslo.)

Fig. 6.12 The dielectric breakdown model generates branching fractal patterns very similar to those seen in diffusion-limited aggregation. They have a fractal dimension of about 1.75. (Image: Luciano Pietronero, University of Rome 'La Sapienza'.)

The dielectric breakdown model can be imported wholesale into a theory of fracture in disordered materials. All you have to do is to regard the discharge as a crack, and the lattice as a network of interconnected atoms or particles joined by bonds of equal strength. At each time step, a new bond breaks from each point along the existing crack, with a probability related to the magnitude of the *stress* field at each of the accessible points. So in this model, the electric field that promotes the discharge is replaced by the stress field that causes bond breaking. It turns out that the two behave in the same way: the stress field is greatest at the tips of a crack, and is smaller in its crevasses.

This is a very simplistic picture of fracture: for one thing, it insists that one bond must always break at each point along the crack with each time step—but in reality there is no reason why this has to be so if the stress isn't large enough. But all the same, the model provides some indication of why cracks might have a fractal branching structure. A better model would make allowance for the fact that bonds can stretch a little without breaking: they are not like rigid rods, but more like springs. This means that, each time a bond breaks, it will release stress in the immediate vicinity and the surrounding bonds can relax somewhat. Fracture models that modify the dielectric breakdown picture to allow for bond stretching and relaxation have been developed by Paul Meakin, Len Sander and others, and they can generate a range of different fracture patterns depending on the assumptions made about bond elasticity and so forth; an example is shown in Fig. 6.13. This crack has a much less dense network of branches than those generated by the 'pure' dielectric-breakdown model, and to my eye it looks much more like the kind of pattern you might finds creeping ominously across the ceiling. The fractal dimension is 1.16, showing that the crack is less like a two-dimensional cluster and more like a wiggly line.

Patterns in the dry season

In all of these examples the crack starts at a single point and spreads from there as the material is stressed. But not all cracks are like that. Think of the fragmented hard mud of a dried-up pond during a drought (Fig. 6.14). What has happened here is that, as the wet mud at the pond bottom has become exposed and dried, the tiny particles have all drawn closer together and aggregated into a compact layer. In effect, the wet mud has been exposed to an internal stress that acts *at all points* as the material contracts. This means that cracks have been initiated at random throughout the system and have propagated to carve up the mud into islands.

This kind of cracking due to uniform shrinkage (or expansion) of a thin layer of material is a common problem in engineering. It might happen to a layer of paint as the material on which it sits expands or contracts because of temperature changes. Surface coatings

Fig. 6.14 When a thin layer of material is stressed as it shrinks, it can fragment into a series of islands of many different size scales. Here this process has occurred in drying mud. (Photo: Stephen Morris, University of Toronto.)

are commonly deposited in a 'wet' form onto an engineering component to protect it or to modify its surface properties (to make it more wear-resistant or less reflective, for instance), and these coatings then shrink as they dry, while the underlying surface retains the same area. Integrated microelectronic devices often incorporate a thin film of one material (an insulator perhaps) laid down on top of another (a semiconductor, say) in which the spacing between atoms is slightly different—so to maintain atom-to-atom bonding at the interface, the overlayer has to be slightly expanded or compressed, and the film is uniformly stressed and liable to crack. Thus there are many very practical reasons for wanting to understand the fracture patterns produced in thin layers of material that are uniformly stressed by expansion or shrinkage.

Arne Skjeltorp from the Institute for Energy Technology in Norway has explored a model experimental system for this type of fracture, consisting of a single layer of microscopic, equal-sized spheres of polystyrene, just a few thousandths of a millimetre in diameter, confined between two sheets of glass. This is an excellent model for the shrinkage of dried mud in a pond bed, because the interactions between the particles are directly analogous to those between silt particles, and because the layer of microspheres, deposited from a suspension in water, likewise contracts and cracks as the water evaporates.

Skjeltorp found that these layers of spheres fracture into complex 'crazy paving' patterns, highly reminiscent

of dried-up river or lake beds, as drying progresses. Figure 6.15a shows the early stages of the process, and Fig. 6.15b and c show the final pattern at two different scales of magnification. The first thing to notice is that the cracks have preferred directions, at angles of 120° to one another (this is particularly evident in Fig. 6.15a). This reflects the symmetry of the underlying lattice of particles, in which they are packed in a hexagonal array. The cracks tend to propagate along the lines between rows of particles, as can be seen clearly in c. The particles in mud are likely to be packed together in a much more disorderly fashion, and so the shapes of the final islands are less regular (Fig. 6.14).

The second thing to note is that the pattern looks similar at different scales of magnification (this can be seen to some degree by comparing Fig. 6.15b and c, except that in the latter we lose the smallest scales because we are reaching scales comparable to the size of the particles themselves). This property is, as we now know, a characteristic of fractal patterns. And indeed these fracture patterns are fractal over the appropriate range of scales—Skjeltorp found that they have a fractal dimension of about 1.68, slightly lower than that of DLA clusters.

Can we reproduce these patterns using the sort of simple probabilistic models of fracture described above? We can indeed. Paul Meakin has adapted the 'elastic' dielectric breakdown model so that it is an appropriate description of Skjeltorp's thin layers of polymer microspheres uniformly stressed by shrinkage. It was important in this model to include the fact that the microspheres are attracted weakly to the confining glass plates—this, Skjeltorp points out, means that the cracks propagate further than they would do otherwise because a crack shifts the spheres away from their initial point of binding to the glass and so sets up additional stresses that drive the crack onward. Allowing for this effect, Meakin found that the model produces crack patterns similar to those observed in the experiments (Fig. 6.16).

What should we conclude from all of this about the web-like branches of cracks? The detailed investigations of the stresses around a rapidly propagating crack tip performed in recent years have enabled us to understand why it is that these fast cracks tend to split into branches: there is a dynamical instability which makes simple forward movement of the tip untenable. Beyond this threshold there is an underlying unpredictability in the motion of the crack tip, so that the crack carves out a jagged path that splits the material into rugged (and

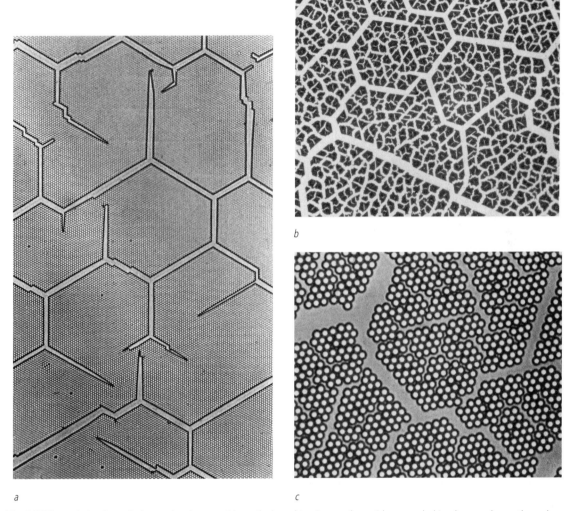

a

b

c

Fig. 6.15 The cracks in a layer of microscopic polymer particles as the layer dries. Because the particles are packed in a hexagonal array, the cracks tend to follow the lines between rows of particles and so diverge at angles close to 120°. This is particularly evident in the early stages of cracking (*a*). The final crack pattern (*b*, *c*) looks similar at different scales, until we reach a scale at which the discrete nature of the particles makes itself evident (*c*). The region in frame *b* is about one millimetre across; that in *c* is ten times smaller. (Images: Arne Skjeltorp, Institute for Energy Technology, Kjeller.)

generally fractal) fracture surfaces. Randomness and disorder in a material's structure provide a background 'noise' that can accentuate the pattern. While in some ways fracture remains a unique and immensely challenging (not to mention practically important) problem, it is nonetheless possible to develop models that seem capable of describing at least some kinds of breakdown process while establishing a connection to other types of branching pattern formation.

A river runs through it

When biologist Richard Dawkins, in his book *River Out of Eden*, compared evolution to a river, his metaphor was based on pattern. Like a river, evolution has its luxuriant branches (Fig. 6.17), a host of tributaries arrayed through time and converging to the broad primary channels of life in the distant past. (Don't look at the analogy too closely, however. It has its strong

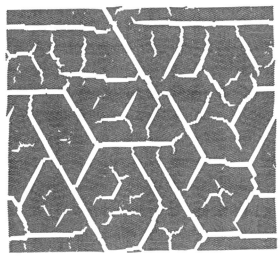

Fig. 6.16 A modified form of the dielectric breakdown model is able to reproduce the fracture patterns seen in contracting thin films. (Image: Paul Meakin.)

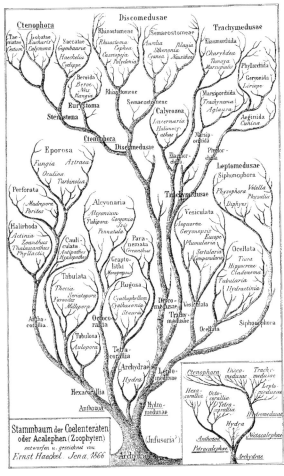

Fig. 6.17 The phylogenetic trees that trace out evolutionary relationships have something of the branching structure of a river delta. Older phylogenies, such as that shown here due to Ernst Haeckel, tended to over-emphasize this pattern, however; Stephen Jay Gould cautions against regarding evolution as a force of increasing diversification.

points, but a river branches upstream, whereas if time is evolution's directional arrow then its bifurcations are distinctly downstream. And some biologists, like Stephen Jay Gould, have spent their lives arguing vigorously that evolution has no 'direction' at all.)

The curious thing about a river network is that it generally grows in the opposite direction to the way the water flows—from the tips of the tributaries into the surrounding rock. There is a very real sense in which we can regard it as a crack, propagating slowly (quasi-statically) through the rock of a hill or mountain range. Yet the physics of this growth process are at face value very different from those of a crack spreading through

Fig. 6.18 River networks—geomorphological cracks on a grand scale? (Photo: Jim Kirchner, University of California at Berkeley.)

stone. Streams grow back from their tips as water from the surrounding slopes flows down into the channel, wearing the rock away little by little. All the same, the result (Fig. 6.18) is a pattern that looks strikingly like a crack, or for that matter like a fractal aggregate or an electrical discharge—but on scales perhaps a million times greater. Already we can smell universality afoot. To what extent is it really so?

For geomorphologists—those who study the shapes of landscapes—many decades ago, there was none of the modern language for describing or conceptualizing branched patterns like this, and they struggled to invent one. The first attempt to do so was made by the

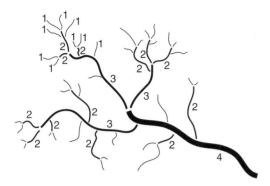

Fig. 6.19 The hierarchy of river network elements in Strahler's modification of Horton's classification scheme. Each branch is assigned an order that increases downstream.

American engineer Robert E. Horton in the 1930s. He formulated a series of 'laws of drainage network composition' which were held to be universal for stream networks. Horton's scheme was modified by A.N. Strahler in 1952, who classified the elements of a network by assigning them an 'order' that signifies their position in the hierarchy of branches. The outermost streams, which themselves have no tributaries, are first-order. Where two first-order streams join, the resulting stream is second-order; and in general, the meeting of two streams of a given order signals the beginning of a stream of next-highest order (Fig. 6.19). If a lower-order stream flows into a higher-order stream, the former terminates but the latter's order is unchanged.

This sensible but somewhat arbitrary classification scheme enabled Horton to identify some general rules governing stream networks. His 'law of stream numbers' states that the number of streams of a particular order decreases with order—there are fewer higher-order streams than lower-order. You could probably guess this rule from Fig. 6.19, but Horton was able to express it with mathematical precision: the number of streams of order n is roughly proportional to the inverse of a constant raised to the power n. In other words, this law of Horton's is a *scaling law*. Another way of expressing this relationship is to say that the number of streams in each order is a constant times the number in the next-highest order. The number of first-order streams in a particular network might, for example, be four times the number of second-order streams, which is itself four times the number of third-order, and so on.

Horton also proposed a law for stream lengths, and this too is a scaling law: the average length of a stream of order n is proportional to a (different) constant raised to the power n. (Or again: the average length for each order is a constant times the average length of the next-lowest order.) Thus, streams of higher order are longer—again what you'd anticipate intuitively from Fig. 6.19. A third scaling law relates the downstream slope of a stream to its order. In 1956 Stanley Schumm proposed a fourth law, in the same spirit as Horton's: the area of the drainage basin feeding a stream with water increases with stream order in the same way as stream length—that is, proportional to a constant raised to the power n. And in 1957, American geologist John Hack proposed a further scaling relationship for river networks: he pointed out that the area of the full drainage basin for a network increases proportionately with the length of the principal river (that is, the highest-order element of the network) raised to the power of about 0.6. Hack's relationship seems to hold some validity for drainage networks ranging in size from those produced in small laboratory experiments to those almost as big as the Amazon. But there is some debate about the precise value of Hack's exponent; other estimates place it closer to 0.5 than to 0.6, and it may be that it does not really have a universal value at all, but varies slightly from place to place.

These scaling laws are really expressions of self-similarity—the networks look the same over a wide range of magnification scales. Benoit Mandelbrot suggested in 1982 that indeed river networks are true fractals, and observations subsequently bore this out. The question is: why? And why, then, do the networks follow these particular scaling laws?

When Horton first reported his laws, they were regarded almost with awe, as though a profound secret of nature's order had been uncovered. But in 1962 Luna Leopold and Walter Langbein showed that randomness alone is enough to ensure that these relationships hold for any branching network. Horton himself suggested that networks emerge as rain falls on a more or less even surface and begins to carve out little gullies or 'rills' wherever the rate of water delivery by the rain exceeds its rate of removal as it filters down through the rock bed. As they grow larger, the rills begin to merge. Leopold and Langbein proposed a model in which rills form at random over a surface and larger channels arise from the merging of smaller ones. The perimeters of rills grow through random walks, constrained only to ensure that the 'walkers' do not recross their own tracks—a property called self-avoidance. This model generates networks that obey Horton's laws as if by magic, even though its ingredients reflect only the barest details of the real geological processes.

In 1966 Ronald Shreve put this picture on stronger foundations by showing that Horton's laws are extremely likely to result from *any* process that connects at random a given number of stream sources within a drainage basin into a network. And geomorphologist James Kirchner demonstrated in 1993 that even randomness is not essential: almost *every* kind of branched network conceivable obeys Horton's laws, not just those arising from random processes. In other words, Horton's laws don't really tell us anything at all about the fundamental patterns of stream networks—they are probably instead an inevitable consequence of the scheme that Horton (and subsequently Strahler) used to break down the networks into fundamental units of different order. So consistency of a particular model of river development with Horton's laws is no good measure at all of whether the model is a good one.

But in any case, it is now clear that drainage networks do *not* usually form by random initiation of rills followed by their merging. Instead, a network grows from the heads (tips) of the channels, where erosional processes cut back into the rock. If we want to understand why networks have the form they do, we would be best advised to focus on what is happening here at the stream heads. And by doing so, we can start to see why drainage patterns have much the same kind of fractal structure as cracks and DLA clusters.

Invasion of the highlands

Recall that in both the latter cases, growth of the pattern from the branch tips is more probable than from deeper within the 'tree'. For cracks this is because the stress is greatest at the tips, just as, within the dielectric breakdown model, the electric field around the discharge tips is largest. The energetic driving force for stream network growth, analogous to the stress imposed on a fracturing material or the electrical power fed into a spark discharge, is the kinetic energy of the rainwater flowing down the contours of the landscape. This energy input to the system is greatest where the water flows fastest and most abundantly—that is, where steep slopes converge. They do so at the head of the stream channels, where water flowing across the rock surface becomes funnelled into the channel. It is this focusing effect at the steep stream heads that creates a greater rate of rock erosion there than elsewhere, leading to predominant growth at the branch tips.

Only predominant, mind you, and not exclusive—because all landscapes are 'noisy'. That is, they all have an element of randomness—variations in surface con-

tours, in soil type and drainage behaviour, in rock type, in vegetation cover and so forth. This noise is the equivalent of the random walks of particles in DLA or of variations in bond strengths in models of fracture in disordered materials. It ensures that networks send out new branches, and that there is still a finite chance of tributaries sprouting from higher-order streams rather than growth taking place only at the stream heads. And like the other branching processes that I have discussed, the growth of drainage networks contains an instability that amplifies small perturbations caused by this landscape 'noise': once a new channel begins to form, its focusing effect on surrounding surface-water flow enhances its growth further.

There is one other aspect of stream networks that bears explanation: stream heads hardly ever cut back across other streams to create islands or loops. This is because, as a stream head advances towards an existing channel, the area feeding it with water diminishes because the existing channel starts to cut off the supply from surrounding ground. Stream heads therefore generally run out of steam (or more properly, of water!) before they intersect other streams. Analogously, the tips of a DLA cluster very rarely merge with other branches because new particles can't reach them once the approach becomes too close.

The connection between these processes and those in crack formation can be made explicit by means of a theoretical model called invasion percolation, which is commonly used for modelling cracks. Percolation is the process by which a fluid passes through a porous medium. D. Wilkinson and J.F. Willemsen devised the invasion percolation model in 1983 to describe the process in which one fluid displaces another in such a medium. We saw in Chapter 5 that the displacement of one fluid by another can create branching instabilities that lead to viscous fingering patterns, whose broad branches have a thickness determined by the surface tension at the interface of the fluids. In invasion percolation, however, the pore network of the surrounding medium imposes its own pattern, and the invading fluid advances through this network in a densely interweaving pattern (Fig. 6.20). The probability of the invading fluid displacing the other is dependent on the size of the pore through which the fluid passes, since this modifies the pressure at the displacement front. If the pore network is highly disordered, this probability varies more or less randomly through the system.

In the model of Wilkinson and Willemsen, this randomness in the advance of the invasion front was

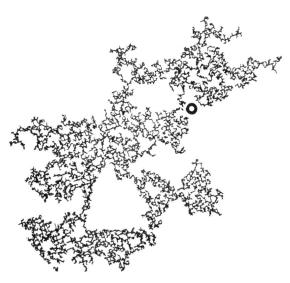

Fig. 6.20 Invasion percolation: the displacement of one fluid by another within a porous medium. The 'invading' fluid is injected here at a single point, and moves forward in a dense, convoluted network. (Image: Roland Lenormand, Institut Français du Petrole, Rueil-Malmaison.)

captured in the following way. The medium being 'invaded' was modelled as a lattice of points linked together by bonds whose strength varies randomly from place to place. Growth of the invasion 'cluster' was initiated at a single point and was assumed to occur in a stepwise manner, with one bond breaking at each step. The next bond to break was always chosen to be the weakest one along the perimeter of the cluster. You can now see that this model describes essentially the same process as the dielectric breakdown model, except that the next bond to break is *always*, rather than most probably, the weakest. It is simply another slight variant on the model of fracture in a disordered solid.

The advance of an invasion percolation cluster occurs mostly at the tips, because as it grows, the cluster 'seeks out' the weakest bonds in its path and leaves behind along its perimeter those bonds that happen to be stronger. The chance of finding at the tips a bond weaker than those still unbroken further inside the cluster is usually pretty good; only rarely will the tips happen all to alight on strong bonds, forcing the breakage of one further back down the cluster's branches. The cluster therefore soon reaches a state in which only bonds with strengths lying in a certain range tend to be broken, and it develops a fractal form.

Colin Stark, working at the University of Leeds, proposed in 1991 that invasion percolation is also much like drainage network evolution. The breaking of bonds mimics the erosion of bedrock by a steady supply of surface water from rainfall; and the randomness in bond strengths reflects the non-uniformity in the landscape. He added only one extra element: the constraint that a stream head could not intersect an existing channel (self-avoidance), included for the reason mentioned earlier.

Stark showed that this model produced stream networks that looked rather realistic at first glance (Fig. 6.21). A trained eye will spot some shortcomings (for example, sometimes three or more tributaries converge at a point, which is not typically seen in real river networks); but Stark went beyond eyeball tests, showing that his model networks obey Hack's scaling law with an exponent of 0.565. Although, as I've said, the 'real' value of this exponent is uncertain, it does seem to lie between 0.5 to 0.6. A related test focuses on the nature of the principal stream—the channel that traces the shortest path through the network (which is what we would normally identify with the 'river' of a particular river network). Observations indicate that this wiggly path has a fractal dimension of around 1.12—it is slightly more wiggly than a simple line. The self-avoiding percolation invasion model predicts a value of 1.13 for this parameter. The 'principal stream' for a branched network formed by DLA, incidentally, has a fractal dimension of 1.0, which isn't really fractal at all but just the same as that of a line. This goes to show why scaling laws are important for distinguishing between network mod-

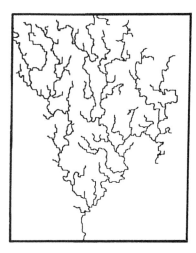

Fig. 6.21 The invasion percolation model, with a slight modification to ensure self-avoidance, produces networks resembling those carved out by rivers as they cut back into the bedrock. (After: Stark 1991.)

els—to the eye, a DLA network doesn't look much different to those of real rivers.

Like all simple models that have been proposed for explaining the form of river networks, the invasion percolation model has its strengths and weaknesses. (For one thing, the physical basis of invasion percolation into a random medium scarcely mimics the processes of dynamic erosion and sediment transport in real rivers.) Most of these physical models include a strong element of randomness, as well as growth instabilities that cause branching and amplify the development of new channels, and they all produce fractal patterns, along with more or less equable agreement with some of the scaling laws seen in the natural networks. The Venezualan scientist Ignacio Rodriguez-Iturbe and co-workers have taken a somewhat broader perspective, by asking whether there is some universal physical principle that underlies the fractal nature of river systems. They have in mind a principle akin to those that physical scientists seek to identify as guiding rules for predicting the course that a system takes when it undergoes a change. For example, we know that objects in the Earth's gravitational field fall downwards because that decreases their gravitational potential energy. But what path does their fall take? The Irish mathematician William Hamilton showed in the nineteenth century that the trajectory of a falling object is that which minimizes a quantity called the *action*, roughly speaking the multiplicative product of the energy change and the time taken for it to happen.

Hamilton's law of least action specifies the parabolic trajectory of a cricket ball as it is thrown and falls in the Earth's gravitational field. Rodriguez-Iturbe and colleagues have made the controversial claim that there is an analogous principle that guides a natural river drainage network into a branched, fractal structure. This principle is that the network evolves in such a way as to minimize the total rate at which the mechanical potential energy of the water flowing through the network is expended. Let me unpack that a little.

As water flows downhill through a river network, it loses potential energy just as does a falling cricket ball. This energy is largely converted into kinetic energy: the water moves. And it is this kinetic energy that ultimately drives the process of erosion that leads the network to expand and rearrange its course. Now, suppose we had a godlike ability to measure everywhere at once the amount of potential energy that all the water was losing each second. (We can't hope to do this in real river systems, but the total can be easily totted up in computer models.) Rodriguez-Iturbe's principle of energy minimization says that the network's shape will change until it finds that for which the total rate of potential-energy dissipation is as small as possible, given the constraint that a certain amount of water must flow through the network each second. This principle says nothing about whether a tributary will or will not appear at a specific location, and it's likely that there will be a large (perhaps huge) number of alternative networks, with broadly similar characteristics, that all come close to satisfying the energy-minimization principle. Rodriguez-Iturbe calls these 'optimal channel networks', and has shown that they have scaling properties that obey Horton's laws, Hack's law and several other empirical laws of river patterns too. In other words, natural drainage networks may be optimal channel networks that have 'sought out' a form that minimizes the rate of energy expenditure. One can show that this optimal form in fact minimizes the average altitude of the drainage basin.

The researchers demonstrated this optimization tendency by conducting computer simulations in which a model network was allowed to alter its channel pattern at random, with the sole constraint that each alteration was more likely to be adopted if it turned out to decrease the rate of energy expenditure. This constraint alone was enough to allow an initial network that looked nothing like a natural drainage pattern to evolve into one that showed all the right scaling laws. Because their model did not include any elements that directly mimicked the geological processes of river drainage (unlike, say the invasion percolation model, which has growth instabilities at the branch tips), the researchers suggested that many other natural, fractal branching patterns might also be optimal channel networks guided by an energy-minimization principle.

But why *should* river networks seek to minimize energy expenditure? Rodriguez-Iturbe merely assumed that they *did*, and showed that this assumption gave realistic branching patterns. They did not attempt to justify this assumption. Kevin Sinclair and Robin Ball of Cambridge University have tried to explain how the energy-minimization principle arises from the fundamental physics of the hydrodynamic processes that govern network evolution. They started with some well-known relationships between quantities, such as the volume and velocity of water discharging through a channel, its width and slope, and used computer simulations to relate these to the rate of erosion of the landscape. They then showed that the resulting relationship between discharge rate and erosion looked mathematically like the expression for Hamilton's law of least action—another minimization principle. In other words, within the very physics of water flow and erosion lies a prescription for the pattern of the drainage networks that these processes will generate. But you'd never guess this pattern by staking out a single channel with any number of flow meters, depth gauges and so forth—the branching pattern is an emergent global property.

The eternal braid

Self-avoidance is the rule for river networks: they do not form closed loops. But all rules are made to be broken. When rivers flow across very flat, broad beds, they often break up into a series of channels that split and rejoin into a series of loops which isolate island after island (Fig. 6.22). These are called braided rivers. They may look familiar—you can see the same kind of braided pattern on a smaller scale when streams run into the sea across a flat, sandy beach. The dried-up imprint of surface flows like this have been seen on Mars too. The pattern appears whenever a broad sheet of water runs over a gently sloping, grainy sediment.

Brad Murray and Chris Paola of the University of Minnesota have proposed that the transport of entrained sediment (something that is ignored in the models described earlier) is crucial to the formation of these braided patterns. Water can scour sediment out of some regions and redeposit it elsewhere to create new bars and islands. In particular, if the scouring rate increases rapidly with increasing flow rate, then an isolated depression in the river bed becomes unstable against deepening. In other words, it captures more of

the flow than the surrounding regions, and so more sediment is washed away from the depression than from its surroundings. The reverse is true for an isolated protrusion: the flow passes around it rather than over it, and so it suffers less erosion and gets higher than its surroundings. As a result, random small protrusions become islands that divert the flow to either side.

It sounds simple enough—but to capture the real dynamics of flow and sediment transport in a theoretical model, Murray and Paola had to include some rather precise rules that related stream flow to sediment flux. In their model the water flows across a checkerboard lattice of square cells, whose heights decrease on average in one direction to define the direction of flow; but superimposed on this smooth slope are small, random variations in height from cell to cell. The amount of water flowing through each cell depends on its height relative to its uphill neighbours: the lower the cell, the greater its share of water from the uphill cells. The height of each cell changes at each computational step, depending on the balance of sediment transport to and from the cell. Because the 'behaviour' (the change in height) of each cell depends on that of its neighbours, this model is a cellular automaton (p. 57).

Murray and Paola found that their model simulations (Fig. 6.23) captured many of the features of real braided rivers. Channels continually form and reform, migrate, split and rejoin: the shape of the river is never steady. Although on average the flow of water and sediment down the river remains constant, it is subject to rather strong fluctuations—more so than in non-braided rivers—because of this constant reorganization of the flow paths. The researchers concluded that it is the processes of sediment scouring, transport and deposition that distinguish braided rivers from branched ones: if the river simply cuts its way by eroding a cohesive, rocky bed to form steep-banked channels, it creates meandering branches rather than braids.

The striking thing about all these river systems, however, is that they are *self-organizing*, in the sense that the flow becomes organized into a stable pattern with properties that remain statistically stable even though the details are constantly changing. This is a hallmark of self-similar growth, which allows an object to preserve its form while it grows indefinitely.

Fig. 6.22 Braided rivers have channels that loop and converge, creating isolated islands that come and go as the river channels change their course. The same pattern can be seen in streams running over flat sand to the sea. (Photo: Chris Paola, University of Minnesota.)

What's left

When we think of river patterns, what usually comes to mind is the plan view: the convergent, branched

Fig. 6.24 Fracture surfaces in brittle materials are commonly highly irregular at high magnification. Shown here is the surface of a fractured hard plastic. (Photo: John Mendenhall, Barbara Goettgens, Jens Hanch and Michael Marder, University of Texas at Austin.)

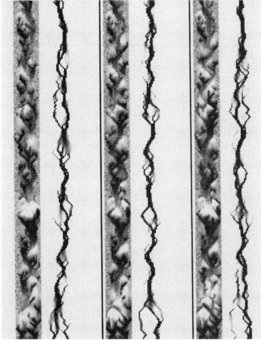

Fig. 6.23 A cellular automaton model of fluid flow and sediment transport captures the essential features of braided rivers: both the instantaneous flow patterns and the way in which these are constantly shifting. Here I show three snapshots of the topography (*left*) and discharge (*right*) produced by the model. (Image: Chris Paola.)

rugged topography. Usually we think of surfaces as two-dimensional objects; but when they become very rough, with peaks and valleys over many size scales, surfaces start to fill up three-dimensional space, and so can be fractals with a dimension greater than two (just as the river network is itself not quite a one-dimensional object but a fractal with a dimension between 1 and 2). You soon find out when a landscape becomes fractal, because it then takes a lot more time and effort to get between two points separated by a given distance as the crow flies, relative to the same journey on a flat plain. Journeys in fractal-land are arduous.

Why the coastal path takes longer

The surface textures that fractures generate are rich and varied. Wood cracks into a spiky array of splinters, reflecting its fibrous texture. Sheets of soft plastics like polyethylene rupture under tension into webs of aligned fibres (Fig. 6.25), a consequence of the fact that the material is made up of entangled chain-like polymer molecules. No single theory can account for all of these textures, since they are generally a consequence of the differing microstructures and atomic-scale structures of the materials. But the idea that many hard materials break to give rough, pitted fracture surfaces like that in Fig. 6.24 is one that seems intuitive—and since, as we've seen, crack networks are typically fractal, we should not be too surprised that the surfaces they leave behind have this character too.

But what does it mean for a surface to be fractal? Simply put, it means that the bumps have no character-istic size scale: they come in all sizes. Put another way, it

network as seen from above. I suppose that this is the perspective we have inherited from map makers, and more recently from aerial photographs and satellite images. But it doesn't much reflect our *experience* of rivers—for what we see instead from our nose-high view of the world is the effect that a river has on the landscape. In other words, we see the topographic *profile* that the river carves into the landscape. Flowing water doesn't just trace sinuous channels through the land; it imposes height variations—hills and valleys, gorges, ravines and lone peaks (Plate 13). There is as much characteristic shape and form in what the river leaves behind—in its profile—as there is in the course it takes.

The river network is, to a first approximation, traced out as a pattern of lines. The topographic profile of the network, meanwhile, is defined in terms of a *surface*—the contoured landscape of the river's hinterland. In just the same way, I discussed cracks earlier from the point of view of a branched network; but what a fracture commonly leaves behind is a rough surface (Fig. 6.24) with a

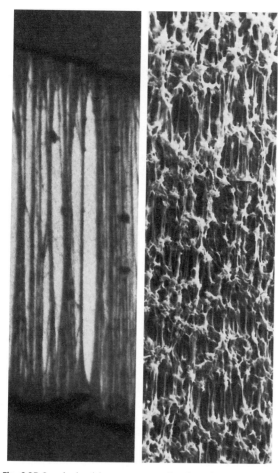

Fig. 6.25 Complex breakdown patterns are found in polyethylene, owing to its fibrous texture. (Image: Paul Meakin, Oslo University.)

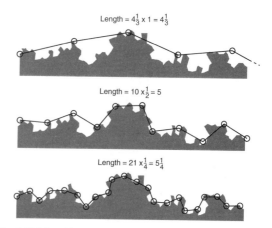

Length = $4\frac{1}{3}$ x 1 = $4\frac{1}{3}$

Length = 10 x $\frac{1}{2}$ = 5

Length = 21 x $\frac{1}{4}$ = $5\frac{1}{4}$

Fig. 6.26 A fractal boundary (like a cross-section through a fractal surface) has a length that depends on the yardstick used to measure it. As the measuring stick becomes smaller, the apparent length seems to increase as we capture more and more of the details. Here the measured length increases slightly each time we reduce the measuring stick by half.

means that the apparent area of the surface depends on the size of the ruler that one uses to measure it. Take a look at a typical cross-section through such a surface (Fig. 6.26). What is the length of this cross-section? That depends on how we measure it. If we use smaller and smaller yardsticks, we capture more and more of the detailed ups and downs and so the overall measured length gets longer. Of course, the real length does not get any longer just by our act of measurement—we just 'see' more of it. But a genuinely fractal boundary has no 'real' length at all: it has ups and downs on all length scales down to the infinitely small, so the apparent length goes right on increasing as we measure it at ever smaller scales. True fractals like this are just mathematical abstractions, however, since the crenelations of any real boundary cannot get any smaller than the sizes of atoms.

It was this apparent dependence of perimeter length on the size of the yardstick that led Benoit Mandelbrot to uncover fractal geometry. In 1961 he came across the attempts of the English physicist Lewis Fry Richardson to specify the length of coastlines and borders, including the west coast of Britain and the border of Spain and Portugal. (Of course, many coastlines can be regarded as fractures on a geological scale, where the Earth's surface has been pulled apart by tectonic forces.) Richardson found that the apparent length of these boundaries depended on the scale of the map that one used to make the measurement: small-scale maps show more detail than large-scale ones, and so capture more of the nooks and crannies, making the total length seem longer. If the logarithm of the length of the boundary is plotted against the logarithm of the length of the yardstick, the points fall on a straight line (Fig. 6.27). Mandelbrot came to appreciate that, for objects like this, length is not a very meaningful parameter since it depends on how it is measured. The form of the object *can* be uniquely specified, however, by the slope of this so-called log-log plot, which is related to the fractal dimension.

When I introduced the concept of fractal dimension in the previous chapter, I did so in a rather different way: by suggesting that it is a measure of how the mass of a fractal object like a DLA cluster (Fig. 5.7) depends on its size. But it is probably not too hard to see from that figure that the DLA cluster has a highly convoluted

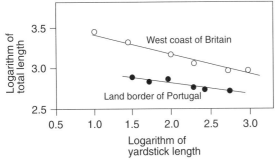

Fig. 6.27 Lewis Fry Richardson found that the lengths of many coastlines and borders depend on the size of the measuring stick, increasing as the stick gets smaller. When the logarithm of the apparent length is plotted against the logarithm of the stick length, the measurements fall onto straight lines that have a characteristic slope for each boundary.

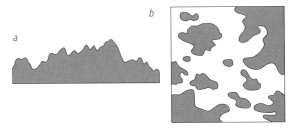

Fig. 6.28 A vertical cut through a rugged landscape reveals an irregular profile of peaks and valleys (*a*). A horizontal cut, meanwhile, isolates islands—cross-sections of the peaks—separated by gaps (*b*).

perimeter which will also have a yardstick-dependent length. The fractal dimension of the perimeter has the same value as that which characterizes the size–mass relationship: about 1.7. There is often more than one way of getting at the fractal dimension of an object, which is an invariant geometrical property of the way it occupies space.

But we must be careful here. Yes, a jagged fracture surface may be a fractal, but it is a fractal of subtly different complexion to the branched structures of DLA clusters or dielectric breakdown patterns. I explained in Chapter 5 that a DLA cluster is *self-similar*, in the sense that if you turn up the magnification at any part of it, you just keep seeing the same sort of delicate web of branches repeated again and again (so long as you don't get to such small scales that the constituent particles themselves start to become evident). More precisely, self-similar objects are composed of copies of themselves scaled down by a constant ratio; and they are *isotropic*: they have the same fractal dimension in all directions.

Self-similar fractals are the easiest sort to understand. But fractal surfaces are, I'm afraid, not like that. Although they have a fractal dimension of between 2 and 3, indicating that they have a tendency to fill up three-dimensional space in a way that a flat or smooth surface does not, this space-filling tendency is *not* isotropic. Imagine taking cross-sectional slices through a rugged mountainous landscape. A vertical cut reveals one thing—a single rising and plunging (but continuous) transect across valleys and peaks (Fig. 6.28*a*)—but a horizontal cut reveals something else entirely—the isolated 'islands' of sections through peaks, separated by

space (Fig. 6.28*b*). In other words, this fractal landscape is not isotropically self-similar. It is instead said to be *self-affine*, which crudely means that the ratio by which the component features are scaled at successive levels of magnification is different in different directions. Notice, however, that the *perimeter* of a vertical cut through a self-affine surface (Fig. 6.28*a*) *is* self-similar—it is a line with a fractal dimension of between 1 and 2 (generally closer to 1, since the line does not tend to bend back on itself so as to more completely fill two-dimensional space).

So it's not quite so straightforward to measure the fractal dimension of a self-affine surface. One way is to look at many cross-sections like Fig. 6.28*a*, by taking cuts through the surface, and to see how their length depends on the length of the ruler (see Fig. 6.26). The fractal dimension of the wiggly cross-sections can then be related to that of the surface as a whole. But in 1984, in one of the first demonstrations that fracture surfaces could be fractal, Benoit Mandelbrot and co-workers took the alternative approach of looking at horizontal cuts through the surface (like that in Fig. 6.28*b*). They examined the nature of fractured steel by shaving down the rough surface in a series of flat cuts, and looking at the rough-edged, flat-topped islands that this left behind. If these islands had had smooth, circular edges, their area would have increased in proportion to the square of their perimeter, and a graph of the logarithm of the area against the logarithm of the perimeter would be a straight line with a slope of 2. Because they (like the surface itself) were rough, fractal objects, however, their areas increased more rapidly with increasing perimeter, and the log–log plot had a slope of 2.28—which is the fractal dimension of the surface.

Mandelbrot realized in the 1970s that the natural topography of the Earth is typically a self-affine fractal. He notes how this aspect of mountain landscapes can be discerned in Edward Whymper's comments from *Scrambles Amongst the Alps in 1860–1869*: 'It is worthy of remark that … fragments of … rock … often present the characteristic forms of the cliffs from which they have been broken'. Fractal geometry has since been used to produce stunning simulated images of imaginary mountainous terrain (Plate 14), and to manufacture computer-generated but realistic-seeming landscapes in Hollywood movies. The crucial point here is that these landscapes are *not* simply random; if you let the computer generate an image in which the ups and downs are merely determined by a random process, the result is a relief pattern that is certainly uneven but that just *looks* wrong. Fractal landscapes are 'noisy' and unpredictable, but are not simply random.

Carried away

Geomorphologists who study landscape formation have embraced the concept of fractals more or less eagerly, but for them this means more than just using some abstract mathematical procedure to churn out endless images of virtual rugged terrain. They want to know how one can understand the evolution of these forms from the fundamental geological processes of nature. This is an old and distinguished field of study,

and I'd be doing it a disservice if I do not make clear that the ideas of fractal form and of self-organization that have become in vogue with physical scientists in recent years provide but a gloss (albeit a very attractive one) on the substantial foundations of geomorphology that were laid down in the nineteenth century, when physical modelling was first attempted. What's more, although much of the work on the spontaneous appearance of geomorphological form focuses on the processes that operate on a daily basis in a geological system to shape it—erosion and other forms of weathering, sediment transport, ground freezing, vegetation growth and so on—these aren't by any means the only or even always the most important influences at work. Sometimes geological forces that operate from outside the system itself—so-called eksystemic influences—come into play in a critical way. Global shifts in climate during ice-age cycles, glacier advances or retreats, and large-scale plate-tectonic events like the collision of tectonic plates, are examples of these. In general, the smaller, shorter-lived features of a landscape—rills, gullies, hillslopes—are self-organized by interactions between them and the other intrinsic elements of the system, whereas larger, long-lived features like mountain ranges come about through external, eksystemic influences.

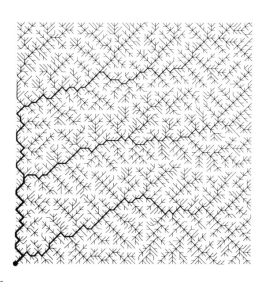

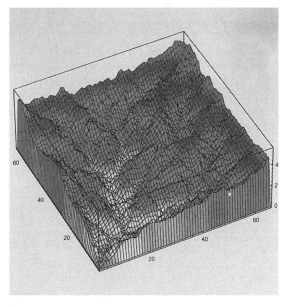

a   *b*

Fig. 6.29 In this model of landscape evolution, water falls on a random landscape and causes erosion as it flows. The streams organize themselves into an 'optimal river network' (*a*), while the topography of the landscape changes from random to fractal, with hills and valleys on all scales (*b*). (Images: from Bak and Paczuski 1993.)

Erosion by flowing water is without a doubt one of the major influences on the Earth's topography, and a great many traditional geomorphological models represent attempts to capture the interactions between rock and sediment removal, transport and deposition. Often these incorporate complicated mathematical expressions for how the water flow properties affect the rate of erosion, the sediment load it bears and so forth. But recently some researchers have suggested that the kind of self-affine relief seen in nature is a robust form that emerges automatically as an erosive river network develops across an initially flat or randomly corrugated (non-fractal) landscape, regardless of the finer points of a particular flow model. Ignacio Rodriguez-Iturbe, Andrea Rinaldo and their co-workers have, for example, studied a model of river evolution that includes the

effects of erosion on the profile of the landscape in a simple way. They began with a plain whose roughness was totally random. A surface of this sort is more like sandpaper than a mountain range—it is uneven, but without scale-invariant self-affinity. In the model, rain falls onto this plain at a uniform rate everywhere, and the resulting flow of water generates an erosion force that depends, at each point, both on the rate of flow (volume of water per second, say) and the steepness of the gradient down which the water flows. This erosive force is assumed to remove and carry away material only when it exceeds some critical threshold value. When this happens, the height of the landscape at that point is reduced.

Notice that there's nothing in this model to ensure that the flow gets channelled into a single, connected

a

b

Fig. 6.30 An experimental scale model of erosion on a bed of sand and clay produces a rugged skyline (a) that resembles those seen in nature at scales thousands of times larger: (b) a mountainscape in the Dolomites. (Photos: Tamás Vicsek.)

river network. Yet as the simulation of landscape erosion proceeds, such a network emerges (Fig. 6.29*a*), in which tributaries feed into higher-order streams (in Horton's sense) that eventually all converge into a single channel. And at the same time, the topography of the plain deepens into a rugged range of hills and valleys with a fractal character (Fig. 6.29*b*). The river network has the properties of an optimal channel network described earlier. This topography looks to the untrained eye much like that of a real landscape, although a geomorphologist might point out that the streams run unusually straight and parallel, and sometimes converge too abundantly at a single junction.

Tamás Vicsek and co-workers in Budapest have been interested in this same process, but with a willingness to get their hands dirty. Their model of landscape erosion is no digital cyberworld but a thing made of real mud and water. They mixed sand and soil (purchased at a Budapest florist's shop) to simulate the grainy but somewhat sticky substance of hillslopes. From this they modelled a flat-topped ridge just over half a metre long, and they sprayed it evenly with water to see what kind of surface would be carved out by erosion.

The running water carries off material through a combination of two processes. The granular substance is worn down quite gradually as it becomes suspended in the flow; but from time to time more profound changes to the model landscape take place through landslides. Both of these processes, of course, can occur in real hill and mountain ranges. The result is a rough, bumpy ridge that one could easily mistake for a rocky hillslope on a scale thousands of times bigger (Fig. 6.30*a*). In fact, Vicsek and colleagues pointed out the similarity to a mountain ridge in the Dolomites (Fig. 6.30*b*), which stretches over kilometres rather than centimetres: a striking example indeed of the scale-invariance of these erosion surfaces.

In some cases, a careful look at the cross-sectional profiles of the model ridges reveals a deeper similarity with mountain ridges than is immediately apparent. The rather flat ridge shown in Fig. 6.31*a* doesn't obviously resemble the jagged section of the Dolomites in Fig. 6.31*b*—until you exaggerate the vertical scale of the ridge's profile, whereupon the two look remarkably alike (Fig. 6.31*c*). You might ask whether it's really a fair comparison to blow up the experimental data in this

a

b

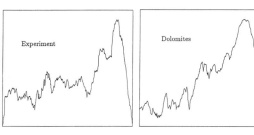

c

Fig. 6.31 (*a*) The profile of a ridge produced in a laboratory model. A section of the the Dolomites (*b*) has the same degree of roughness when the profile in (*a*) is 'scaled' by expanding the vertical scale (*c*). (Photos: Tamás Vicsek.)

way; but the fact is that all this is doing is making more visually apparent the underlying *statistical* similarities of the two profiles. In other words, the two surfaces follow much the same scaling relation for self-affinity (between, say, the degree of height variation and the distance over which it is measured); it's just the *amplitude* of the roughness that differs.

But, you might object, the Dolomites are made of rock, not a soft mixture of soil and sand! This might not, however, be as great a distinction as it appears. Both of these substances are worn away by flowing water—it's just that it happens much faster in the softer medium. And both have an erosion resistance that varies from place to place—the sand and soil were only crudely mixed, and even the rock is highly non-uniform. Finally, both materials suffer from erosion due to the same two processes: gradual removal of suspended small particles, and abrupt landslides. In mountain ranges, the latter can take place over distance scales of up to a mile or so, and become possible when the rock is fragmented by freezing.

So you might, when next walking in the mountains, like to scan the slopes all around for miniature replicas of the giant peaks in the distance: demonstrations of the Earth's scale invariance carved by the elemental forces of nature.

FLUIDS

The general surface grew somewhat more smooth, and the whirlpools, one by one, disappeared, while prodigious streaks of foam became apparent where none had been seen before. These streaks, at length, spreading out to a great distance, and entering into combination, took unto themselves the gyratory motion of the subsided vortices, and seemed to form the germ of another more vast.

Edgar Allan Poe
A Descent into the Maelstrom

If you want to see one of the key differences between Eastern and Western thought, look at the classical art of the two cultures. The West is deeply concerned with static form, with the angle of hand and arm, the tilt of a head, the naturalistic reproduction of shape. The Eastern tradition works differently: not with light and shade, not with a limitless blend of mimetic colour, but with quick, broad strokes, alive with the energy of the artist. It is like the apotheosis of a sketcher's technique, capturing the instant while exclaiming the transience of forms in motion. It is, in short, an art that embraces change—an embodiment of the essential difference between a Platonic and a Taoist tradition.

Traditional Western artists have seldom faced up squarely to the challenge of change. It's not easy to paint something that is never still. Yet to the traditional Chinese artist, that can be the whole point of the exercise—to capture the fundamental forms of motion. This is nowhere more clear than in the ways in which these two traditions have attempted to depict the most challenging of all movements: that of flowing water. The West has relied on the play of light to suggest the froth of wave caps (take a look at George Morland's *The Wreckers* (1791)) or the swirl of mist and sea (take a look at almost any painting by Joseph Turner). Chinese and Japanese artists, meanwhile, have sought to capture the structures of fluid trajectories in a series of lines (Fig. 7.1), which are remarkably close to the *streamlines* that scientists use to depict fluid flows (as we shall see).*

This is not a naturalistic representation, but an artistic response to the same problem that now occupies a great many physical scientists: what are the fundamental forms of turbulent flow?

We like to think that the calculus of Newton and Leibniz gave us a tool to handle the science of change; but for a problem like turbulence, calculus is merely as the brush is to the picture. It provides a formalism with which to frame the problem mathematically, to write an equation for turbulent flow. Then we can stare at this equation and realize that we can't solve it, and in the end we are forced to go back, like the French mathematician Jean Leray in the early twentieth century, and gaze instead at the real thing: the eddies of the Seine as it flows beneath the Pont Neuf in Paris. There are patterns in there, to be sure—we observe the swirling vortices being born and swallowed up—but how can one formulate an exact description of them?

Many of the greatest scientists have bloodied their knuckles against the implacable walls that surround the problem of turbulent fluid flow. David Ruelle, a physicist who has contributed more than many to our understanding of it, points out that 'turbulence is the graveyard of theories'. He notes with glee how the classic text *Fluid Mechanics* by the Russian physicists Lev Landau and Evgeny Lifshitz suddenly dissolves from its charac-

---

* I do not know why Leonardo da Vinci's astonishing sketches of flow patterns along similar lines have not had more influence on Western art. This theme seems to have resurfaced only in the late nineteenth century, notably with van Gogh, Munch, graphic artists like Arthur Rackham, and the art nouveau movement.

Fig. 7.1 In Chinese painting, the flow of water is commonly represented as a series of lines that more or less approximate the paths of suspended particles. This is not a realistic, but a schematic, depiction of flow. These images are taken from a painting instruction manual compiled in the late seventeenth century. (From: M.M. Sze (ed.) (1977), *The Mustard Seed Garden of Painting*. Reprinted with permission of Princeton University Press.)

teristically complicated mathematical formulae into pure narrative description when they come to talk about turbulence. These formidable scientists were forced into the equivalent of the Chinese artists' efforts—to paint pictures in words.

But turbulence is the hard part of fluid mechanics (a discipline also known as hydrodynamics, since through most of its history the fluid of interest has been water). Before a fluid flow is driven so hard that it gives way to turbulence, many things can happen and all manner of interesting and unexpected patterns lurk. In coming to understand these pattern-forming processes in fluid flow, scientists have gradually fashioned tools sharp enough to start to chip away the carapace of turbulence.

On the boil

There can be few more contradictory places on Earth than Iceland. Though not quite the icy wasteland that the name suggests (and which, legend has it, was invented by jealous Viking settlers hoping to deter others from coming to contest their lands), it is nonetheless tucked up against the Arctic Circle, at the same latitude as Fairbanks, Alaska and the barren Siberian tundra. Permanent glaciers nestle in the island's centre. And yet heat and fire are as much a part of the island's culture as ice. Vast solidified lava flows, up to thousands of metres deep, stretch as far as the eye can see. Hot volcanic springs attract bathing tourists. Mounts Hekla and Heimaey frequently spit fire and ash.

Iceland owes its fiery character to fluid movements in the Earth's mantle. The island has the dubious distinction of sitting right on top of a part of the deep Earth where hot rock wells up to the crust and bursts forth through fissures at the surface. The rock is imponderably sluggish—about as viscous as window glass—but the Earth has plenty of time to conduct its internal gyrations. These fluid motions are an example of convection, the movement that arises in a fluid when it varies in temperature from one region to another.

Like so many other pattern-forming processes, convection is a non-equilibrium phenomenon. A fluid at equilibrium must have a uniform temperature throughout; imbalances in temperature will induce a flow of heat from hot to cold. But this heat flow need not in itself involve motion of the bulk fluid: if the temperature differences are only slight or gradual, heat can be redistributed by conduction, in which the excess energy of the hotter region is passed out to cooler parts from molecule to molecule, like a bucket brigade.

Convection is a fluid flow brought about by the fact that a warmer fluid is generally less dense than a cooler one. If a layer of fluid is heated from below, the lower parts of the fluid become warmer and less dense than the upper parts. This then gives the warmer fluid more buoyancy: like a bubble, it will have a tendency to rise. By the same token, the cooler, denser fluid on top will tend to sink. If you hold a thin metal dish of water above a candle, you will see how this imbalance creates convection currents. The hot water in the centre of the dish, directly above the candle, will rise up in a plume, while the cooler water at the top will sink back down at the edges. (The flow may be visible owing to the differences in refractive index of hot and cold water, but it can be made more evident by dispersing small particles, like powdered metal, in the water.) One can watch convection currents carry dust aloft above radiators in a heated room—the dust traces out the otherwise invisible motions of the air.

But if a fluid in a shallow pan is heated *uniformly* from below, then there is a conundrum. All of the lower layer has the same temperature and so the same tendency to rise up through buoyancy. And all of the fluid at the top has the same sinking tendency. But clearly the two parcels of fluid cannot merely pass through one another. The sheer symmetry of the system poses an obstacle to convection. This was the situation studied by the Frenchman Henri Bénard at the start of the century. The outcome that he observed should not by now surprise us. Driven away from equilibrium by the heating

from below, the system is forced to break its symmetry. And as we know, that is when patterns start to appear.

What Bénard saw was that the uniform fluid breaks up into cells in which the liquid circulates from top to bottom (Fig. 7.2; see also Plate 1). These are now called Bénard cells, and in Appendix 6 I say more about how to manufacture them. For a heating rate just sufficient to start convection the cells are generally sausage-like rolls, which, when seen from above, give the fluid a striped appearance (Figs 7.3 and 7.4*a*). Neighbouring roll cells circulate in opposite directions, so that the fluid at their boundaries is alternately sinking and rising. Clearly, the symmetry of the fluid is broken when these cells appear. Before, every point in the fluid was the same as any

Fig. 7.2 When heated uniformly from below, a layer of fluid will develop convection cells, within which warm, less dense fluid rises and cool, denser fluid sinks. (Photo: Manuel Velarde, Universidad Complutense, Madrid.)

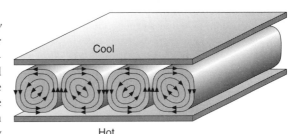

Fig. 7.3 Convection roll cells, which appear in a fluid confined between a hot bottom plate and a cooler top plate. The cells are roughly square in cross-section, and adjacent cells rotate in opposite directions.

a

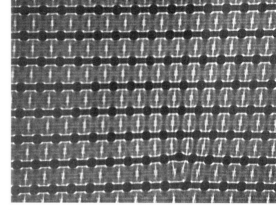

b

c

Fig. 7.4 The complexity of convection patterns increases as the driving force—the temperature gradient from the bottom to the top of the vessel, measured as a quantity called the Rayleigh number—is increased. The convection cells that first appear are roll cells (*a*); at higher Rayleigh numbers, rolls develop in the perpendicular direction too, and the pattern consists of roughly square cells (*b*). This is called bimodal flow. At still higher Rayleigh numbers, the pattern becomes irregular and changes with time (*c*). This 'spoke pattern' is turbulent. (From: Tritton 1988.)

other, whereas after, a microscopic swimmer would find himself in a different predicament in different locations—either buoyed up by the liquid rising from below, carried along by the flow at the top of a cell, or dragged down by the sinking liquid at its edge. But most strikingly, this roll pattern has a characteristic scale that seems to have come out of nowhere. The cells are about as wide as the fluid is deep (typically a few millimetres in experiments like Bénard's), whereas the scale of the interactions between water molecules is about a million times shorter. As in the case of Turing patterns (Chapter 4) or viscous fingering (Chapter 5), a particular pattern with a particular size has been selected; yet, a moment before its appearance, there was nothing in the system to give any clue of its imminent arrival or its scale.

In 1916 Lord Rayleigh tried to understand what triggered the sudden appearance of this convection pattern.

It does not arise as soon as there is a gradient in the fluid's temperature from warm at the bottom to cool at the top, even though the lower layer becomes buoyant as soon as this imbalance is set up. Rather, a certain threshold in temperature difference has to be reached before convection starts, and this threshold depends on the composition and the depth of the fluid. At first sight, this dependence on the experimental set-up seems to spell doom for any attempts to establish a general criterion for the onset of convection. But Rayleigh showed that the various controlling factors can be combined to define a single parameter, called the Rayleigh number, whose value provides a universal criterion for whether or not convection occurs. The Rayleigh number is basically a measure of the balance between the forces that promote convection (the buoyancy of the fluid, which is determined in part by the temperature difference) and those that oppose it (the frictional forces that arise from the fluid's viscosity, and the thermal diffusivity, which tends to even out the temperature imbalance by allowing heat to diffuse from hot to cool regions). Convection does not arise as soon as the bottom becomes warmer than the top because the fluid motion is opposed by friction. Only when the driving force (the temperature gradient) becomes big enough to overcome this resistance do the convection cells appear. The Rayleigh number *Ra* is dimensionless—it is just a 'bare' number, without units, like a percentage or a probability. This is because all of the units in the two opposing forces cancel out when one takes their ratio.

The beauty of treating the problem this way is that all that matters (well, nearly all, as we'll see) is the Rayleigh number—two different fluids in vessels of different dimensions will behave in the same way when their Rayleigh number is the same. This means that one can map out the *generic* behaviour of convecting fluids as a function of Rayleigh number, without having to worry about whether the fluid is water, oil or glycerine. For what it is worth, the critical Rayleigh number for the onset of convection is 1708. Rayleigh showed that there is also a characteristic width for the Bénard rolls that appear at the onset of convection: it turns out that this 'critical width' is very nearly equal to the depth of the fluid, so that the rolls are approximately square (Fig. 7.3). The critical width can be expressed most conveniently in the form of a 'wave vector', a dimensionless measure of the ratio of the roll-cell width to the fluid depth. In other words, while the critical width differs for fluids of different depth, the critical wave vector is the same for all fluids at the onset of convection: its value is 3.12.

If the Rayleigh number is increased beyond its critical value of 1708 to a value of several tens of thousands, the convection pattern can abruptly switch to a so-called bimodal form, in which there are essentially two sets of perpendicular rolls (Fig. 7.4*b*). At still higher values of *Ra*, the roll pattern breaks down altogether and the cells take on a random polygonal appearance called a spoke pattern (Fig. 7.4*c*). Unlike the rolls, this pattern is not steady: the cells continually change shape over time. It is, in fact, a turbulent form of convection.

The master equation

Rayleigh's analysis began with the standard theory of fluid dynamics, enshrined in a single equation that describes how the flow pattern changes over time as a result of the forces that the fluid experiences. In principle this theory is straightforward, since it invokes nothing but the basic laws of motion formulated by Isaac Newton in the seventeenth century. What one wants to know is how, when acted on by all the various forces they experience, each infinitesimally small parcel of fluid moves. Newton's second law tells us how a force causes a change in motion: a constant force on a particle of a certain mass brings about a constant rate of change in velocity (a constant acceleration): the rate of change is equal to the force divided by the mass. Simple.

In the middle of the nineteenth century, George Gabriel Stokes wrote down an equation for fluid motion based on Newton's second law. Stokes's equation was really just a more rigorous restatement of a formula derived by the French engineer Claude Navier in 1821, and so it bears the name of the Navier–Stokes equation. It says that the rate of change of velocity at all points in a fluid is proportional to the sum of the 'inertial' forces—those promoting movement, such as pressure and gravity, plus the retarding force of viscous drag. OK, so that *still* sounds simple.

The catch is that the Navier–Stokes equation is often exceedingly difficult to solve without making several simplifications and assumptions about the nature of the fluid. One prime reason for this is that every 'parcel' of fluid exerts a viscous drag on all the particles around it as soon as their velocities differ—so in general, the behaviour of the fluid at any one point in time and space depends on the behaviour all around it. These complications meant that Stokes's equation was not put to any rigorous test until over a century after Navier derived its initial form.

Much of the work on fluid mechanics even today revolves around the issue of how to introduce appropriate simplifications into the Navier–Stokes equation for particular types of flow so that it can be solved without, in the process, losing the essential features of that flow. Rayleigh's analysis of convection made several assumptions of this sort. He considered a fluid trapped between two parallel plates such that it filled the gap entirely—the fluid had no free surface (as depicted in Fig. 7.3). Convection that takes place under this circumstance is called Rayleigh–Bénard convection. Rayleigh assumed that the fluid was incompressible—the weight of the overlying fluid did not itself alter the density of the lower layer. This is a reasonable assumption for convection in thin fluid layers. Less easy to justify was the assumption that only the density of the fluid changes with temperature, other properties (such as the viscosity) staying the same. We know that for most fluids this is not true—they get less viscous and more runny when they are heated. And most importantly of all, Rayleigh assumed that the temperature gradient—the rate at which temperature changes from bottom to top of the fluid layer—stays constant and uniform. This might seem reasonable enough until you realize that a rising parcel of hot fluid carries heat *up* with it, and in the same way, a cooler sinking parcel can cool down the lower regions. In other words, the motions of the fluid alter their very driving force (the temperature gradient). Rayleigh could not take this into account, and so strictly speaking his theory applied only to infinitesimally small displacements of the fluid, which did not alter the temperature distribution.

Rayleigh's calculations predicted, as I've said, that the fluid in this model system becomes unstable to convection at the critical Rayleigh number, and that the convection rolls have characteristic proportions. But he also found that, as the Rayleigh number is increased beyond the critical value, there is no longer a uniquely allowed wave vector for the convection pattern; rolls both wider and thinner than the critical wave vector of 3.12 (which have, respectively, smaller and larger wave vectors) may also be created. Physicists call these different wave vectors 'modes'—they are rather like the different acoustic oscillations (sound frequencies) that can be excited in an organ pipe or a saxophone's horn. Typically, the harder you blow into a saxophone, the more acoustic modes become excited and the more harmonically rich the note becomes. Rayleigh's treatment of convection shows how to calculate the modes that may be excited for a particular value of *Ra*. If a given mode (with a characteristic wave vector and corresponding length scale) is permitted, this means that applying the slightest of perturbations to the initially uniform fluid on that length scale (by, say, hypothetically moving small parcels of hot fluid infinitesimally upwards at regular intervals of that length) will trigger the appearance of convective roll cells. This kind of mathematical analysis yields a stability boundary for convection which borders an ever widening region beyond the critical Rayleigh number of 1708 (Fig. 7.5). Notice that precisely *at* the critical Rayleigh number, *only* the roll pattern with the critical wave vector of 3.12 is stable.

Given the simplifying assumptions on which it is based, it's perhaps surprising that Rayleigh's theory does so well. It is able to predict not only under what conditions convection starts, but also what the proportions of the convection cells are (or at least, what is their maximum and minimum wave vector). But it cannot tell us

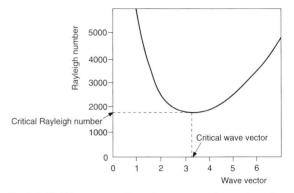

Fig. 7.5 A fluid becomes unstable to the appearance of convection cells at a critical Rayleigh number of 1708, at which point the wave vector (which determines the ratio of height to width) of the convection rolls takes the unique value of 3.12. At the critical Rayleigh number, the roll pattern will grow from the slightest perturbation to the fluid's uniformity. Above this point, other wave vectors become stable too. Here the solid line shows the boundary of stability of the roll cells: only wave vectors inside (*above*) the boundary are stable.

anything about the *shape* of the cells—only a more sophisticated analysis will show that they are roll-like under the conditions that Rayleigh assumed. Moreover, to know whether a particular convection mode is truly stable, it is not enough to deduce whether it will be sustained if stimulated by an infinitesimal perturbation to the fluid; one also needs to know if all other imaginable disturbances, such as a snake-like 'shudder' of the sausage-like rolls, will die out or grow into something catastrophic. Figuring out the stability of the various allowed modes in the face of all such disturbances is no mean task, involving mathematical analysis considerably more complicated than that employed by Rayleigh. During the 1960s and 1970s the German physicist F.H. Busse and co-workers embarked on these difficult calculations. They discovered all manner of instabilities in the parallel sets of rolls that would become manifest under different conditions, and found that the onset of these instabilities depended on the Rayleigh number and the wave vector of the rolls. Busse gave the instabilities graphic names, such as zigzag, skewed varicose and knot. The boundaries of the instabilities cross each other in such a way as to create an enclosed area on a graph of *Ra* against wave vector, called the Busse balloon (Fig. 7.6). At all points within this enclosed field, parallel rolls are stable—at least in theory.

But straight rolls are actually the exception rather than the rule in experiments—generally they are seen only in long, narrow vessels. Even here the rolls can become mildly deformed, and strange things happen at

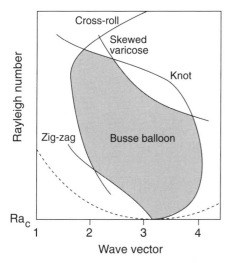

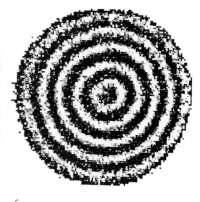

Fig. 7.6 Parallel convection rolls are susceptible to instabilities that distort the pattern. At some wave vectors, for example, roll cells may spontaneously develop a zigzag deformation. The boundaries at which various instabilities set in restrict the region in which parallel roll cells can persist to the so-called Busse balloon (*shaded region*). Rayleigh's original stability boundary is shown as a dashed line.

Fig. 7.7 Convection rolls in a rectangular vessel. Ideal, parallel rolls are frequently distorted by the effect of the vessel's edges. Here the rolls acquire a wavy undulation, and at the ends of the vessel the pattern breaks up into square cells. (From: Cross and Hohenberg 1993, after LeGal 1986.)

the ends (Fig. 7.7). The fact is that all the theories that predict stable rolls begin with Rayleigh's model of a fluid between two parallel plates of infinite extent. But no apparatus is of infinite extent; the vessel holding the fluid must obviously have edges. You might not imagine that this would matter very much, except perhaps close to the edges, but in fact edge effects can have a profound influence on the patterns generated by convection. This

makes life all the harder for theorists trying to predict how a convecting fluid will behave, but it has the attraction of adding a whole palette of new patterns.

Rayleigh–Bénard convection is commonly studied in a circular cell containing a shallow layer of fluid confined between plates. Parallel roll patterns can occasionally be seen in this geometry (Fig. 7.8*a*), but often these become distorted into a pattern that resembles the

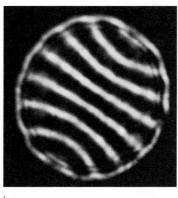

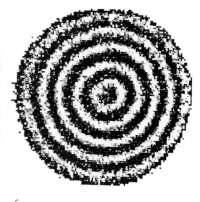

a   *b*   *c*

Fig. 7.8 In a circular dish, roll cells take a variety of shapes. They may remain parallel (*a*), or become curved into a pattern resembling the Pan-Am logo (*b*) to lessen the angle at which peripheral rolls meet the vessel walls. There are no intersections with the walls at all if the cells take on the form of concentric circles (*c*). In (*a*) the fluid is carbon dioxide gas, in (*b*) it is argon gas, and in (*c*) water. (Images: (*a*) and (*c*), David Cannell, University of California at Santa Barbara; (*b*) from Cross and Hohenberg 1993, after Croquette 1989.)

old Pan-Am logo (Fig. 7.8*b*). This is because rolls are generally more stable when they meet a boundary at right angles; so the rolls bend at their ends to try to satisfy this condition. Another option is for the rolls to adapt themselves to the shape of their environment: by curling up into concentric circles, they can avoid having to meet any boundaries at all (Fig. 7.8*c*).

Near the onset of convection, rolls can sometimes break up into polygonal cells, which can be regarded as a combination of two or more roll arrays crossing one another at an angle. Square, triangular and hexagonal patterns (Fig. 7.9) have all been observed, the latter being particularly common. These patterns are all predicted by Busse's complicated calculations.

Because of this rich diversity of patterns that are accessible to the convecting fluid, it is not easy to predict which will be observed in any given experiment. When several alternative patterns are possible in principle for a particular geometry or Rayleigh number, which is selected may depend on the way in which the system is prepared—that is, on the initial conditions and the way in which these are changed to reach a specific set of experimental parameters. Pattern formation is then dependent on the past history of the system.

Moreover, not all convection patterns are unchanging over time. In cylindrical dishes, the regular patterns

Fig. 7.10 Convection rolls can become twisted and fragmented into disordered patterns that are constantly changing in time. (Image: David Cannell.)

described above are unusual; more often the convection cells form an irregular network of worm-like stripes which constantly shift position (Fig. 7.10). Although these patterns are disordered, and might even be considered turbulent after a fashion, nonetheless they clearly retain some vestiges of a pattern with identifiable features. For one thing, all of the wavy rolls tend to intersect the boundaries more or less at right angles. The pattern is reminiscent of the stripe phase of surfactant films (Fig. 2.21*c*) and of Turing structures (Fig. 4.3). It is in fact simply a roll pattern containing a high density of 'defects', characteristic misalignments of the linear cells. These defects can be classified into several types (Fig. 7.11*a*; you should be able to spot most of these in Fig. 7.10), all of which have direct analogies in crystal physics. That is to say, similar misalignments can be seen between rows of atoms in crystalline materials. Defects such as dislocations (Fig. 7.11*b*) in metals are responsible for their ductility—without these defects, metals would be harder but also more brittle. Analogous defects can also be found in the peculiar materials called liquid crystals, where they arise from misalignments between layers of oriented, rod-like molecules (Fig. 7.12).

While investigating the transitions between concentric roll and hexagonal patterns in 1991, Eberhard Bodenschatz and co-workers at the University of

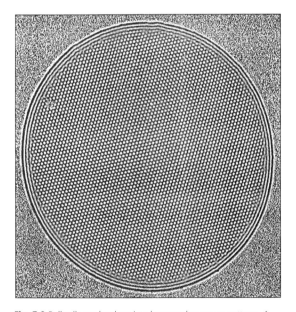

Fig. 7.9 Roll cells can break up into hexagonal or square patterns. A highly ordered hexagonal pattern is seen here in carbon dioxide gas. (But notice that the fluid keeps a couple of circular rolls right at the edge.) (Image: David Cannell, University of California at Santa Barbara.)

a

Focus singularity   Dislocation

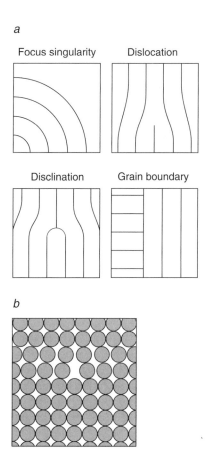

Disclination   Grain boundary

b

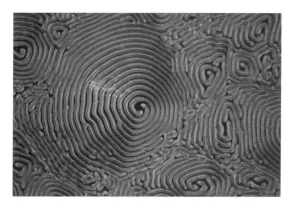

Fig. 7.12 Dislocations in the domain structure of liquid crystals that form spiral domains. The orientation of the rod-like molecules differs in adjacent domains. As the orientation changes, so to does the way in which the material scatters polarized light. So the pattern shows up under illumination with polarized light. The domains are a few micrometres in width. (Photo: Michel Mitov, CEMES, Toulouse.)

Fig. 7.11 (*a*) Several types of defect can be identified in the disordered patterns like those in Fig. 7.10. (*b*) Dislocations between rolls—where two run into one, for example—are analogous to those that appear in the atomic lattices of crystalline materials such as metals, where the regular rows of atoms are disrupted.

California in Santa Barbara saw a new pattern emerge: a spiral, in which one or more roll-like arms twist their way to the circular edge of the cell (Fig. 7.13). The researchers saw spirals with up to 13 arms; that shown here has only two (you can see this by looking at its centre). Israeli physicists Michel Assenheimer and Victor Steinberg showed in 1994 that both concentric (target-like) and spiral convection patterns could exist at the same time in a fluid undergoing Rayleigh–Bénard convection (Fig. 7.14). They were able to induce switches from one pattern to another by changing the balance between the viscous and heat diffusion effects that oppose fluid flow. This balance is characterized by another dimensionless number, called the Prandtl number, which is simply the ratio of the viscosity to the heat diffusivity. In Rayleigh's theory this number stays

constant for a given fluid; but Assenheimer and Steinberg were able to vary the Prandtl number of their fluid (sulphur hexafluoride) with great sensitivity. This was possible because they conducted the experiments at a temperature close to the so-called critical temperature

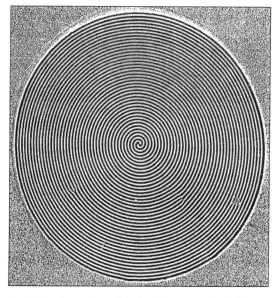

Fig. 7.13 Spiral convection rolls look a lot like concentric rolls (Fig. 7.8*c*), except for a 'defect' at the pattern's centre where the cells meet. Notice that the spirals contain other defects—one is evident towards the *bottom left*, and another towards the *bottom right*. (Compare these with the defects in the spirals of Fig. 7.12.) The spiral is not a stationary structure, but rotates slowly. (Image: David Cannell, University of California at Santa Barbara.)

Fig. 7.14 Coexisting spiral and concentric convection patterns in sulphur hexafluoride. (Photo: Michel Assenheimer, Weizmann Institute of Science, Rehovot.)

of the fluid, where the distinction between a liquid and a gas vanishes (see p. 213). Near the critical point, properties like the viscosity and heat diffusivity vary sharply with temperature. The researchers found that for Prandtl numbers of around 3, spiral patterns were stable, whereas for values close to 6, target patterns were preferred. In between these two values, the two could coexist. There is an analogy to be drawn here with the spiral and target patterns of the BZ reaction (Chapter 3), which can also coexist.

Surface matters

In his convection experiments in 1900, Bénard himself saw polygonal patterns (Plate 1 and Fig. 7.2). The cells here are roughly hexagonal, with an average of six sides each; they become more uniformly hexagonal, resembling those in Fig. 7.9, as the pattern 'matures' after the onset of convection. As we have seen, Rayleigh developed his theory to try to account for Bénard's observations, and on the whole it was very successful: in particular, hexagonal patterns in Rayleigh–Bénard convection can be regarded as superpositions of roll-like patterns. But we now know that Bénard's hexagonal cells did *not* have the same origin as those seen in Rayleigh–Bénard convection, because Bénard's experiments did not correspond to the situation considered by Rayleigh. The crucial difference is that, while Rayleigh dealt with a fluid filling the space between two plates (Fig. 7.3), Bénard's fluid was a shallow layer with a free surface exposed to air. This free surface has a surface tension (Chapter 2), and this can come to exert a dominant influence on the pattern-forming convection process. Yet Rayleigh's theory takes no account of surface tension.

The surface tension of a liquid changes with temperature, generally becoming larger as the liquid gets cooler. This means that if the temperature of a liquid surface varies from place to place, a surface flow may be set up because the higher surface tension in the cooler regions pulls warmer liquid towards it—remember that surface tension can be considered to be a force acting on the surface. Now, upwelling of hot fluid due to buoyancy-driven convection can set up precisely such a non-uniform temperature distribution at the free surface of a fluid: the temperature is higher over the centre of a rising plume of warm fluid than to all sides. If the imbalance in surface tension is the same in all directions around the plume, no surface-tension-driven flow is created because the forces pull equally in all directions. But just as a tiny non-uniformity can give rise to the symmetry-breaking transition of buoyancy-driven convection, so can a tiny heterogeneity in the horizontal balance of surface tensions trigger a symmetry-breaking transition to a pattern of surface flow. When such a flow is established, fluid is pulled up from below to replace that which is pulled laterally across the surface to regions of higher surface tension—and an overturning circulation is induced, just like that of buoyancy-driven convection.

Fluid flows induced by surface-tension gradients were studied in the nineteenth century by the Italian C.G.M. Marangoni, and bear the name of Marangoni effects. Whether or not a flow will be set up by such a gradient depends on the balance between the pull of the surface-tension gradient and the resisting influences of viscous drag and of heat diffusion, which serve to even out the imbalance in temperature that gives rise to the gradient. So just as a similar balance of forces determines when Rayleigh–Bénard convection will begin, so too does a Marangoni flow commence only when a critical threshold of surface-tension difference is exceeded. This threshold is defined by another dimensionless number called the Marangoni number: the ratio of the transverse surface-tension gradient (owing to its temperature dependence) to viscous drag and heat diffusion.

So convection in a Bénard-type experiment is dominated by the Marangoni effect, which sustains the flow and determines the pattern of the convective cells. This means that the onset of convection cannot be predicted in this case by Rayleigh's theory; an alternative theory that takes into account surface-tension effects is required. In addition, the stable pattern is not that of convective rolls but of hexagons in which warm fluid

In the depths, something stirs ...

The Earth is one vast convecting vessel, because it is filled with a fluid that is hotter at the bottom than at the top. The planet's core of mostly molten iron creates temperature of something like 4000°C at the base of the Earth's mantle, nearly 3000 km beneath our feet. The top of the mantle varies in depth from 100 to just over 10 km, and the temperatures there are just several hundred degrees. In addition, the mantle contains many radioactive substances that are gradually decaying and releasing their nuclear energy, thereby heating the fluid mantle from within. As the mantle cools only from the top, this internal heating also contributes appreciably to a bottom-to-top temperature gradient. Even though the mantle is extremely sluggish and viscous, it has a Rayleigh number of several tens of millions, and so is well into the region of turbulent convection. We can expect no well-ordered roll-like convection cells here.

There *are* surely convection cells in the mantle, but they have no regular pattern to them—they most probably shift around over geological time. This is what makes geophysics so interesting, and so hard to unravel—the patterns of mantle circulation are hard to predict. And there are other factors that complicate the matter. First of all, the Earth's continental plates—partly the Earth's crust, partly a cooler, rigid section of the mantle called the lithosphere—ride like a cracked and stony scum on the top of the fluid mantle. It is the slowly overturning convection in the mantle that pulls

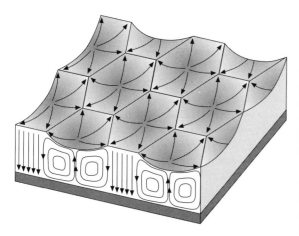

Fig. 7.15 Marangoni convection takes place in liquids that have a free top surface. Although it gives rise to hexagonal cells like those that can be seen in Rayleigh–Bénard convection (where the fluid is confined between two plates), the origin of the pattern is different. It results from imbalances in surface tension, owing to variations in temperature at the liquid surface. This causes the surface to pucker up into hexagonal cells, in which the liquid is pulled from the centre to the edges at the surface.

rises in the centre, is pulled outwards over the surface by the Marangoni effect, and sinks again at the hexagon's edges (Fig. 7.15). What is more, the surface of the fluid becomes deformed and puckered by the imbalance of surface tension, being pulled upwards where this is greatest (at the edges of the cells). This has the counter-intuitive consequence that the hexagonal cells are depressed in the middle (where the fluid is rising) and raised at their edges (where the fluid sinks).

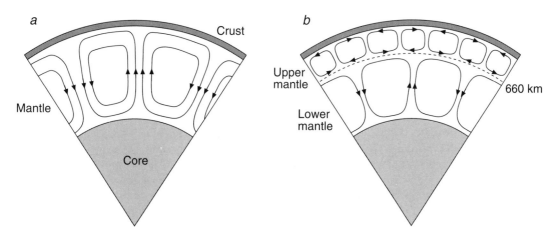

Fig. 7.16 Convection in the Earth's mantle may occur either in one layer or in two. In the former case, convection cells rotate throughout the whole mantle (*a*); in the latter, there are two layers of independent cells separated at a depth of about 660 km (*b*).

these plates around, causing continental drift and plate tectonics. When upwelling convection starts up in the middle of a plate (as it is doing in modern East Africa), the plate is pulled apart and great rift valleys form in the divide. Elsewhere plates collide and mountain ranges like the Himalayas are pushed up; and in other places (such as North America's western seaboard) one plate plunges down beneath another at so-called subduction zones, and the groans of the sinking plate are felt at the surface as earthquakes. All of this movement on top of the convection cells is likely to influence their shape and disposition.

Second, the Earth is not a set of parallel plates or a cylindrical dish, but a sphere. The patterns of convecting fluids on the surface of a sphere are not well studied for low Rayleigh numbers, let alone in a turbulent regime. Finally, and most compromising of all, the structure of the mantle is itself the subject of hot debate amongst geologists. Some seismic waves from earthquakes come bouncing right back towards the surface from a boundary at a depth of 660–670 km, which appears to split the mantle into two layers, like an onion. Most geologists believe that this boundary defines a change in the crystal structure of the mantle material, brought about by the intense pressures and temperatures at these depths. The question is then: does mantle convection punch its way straight through this boundary? Or does it occur in both layers simultaneously but independently? Does the mantle convect as a whole or in layers (Fig. 7.16)?

You might imagine that all of this would leave geophysicists throwing up their hands in despair. But they now have a new experimental tool to enable them to tackle these difficult questions: the computer. Over the past few decades, the computer has advanced to a stage where one can perform billions of calculations per second, and simulate complex phenomena like fluid flow, which could never be solved mathematically by hand. These complex flows are modelled on a computer by dividing up the system into a grid of lots of tiny compartments and then getting the computer to calculate numerically the way in which the fluid in each of the compartments evolves in time. Generally what this means is that the computer deduces how the system will evolve in distinct time steps. For each step it sweeps through the grid, calculating (using the known equations of fluid mechanics) how the fluid within each compartment will move instantaneously in response to the forces acting from all sides. Then it advances one time step so that the initial conditions for the next cal-

culation are the results of the last calculation—and does the whole thing again.

In this way, researchers can investigate what mantle convection would look like under a variety of assumptions—whole-mantle convection, layered convection, convection in which the layers can exchange material weakly, convection with rigid tectonic plates on top, and so on. The calculations are hugely expensive in computer time, and still require that we make some simplifications. But they are now starting to help us make sense of the planet's turbulent insides.

One thing that has become more and more clear to geologists is that the rising and sinking components of mantle convection cells are not equivalent. The latter are sheet-like structures called mantle slabs, which plunge back into the Earth's depths at subduction zones. But the oceanic fissures where hot magma wells up to form new ocean crust—like the Mid-Atlantic Ridge that cleaves the Atlantic almost from pole to pole, or the East Pacific Rise off the west coast of South America—are *not* the corresponding parts of convection cells in which the mantle fluid is driven up by buoyancy. Rather, these linear flaws in the Earth's surface are merely regions of passive upwelling, where hot rock is drawn up from rather shallow depths to sustain the vast tectonic conveyor belts as they head towards subduction zones. The fundamental upwelling structures of mantle convection are instead plumes, cylindrical columns of rising magma. (Iceland has the curious distinction of sitting over a mantle plume as well as over a passive mid-ocean ridge.)

The nature of mantle plumes has been investigated experimentally by simulating mantle convection in tanks of shallow viscous fluids such as silicone oil and glycerine. These experiments show that convection plumes have a mushroom shape (Fig. 7.17), with a broad head whose edges twists into a scroll-like spiral that captures ('entrains') fluid within it. This shape was known to D'Arcy Thompson, who saw reflected in it the form of jellyfish and other soft marine invertebrates (Fig. 7.18). Thompson wondered whether the forms of these creatures might be dictated by some process akin to the rising of a buoyant fluid from the depths.

The diameter of a mantle plume's mushroom head should depend on how far it has travelled from the depths: if the plumes begin close to the base of the lower mantle, as proponents of whole-mantle convection believe, the head can be around 2000 km across by the time it reaches the top of the mantle. There it might then burst forth in a huge outpouring of molten rock,

Fig. 7.17 Convection in viscous fluids at high Rayleigh number creates mushroom-shaped rising plumes. Such features are thought to exist in the Earth's mantle. Where a plume breaks the crust, there is volcanism and an outflow of molten rock. (Photo: Ross Griffiths, Australian National University, Canberra.)

Fig. 7.18 The jellyfish *Syncoryme* has a shape much like a convection plume head. (After: Thompson 1961.)

Why are the rising and sinking features of mantle convection so different? Australian geophysicist Greg Houseman provided a clue when in 1988 he conducted computer simulations of convection in a flat layer of fluid for a Rayleigh number of about 590 000—less than that of the mantle, but well into the regime of turbulent convection. He showed that when the fluid was heated half by internal generation throughout the fluid (like the radioactive heating of the mantle) and half from the base (like the heating from the Earth's core), hot rising plumes and cold sinking sheets appeared—this seems to be the natural structure of convective circulation under such conditions.

The question of whether the mantle convects in one layer or two is still unresolved, but most geophysicists now think that the answer is probably: both. French geophysicists Philippe Machetel and Patrice Weber have carried out simulations of mantle convection in a spherical shell with the proportions of the mantle, in which they allowed for the effects of the change in crystal structure of the mantle rock thought to occur at 660 km depth. (If you are worried about the idea of a convecting fluid with a crystalline structure, remember that the rock is *very* sluggish, moving over geological time scales, and is to all intents and purposes a crystalline material at any instant.) The French researchers found that they could obtain both layered and whole-mantle convection in their model, depending on the character of the structural transition (specifically, on how steeply the transition pressure changed with

laying down vast 'flood plains' of basaltic rock. The 'flood basalt' provinces found in some parts of the world, such as the Deccan Traps in western India, have a comparable width, and might bear testament to the surfacing of a deep mantle plume. Plumes that rise from shallower depths have much smaller heads. Where they break the surface, mantle plumes give rise to so-called hot spots, localized regions of intense volcanic activity. As the tectonic plates are dragged across oceanic hot spots by the pull of subduction, episodic outbursts of magma through the ocean crust create linear chains of islands like those of the Hawaiian chain (Fig. 7.19).

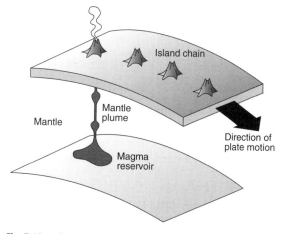

Fig. 7.19 As the Earth's tectonic plates pass over the top of a mantle plume (hotspot), episodic releases of magma from the plume create a chain of volcanic islands. The Hawaiian chain is one such.

temperature). Assuming parameters close to those found experimentally for this transition, they found that convection was a mixture of these two processes. Sinking currents in the upper mantle generally did not penetrate the boundary at 660 km, but if two such currents were pushed together by one of the broad upwelling flows, together they might achieve a threshold size that would enable them to punch through the boundary and flush into the lower mantle. A similar picture of both layered and whole-mantle convection was painted by Paul Tackley of the California Institute of Technology and colleagues when they conducted the same kind of computer simulation, but with a somewhat more realistic model, in 1993. They found that again the flow pattern organized itself into hot rising plumes and cold sinking sheets. The plumes were able to punch their way from the base of the mantle straight through the 660-km boundary to the top; but the cold sinking sheets, the analogue of mantle slabs, generally stopped at this boundary, where the cold, dense fluid accumulated in spreading puddles. When these cold pools became large enough, they would suddenly flush through to the lower mantle in an avalanche, creating a broad sinking column that then spread in a vast pool at the core–mantle boundary (Plate 15).

What limited direct evidence we have of the nature of mantle convection seems to support this picture: sometimes slabs of cool material borne downwards at subduction zones seem to be stopped or deflected at the 660-km boundary, but others appear to pass right through. Little by little, we are starting to piece together the mysteries of the Earth's bowels.

Air, water, earth and fire

There are ample examples of natural convection patterns to keep us diverted at the Earth's surface too. The canvas of the sky is streaked with their imprint. The towering piles of cumulus clouds are erected by convective updrafts as warm air, locally heated by the Sun, rises and bears water vapour with it (Fig. 7.20). As the air cools, the water vapour condenses out into tiny droplets that, by reflecting light, provide the cloud's white billows.

The atmosphere loses its heat primarily by radiation from the uppermost layers, while it is warmed not only by direct sunlight but by heat radiated from the ground. So there is a perpetual imbalance set up between warmer, lower air masses and cooler air higher up—with the consequence that air is always on the move

Fig. 7.20 Clouds trace out the convection patterns of the atmosphere. towering cumulus stacks form around updrafts, where warm air rises. (Photo: Jackie Cohen.)

somewhere, bringing winds, storms and sometimes the violence of hurricanes. When this imbalance is suppressed—when, for example, cold dense air gets trapped in a valley—the result is a temperature inversion, a stagnation of the atmosphere that can allow smog to accumulate.

Convection in the atmosphere cannot be accurately described by Rayleigh's model, because many of the assumptions he made—that the fluid is incompressible, that the viscosity does not vary significantly with temperature—just aren't good ones for air. All the same, many of the general features of convection patterns still apply, and in particular convective motion can become organized into roll-like cells of more or less equal width. These can give rise to banded cloud formations called cloud streets or mare's tails (Fig. 7.21), which mark out the boundaries of the roll cells. These rolls are typically wider than they are deep, unlike the roughly square profile of Rayleigh–Bénard rolls. Approximately hexagonal cells can also be seen in satellite images of cloud convective patterns.

On much larger scales, vast atmospheric convection cells are set up by the differences in temperature between the tropics and the polar regions. These cells don't have a simple, constant structure, and moreover they are distorted by the Earth's rotation; but nevertheless they do create characteristic circulation features, such as the tropical trade winds and the prevailing westerly winds of temperature latitudes. Edmund Halley first proposed in the seventeenth century that convection owing to tropical heating drives atmospheric circulation, and for some time after it was believed that a single convection cell in each hemisphere

Fig. 7.21 Convective roll cells in the atmosphere can create regular cloud streets, as water vapour condenses at the tops of the cells. (Photo: Wen-Chau Lee, NCAR, Boulder, Colorado.)

carried warm air aloft in the tropics and bore it to the poles where it cooled and sank. We now know that this picture is too simplified, and that there are in fact three identifiable cells in the mean hemispheric circulation of the lower atmosphere: one (called the Hadley cell) that circulates between the equator and a latitude of about 30°, one (called the Ferrel cell) that rotates in the opposite direction at mid-latitudes, and one (called the polar cell) that rotates in the same sense at the pole (Fig. 7.22). The polar and Ferrel cells are both weaker than the Hadley cell, and are not clearly defined throughout all the seasons. Where the northern Hadley

and Ferrel cells meet, the effect of the Earth's rotation drives the strong westerly jet stream.

The oceans too exhibit convection patterns over several size scales. Like the atmosphere, the oceans are warmed in the tropics and cooled in the polar regions, and so cool, dense water sinks around the poles. This helps to establish a vast conveyor-belt circulation from the tropics to high latitudes, and the warm water carried polewards at the top of the North Atlantic convection cell brings with it heat that keeps Northern Europe and the eastern North American seaboard temperate. This circulation pattern is modulated, however, by the fact that the density of sea water is also determined by the amount of dissolved salt it contains—the more saline the water, the denser it is. The salinity can be altered by evaporation, which removes water vapour and leaves behind saltier water. Freezing also affects salinity, since ice tends to leave salt behind and so the unfrozen water gets increasingly saline as ice develops. Thus the large-scale pattern of ocean convection is influenced by evaporation in the tropics and freezing at the poles: together, these processes give rise to the so-called ocean thermohaline ('heat–salt') circulation. On smaller spatial scales, the interplay between salinity and thermal convection can create diverse circulation effects in the upper few metres of the oceans, such as oscillatory rising and falling of water parcels or finger-like protrusions of salty water into fresher water below, called salt fingers (Fig. 7.23).

Once you start to spot convection patterns in the world around you, they crop up in the most unlikely places. You can find their polygonal imprint petrified

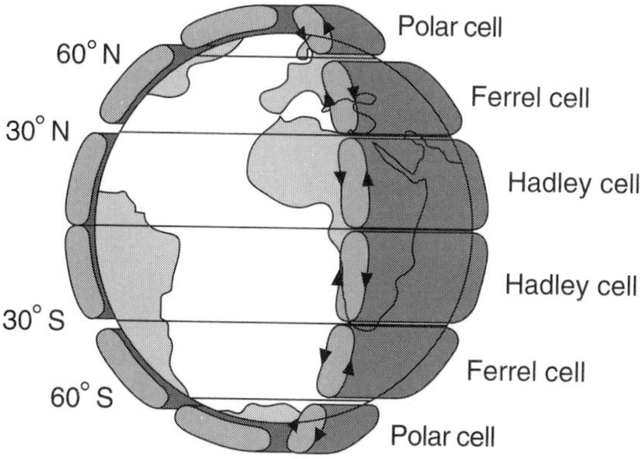

Fig. 7.22 Large-scale convection in the Earth's atmosphere traces out three hemispheric convection cells: the Hadley cell between the equator and about 30° latitude, the Ferrel cell at mid-latitudes and the polar cell over the pole.

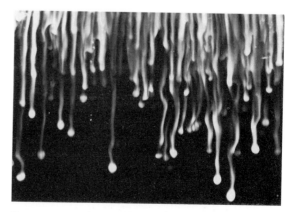

Fig. 7.23 Convection in the surface layer of the oceans due to differences in salinity (and therefore in density) produces forests of sinking 'salt fingers'. Here a laboratory model makes them visible. (From: Tritton 1988.)

Fig. 7.24 The freezing and thawing of water in the soils of northern tundra sets up convective circulation owing to the unique density changes that water undergoes close to its freezing point. The imprint of this circulation can be seen as polygonal cells of stones at the ground surface. Shown here are stone polygons on the Broggerhalvoya peninsula in western Spitsbergen, Norway. (Photo: Bill Krantz, University of Colorado.)

these formations as the water undergoes seasonal cycles of freezing and thawing. The idea that these cases of 'patterned ground' are caused by convection in fact dates back to the Swedish geologist Otto Nordenskjold in 1907, but Krantz and his co-workers were the first to place this idea on a firm theoretical basis. In these cold northern regions, water in the soil spends much of its time frozen. But when the ground warms and the ice thaws, it does so from the surface downwards, so the liquid water gets cooler the deeper it is.

For most liquids this would correspond to a situation in which the density increases with depth in similar fashion. This is a stable arrangement, for which no convection would take place. But water is not like other liquids; perversely, it is densest at 4°C *above* freezing. So when it is warmed by about this amount at the surface, the water closest to the surface is denser than the colder water below it, and convection will begin through the porous soil (Fig. 7.25). Where warmer water sinks, the ice at the top of the frozen zone (the so-called thaw front) will melt, while the rising of cold water in the ascending part of the convection cells will raise the thaw front. In this way, the pattern of convection becomes imprinted into the underground thaw front.

But how does this find its surface expression in mounds of stones? Krantz and colleagues proposed that sub-surface stones are concentrated in the troughs of the corrugated thaw front and then brought to the surface by subsoil processes that are known to shift stones around when soil freezes. This raising up of stones to the surface is known to farmers as 'frost heaving', and results in the littering of a field with stones when it freezes during a frost and then thaws. Polygonal patterns formed in this way can also be found on the beds of northern lakes when the water is shallow enough to freeze down into the lake bed (Plate 16). Krantz and

into stone and rock in some of the frozen wastes of the world in Alaska and Norway (Fig. 7.24). Now here's a real puzzle—it's natural enough to find these patterns in the fluid media of air and sea, and even in hot, sluggishly molten rock, but how do they find their way into frozen, stony ground?

The answer, according to William Krantz and colleagues at the University of Colorado at Boulder, is that convection takes place in the water-laden soils beneath

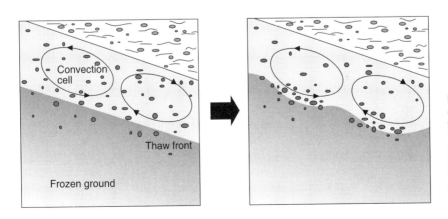

Convection cell

Thaw front

Frozen ground

Fig. 7.25 As water circulates in convection cells through the soil, the pattern is transferred to the 'thaw front', below which the ground remains frozen. Stones gather in the troughs of the thaw front, and are brought to the surface by frost heaving in the soil.

colleagues have developed a theoretical model of the convection patterns that can arise as water circulates through porous soils. They found that polygonal (particularly hexagonal) patterns are favoured on flat ground, but that the convection cells are roll-like on sloping ground, giving rise to striped formations at the surface (Fig. 7.26).

If you want to see convection on a grand scale, look to the Sun (though not literally, I hasten to add). The Sun's visible brightness comes from a 500-km thick layer of hydrogen gas close to its surface, called the photosphere, which is heated to a temperature of about 5500°C. This gas is heated from below and within, and radiates its heat outwards from the surface into space—so that, although it is about a thousand times less dense than the air around us, it is a convecting fluid. The Rayleigh number of this fluid is so high that it should be utterly chaotic and unstructured. But photographs of the Sun's surface show that, on the contrary, the photosphere is pock-marked with bright regions called solar granules, surrounded by darker regions (Fig. 7.27). These granules are convection cells, whose bright centres are regions of upwelling and whose dark edges are regions of cooler, sinking fluid. Each granule is between 500 and 5000 km across, making the largest about half the diameter of the Earth. The pattern is constantly changing, each cell lasting only a few minutes. The very existence of these cells in such a turbulent fluid shows that we still have a lot to learn about convecting fluids and their patterns.

Fig. 7.26 On sloping ground, the convection cells in freezing porous soils can become roll-like. The pattern traced out by stones at the surface is then a series of parallel stripes, seen here in the Rocky Mountains in Colorado. (Photo: Bill Krantz, University of Colorado.)

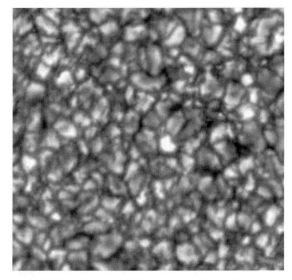

Fig. 7.27 Solar granules are highly turbulent convection cells in the Sun's photosphere. (Photo: The Swedish Vacuum Telescope, La Palma Obsevatory, Canary Islands.)

Riverrun

Let's now return to Jean Leray gazing into the Seine at the Pont Neuf. As the water flows around the columns of the bridge, swirling eddies disturb the surface in the wake downstream. Can we make sense of this flow pattern?

My description of studies of convection will, I hope, have provided an indication of how a scientist might approach this question. First, make an idealized experimental model that captures the essential features of the problem in their simplest form. Then, look at what happens as a single parameter of the experiment is gradually altered while all others are kept constant. The Seine can be a body of water flowing smoothly through a channel. Let's forget about the air/water or water/wall interfaces—as we saw above, they can just complicate things—and just focus on what happens within the body of the water away from any boundary surfaces or edges. We can model a column of the Pont Neuf by a cylinder placed with its axis perpendicular to the direction of flow (Fig. 7.28a). Of course, the column is not really a cylinder, but that seems like a nice simple shape to begin with. If we ignore what happens at the ends of the cylinder, this experiment has translational symmetry along the cylinder's axis—that is to say, the initial flow and the obstacle it encounters are identical for all two-dimensional slices parallel to the flow. This means that we can consider the problem to be a two-dimensional one: we need consider only what happens in a single layer of fluid, and assume that the same thing

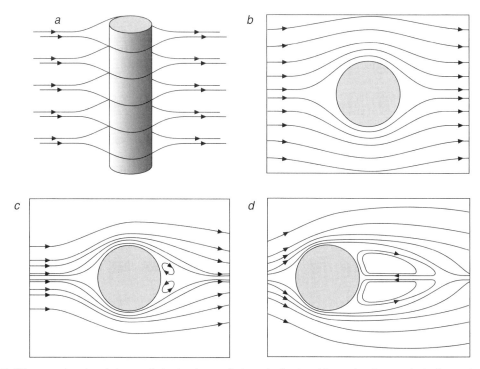

Fig. 7.28 Fluid flow around an obstacle, here a cylinder placed perpendicular to the direction of flow. At low flow speeds, the flow can be regarded as two-dimensional, with the flow pattern being identical in all layers perpendicular to the cylinder (a). The flow can be represented by streamlines, which are essentially the paths taken by tracer particles borne along by the fluid. Far upstream of the obstacle, the streamlines are parallel lines (a and its plan view in b). The flow pattern depends on the Reynolds number (Re), which is a measure of the speed of the flow and the width of the cylinder. At low Reynolds number, the streamlines simply bend around the obstacle (b). At higher Re, circulating vortices appear behind the cylinder (c). These grow with increasing Re, until they become highly elongated (d).

happens to all the other parallel layers too. This assumption is not perfect, and indeed we will find that the flow of the fluid quite readily ceases to be invariant along the entire vertical direction (parallel to the cylinder)—but it will serve adequately for much of what I shall say.

So long as the edges of the channel remain sufficiently distant, the fluid flow towards the cylinder is said to be laminar. This means that if we divide the fluid up into many tiny parcels (which are nonetheless sufficiently large relative to the fluid's constituent molecules that we can regard it as a continuous medium), each parcel travels smoothly on a well-defined path which remains more or less parallel to the direction of overall flow. In the flow upstream of the cylinder, all of these paths are steady, smooth lines. We can depict the flow pattern using the concept of streamlines. Roughly speaking, a streamline shows the path that a given fluid parcel takes in the flow. The streamlines can effectively be made visible by adding tiny solid particles (commonly aluminium flakes, which reflect light) to the fluid, which

are suspended and carried along by the flow. (More properly, a streamline is defined by the condition that a tangent to it at any point shows the direction of velocity of the fluid at that point.)

The cylinder will deflect the flow, which must pass to either side around it. Intuitively, we would expect the disturbance of the flow to be increasingly pronounced as the flow gets faster, or as the cylinder gets bigger. As with convection, it would be useful to have some way of characterizing this effect in a way that does not require us to specify the size of the cylinder, the velocity and viscosity of the fluid and so forth. We would like to identify a dimensionless number, like the Rayleigh number, that allows us to say simply 'For a flow with dimensionless number *x*, the flow around the cylinder has such and such a form'.

Needless to say, there exists such a parameter, and it is called the Reynolds number after the nineteenth-century British scientist Osborne Reynolds, who made an extensive study of fluid flows. The Reynolds number

(*Re*) is essentially the ratio of the forces driving the flow to the forces retarding it (the viscous drag). In its most general form it is given by the product of the velocity of the flow and the characteristic size of the system confining or deflecting the flow, divided by the viscosity. For flow down a narrow channel, the 'size' is the width of the channel; for the experiment described above, where we ignore edge effects by assuming that the channel is indefinitely wide, it is the size (width) of the cylinder that features in the Reynolds number. In a given experiment this stays constant, and so the Reynolds number increases in direct proportion to the increase in the velocity of the flow.

With these necessary preliminaries, we are now ready to see what happens in our model of the Seine passing beneath the Pont Neuf. For Reynolds numbers below four, nothing much happens at all. The streamlines simply bend around the cylinder and then become parallel again on the far side (Fig. 7.28*a*, *b*). As *Re* is increased above four, however, a new flow pattern appears. Immediately behind the cylinder we can now find two little vortices of circulating fluid, called eddies (Fig. 7.28*c*). The eddies circulate in opposite directions,

and they represent little pockets of trapped fluid which have become detached from the main flow and remain in place behind the cylinder. As the Reynolds number is increased, these eddies get bigger; by the time *Re* is about 40, they are highly elongated (Fig. 7.28*d*). But the wake downstream of the cylinder remains laminar: the deflected streamlines outside the eddies converge again until they resume their parallel paths.

Beyond a Reynolds number of about 40, something dramatic starts to happen to the wake. It acquires a wavy disturbance, which becomes more and more pronounced as *Re* increases (Fig. 7.29). This patterning of the wake can be made evident either by suspending tracer particles in the flow or by injecting a coloured dye into the flow from the rear of the cylinder, as shown here; the dye is carried along in a narrow jet, since the rate at which it diffuses and disperses is slow compared with the speed of the flow. Around *Re* = 50, the waves break, their pinnacles curling over into little vortices that leave the wake looking like a swirling art nouveau design (Fig. 7.29*d*). This pattern is called a Kármán vortex street, after the Hungarian physicist Theodore von Kármán. The vortices are carried along with the

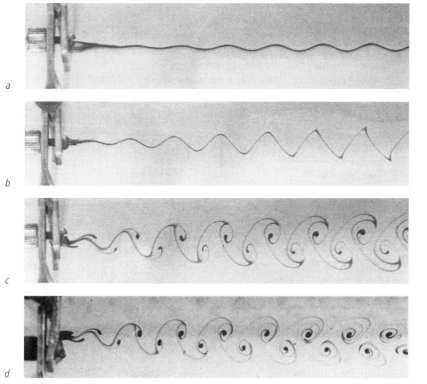

a

b

c

d

Fig. 7.29 At a Reynolds number of about 40, the wake of the flow past a cylinder develops a wavy instability, revealed here by the injection of a dye into the flow from the rear of the cylinder. This wavy disturbance becomes a train of vortices at higher flow speeds (*d*), called a Kármán vortex street. (From: Tritton 1988.)

flow, but more slowly than the average speed of the flow. They slowly dissipate their energy through viscous drag and vanish further downstream.

Like the onset of convection or the appearance of Turing structures in chemical reaction–diffusion systems (Chapter 4), the development of a wavy structure in the wake of the cylinder is an example of pattern formation triggered by a spontaneous instability of a nonequilibrium system (remember that the system *has* to be out of equilibrium for flow to occur at all). Where does the instability come from in this case?

If we measure the velocity of the fluid through a cross-section of the wake for *Re* below 40, it has a profile like that in Fig. 7.30*a*. There is a dip in velocity along the path through the cylinder, so that more or less parallel layers of fluid move past each other at different velocities. Flows with this character are called shear flows, and they are susceptible to pattern-forming processes called *shear instabilities.*

To be sustained indefinitely, any sort of flow structure has to be mechanically stable in the face of small perturbations. That is to say, if we imagine applying a small disturbance to the flow pattern, it is stable if there are restoring forces that return the flow to its initial state. An instability sets in, on the other hand, when a perturbation creates forces that serve to enhance the perturbation still further. This is the case at the threshold of convection, where an infinitesimal upwards displacement of a warm parcel of fluid brings it amongst cooler, denser fluid and so enhances its buoyancy. Shear instabilities involve a similar kind of self-amplification of a perturbation.

One of the best studied is the Kelvin–Helmholtz instability, after the two great nineteenth-century physicists who identified it. It arises in shear flows in which there is an abrupt change in the velocity between adjacent layers of fluid (Fig. 7.30*b*). An extreme case of such a flow is one in which the two layers of fluid flow in opposite directions (Fig. 7.30*c*). Imagine imposing a wavy disturbance on this flow (Fig. 7.30*d*). This pushes together streamlines on the convex side of the disturbance—over the 'peaks'—and pulls them apart on the concave side, in the dips. What this means is that the fluid flows slightly faster (in opposite directions on each side) over the peaks and slower in the dips. (Think of a similar squeezing-together of streamlines when a river flows through a narrow gorge—the flow gets faster.)

Now, the key to the instability is this: along any particular streamline in a flow, the pressure of the fluid decreases as its velocity increases. This fact was demonstrated in 1738 by the Swiss mathematician Daniel Bernoulli, and it is known as Bernoulli's law. It means that the pressure of the fluid against the dips of the wavy interface increases (because the velocity decreases there), and vice versa for the peaks. In other words, the undulations of the wave are pushed outwards—the wave becomes more pronounced (Fig. 7.30*e*). The same principle provides the lift under the wings of an aircraft, since the aerofoils are curved in the same manner—convex on the upper side, concave below.

The undulations are eventually deformed into a train of vortices (Fig. 7.31), whose graceful regularity is fleeting: they subsequently collide and degenerate into

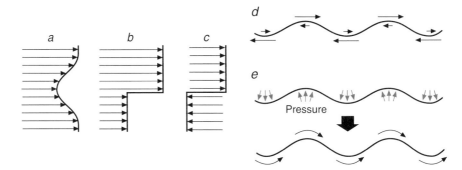

Fig. 7.30 In the wake of the flow around an obstacle, the velocity profile has a dip in the middle (*a*). This is an example of a shear flow, in which layers of fluid move past one another at different speeds. A more extreme shear flow is one in which there is an abrupt discontinuity in the velocity profile (*b*), which corresponds in the extreme case to two layers of fluid moving past one another in opposite directions (*c*). This kind of flow is susceptible to a shear instability called the Kelvin–Helmholtz instability. Any deviation from linearity of the boundary in (*c*)—such as an undulating displacement (*d*)—gets amplified. The sideways pressure on the boundary becomes unequal on either side at the peaks and troughs, because the fluid in the troughs slows down and that at the peaks speeds up (*d*): because of Bernoulli's principle, this sets up a pressure imbalance (*e*, *grey arrows*) at these points, which pushes the peaks outwards. These peaks develop into vortices.

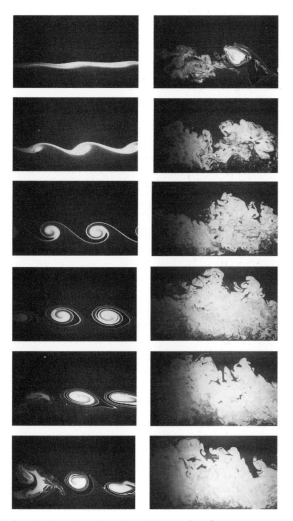

The wavy disturbance in Fig. 7.29*b* is a shear-flow instability akin to the Kelvin–Helmholtz instability, and so it is tempting to identify the vortex street that develops subsequently with the vortices that appear as a result of the latter (Fig. 7.31). But this is not so; the Kármán vortices have a different origin. They are provided 'ready-made' from the flow field immediately behind the cylinder, where the sheared fluid layers acquire 'vorticity'—a rotating tendency—as a consequence of the disturbance that the cylinder imposes on the flow. The instability in the flow behind the cylinder sets up a process of 'vortex shedding', in which vortices break away from the disturbed region on alternating sides of the 'street' and are entrained in the wake. So vortex creation takes place immediately behind the cylinder, not all along the shear flow as in the Kelvin–Helmholtz instability. The vortex-shedding process is highly organized: at the same time as the vortex on one side is being shed, that on the other is in the process of reforming (Fig. 7.32). Such periodic vortex shedding occurring from alternate sides of a bubble rising through water accounts for why bubbles often dance along a zigzag path as they rise.

Above a Reynolds number of 200, vortex streets are still formed in the wake, but instead of remaining coherent structures until they slowly dissipate downstream the vortices now break up downstream into a turbulent wake with an apparently chaotic structure (Fig. 7.33). This break-up of the regular structure is brought about by an instability that breaks the symmetry in the third dimension, parallel to the cylinder's axis—the flow then becomes fully three-dimensional. It is a curious kind of instability, appearing *intermittently* downstream as the Reynolds number of 200 is approached. That is to say, an observer stationed a certain

Fig. 7.31 The Kelvin–Helmholtz instability in a shear flow between two streams moving in opposite directions. The sheared region is made visible by entraining a fluorescent dye. The images show the flow at regular time intervals, beginning with the top left. The wavy instability rolls up into vortex structures, which then interact and lose their identity as the flow becomes turbulent. Structures like those in the second and third frames have been seen in cloud formations, owing to the Kelvin–Helmholtz instability in atmospheric flows. (Photo: Katepalli Sreenivasan, Yale University.)

turbulence. You can see that the Kelvin–Helmholtz instability should apply to *any* flow of the type shown in Fig. 7.30*c*, no matter how slowly it is going. But just like the instability that gives rise to Rayleigh–Bénard convection, it is counteracted by the damping effect of viscosity, which resists fluid motion; and so it is only at some critical shear (or equivalently, Reynolds number) that the instability becomes manifest.

Fig. 7.32 The Kármán vortex street arises from eddy shedding, wherein the circulating eddies behind the obstacle are shed from alternate sides and borne along in the wake. Here one eddy is in the process of forming just after that on the opposite side has been shed. (From: Tritton 1988.)

Fig. 7.33 At a Reynolds number above about 200, the Kármán vortices break up into a turbulent, three-dimensional flow in the downstream wake. (From: Tritton 1988.)

distance downstream would see a regular passage of vortices passing by, interrupted now and then by a more disorganized flow pattern. The disorganization gets more frequent the further downstream you go, so that an observer farther down the line would see more turbulent bursts and fewer regular vortex sequences. Above $Re = 200$, these regular sequences disappear altogether for an observer far enough downstream.

Then at Re greater than 400, a second instability sets in which causes turbulent break-up of the vortex street much closer to the cylinder itself, so that no real 'street' remains at all—just a wild, swirling wake. Closer inspection reveals that this too is a shear instability, which occurs in the fluid just after it moves away from the cylinder's surface to form the eddies that are then shed into the wake—with the consequence that the eddies are themselves turbulent instead of coherent circulating cells. This turbulent wake remains much the same up to a Re of around 300 000, at which point even the flow right next to the cylinder's surface (in the so-called boundary layer) becomes turbulent and the wake narrows into something like a turbulent jet.

Somewhere in these high-Re flows we can see the kind of chaotic billows that Leray must have seen in the Seine (rivers typically have Reynolds numbers of well over a million). But they seem now perhaps less daunting, because we can recognize in them not simply a disorganized mess but a flow pattern that, although undoubtedly messy, results from a series of well-defined instabilities occurring under well-defined conditions in an otherwise regularly structured flow. The precise Reynolds numbers at which these instabilities manifest themselves will depend on the exact shape of the bridge's columns, but their basic character remains the same.

That the structures created by these instabilities are generic is demonstrated by their appearance in real-world systems that are far removed from the idealized case of shear flow past a cylinder. Vortex streets, for instance, may be seen in satellite images of atmospheric flows, such as that in Fig. 7.34. Here the vortex street is superimposed on a cloud street, a series of stripes caused by convection.

Fig. 7.34 A vortex street in clouds, photographed by satellite off the edge of Jan Mayen island on the southeastern edge of the Arctic ice cap. The regular bands are cloud streets caused by convection roll cells. The wind is passing over the island from the top left to the bottom right, so that the convection rolls are aligned parallel to it. Deflection of air around a temperature inversion over the edge of the island has established a shear flow which deforms the convenient markers of the convection rolls into a regular series of vortices. (Photo: Satellite Observing Centre, University of Dundee.)

Canned rolls

Flow through a narrow channel is a kind of shear flow: the fluid near the edges is slowed down by frictional forces against the sides of the channel, and so, while the flow remains laminar, the velocity increases gradually towards the centre of the channel—in effect, the fluid can be thought of as a series of thin layers parallel to the flow, each sliding past its neighbours. This kind of flow is extremely important in nautical and aerodynamic engineering, and can be conveniently studied by confining a fluid between two concentric cylinders that are rotating at different rates of revolution (Fig. 7.35a). This might seem rather different from the case of a fluid flowing down a channel, but you can soon see that it is really a similar kind of shear flow if you imagine looking

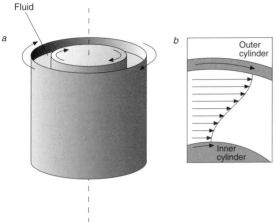

Fig. 7.35 In the Taylor–Couette apparatus, a fluid is sheared between two concentric cylinders rotating at different speeds (*a*). As the velocity of the fluid at the walls is determined by the rotation speed of the cylinders, the velocity profile of the flow in the radial direction has the character of a shear flow (*b*).

at the velocity profile of the fluid from a point located on the edge of the outer cylinder (Fig. 7.35*b*).

Let's first keep the outer cylinder fixed, and just rotate the inner one. This is what the Frenchman Maurice Couette did when he pioneered studies of this kind in 1888. At low rotation speeds of the inner cylinder, the fluid simply tracks the rotation—all the motion is in circles around the axis of rotation. This is called Couette flow. But what happens when you turn up the speed? Well, one crucial thing that *does* distinguish this kind of shear flow from that down a straight channel is that a rotating object experiences a centrifugal force—the force that pulls tight the string on which a threaded conker is spun in a circle. Not only is the fluid carried around in circles, but it is simultaneously forced outwards. As ever, viscous drag resists this outwards force, so that for low rotation speeds the centrifugal force does not appear to affect the flow.

But the British mathematician Geoffrey Taylor (one of the central figures in the development of fluid mechanics and, incidentally, the Taylor of the Saffman–Taylor instability of Chapter 5) found in 1923 that once the centrifugal force can no longer be resisted, patterns start to appear. First, the column of fluid develops stripes (Fig. 7.36*a*). These are in fact roll-like vortices in which the fluid circulates in alternate directions, as if around the surfaces of a stack of doughnuts. The symmetry of the Couette flow is broken, and in a manner that selects a particular pattern of a well-defined size.

Fig. 7.36 As the Reynolds number of the shear flow in the Taylor–Couette apparatus increases, the fluid becomes structured into increasingly complex patterns. First, doughnut-like roll cells appear, partitioning the fluid into a stack of bands (*a*). These then develop wavy undulations (*b*). At higher Reynolds number the roll cells reappear with turbulence amidst their folds (*c*); and ultimately, unstructured turbulence fills the cell. Even a well-developed turbulent state may preserve some structure, however: in *d*, a region of laminar (smooth) flow spirals through the turbulence. (From: Tritton 1988.)

It isn't too hard to see that this situation is directly analogous to convection, which is why just the same kind of symmetry-breaking structure—roll cells—is created. All of the fluid in the inner part of the Couette flow wants to move outwards, due to the centrifugal force, at the same time. But it cannot all move through the outer layers at once. So at the critical rotation speed at which viscous drag is overcome, the system becomes unstable to small perturbations, and roll vortices transport part of the inner fluid to the outer edge, while a return flow replenishes the inner layer. Not only is the instability of the same basic nature as that in convection, but the shape of the rolls is the same: the critical wave vector of the roll pattern is again 3.12, so the rolls are again roughly square, as wide as the gap between the inner and outer cylinders.

But what is the equivalent of the critical Rayleigh number for convection? As with all shear flows, the important dimensionless parameter is again the Reynolds number, this time defined such that the relevant velocity is that at the surface of the inner (rotating) cylinder and the characteristic dimension of the system is the width of the gap between the two cylinders. One can again calculate a stability boundary on a graph of roll wave vector against Reynolds number (Fig. 7.37), just like that for convection (Fig. 7.5). Although other wave vectors (rolls of different widths) can be stable within the limits of the boundary for *Re* greater than the critical value, the 'square' rolls remain if the rotation speed is increased only slowly. While it is dimensionless, the critical value of *Re* for the formation of Taylor vortices does in this case depend on the geometric details of the apparatus, specifically on the ratio of the radius of the inner cylinder to the width of the gap.

Having observed this much, we can be fairly confident that there are riches to be had by increasing the Reynolds number further. And sure enough, this brings about a series of instabilities to the Taylor vortices: first they go wavy, undulating up and down around the cylinders (Fig. 7.36*b*), then the waves get more complex

before becoming more or less turbulent, then the stacked stripes reappear with turbulence inside them (Fig. 7.36*c*) and finally (when *Re* is about a thousand times the critical value) the whole column of fluid goes turbulent (Fig. 7.36*d*). Here, then, is another well-defined sequence of instabilities leading from smooth, featureless flow, through patterns, to turbulence.

But there is more. Taylor realized that the game changes if, instead of keeping the outer cylinder fixed, we let that rotate too. Then the fluid can experience significant centrifugal forces even when the *relative* rotation speed of the inner layer with respect to the outer is small, and so a different balance of forces can be established. Experiments on a system like this have revealed a menagerie of patterns, too numerous to show here but summarized (as far as they are yet known) in Fig. 7.38, which shows the stability boundaries as a function of the Reynolds numbers at the surfaces of the inner and outer cylinders.

Into the whirlpool

It is evident from these examples that if you drive a fluid flow hard enough, you will always end up with turbulence—with chaos. But we can also see that the patterns that appear on the route to turbulence get richer the closer we approach it. The Russian mathematical physicist George Zaslavsky and his co-workers have provided some of the most extreme demonstrations of this. They have found fluid flows poised on the brink of turbulence in which chaos is delicately interwoven with symmetrical patterns of the most extraordinary complexity.

In the flows that I have considered so far, the driving force of patterning has been constant through time. For convection it was the buoyancy force created by a temperature gradient; for shear flows, it was a shear created either by the frictional drag experienced by a constant-velocity flow as it passed over a solid body or by the movement of one confining surface relative to another. Zaslavsky has looked at flows driven by a force that varies periodically in both time and space—that is to say, at each point in the flow field, the inertial force on the fluid includes a component that waxes and wanes regularly as time progresses. You might wonder where such strange flows could possibly be found, but they are not quite so contrived as they may at first seem: they crop up, for instance, in the behaviour of charged fluids called plasmas.

Setting up experiments to study the characteristics of these flows is not easy, but Zaslavsky chose instead

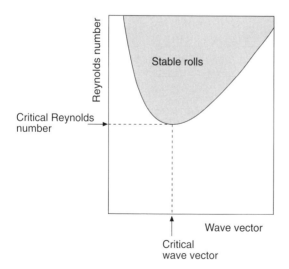

Fig. 7.37 The onset of roll patterns in the Taylor–Couette cell occurs at a critical threshold of Reynolds number, just as convection roll cells appear for a critical Rayleigh number. Above this point, an increasing range of wave vectors of the rolls can be supported.

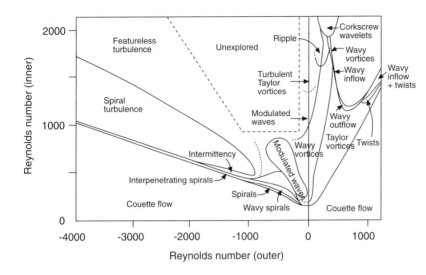

Fig. 7.38 There is a whole zoo of patterns that the Taylor–Couette cell will support. Which pattern is selected depends on the relative rotation speeds of the two cylinders. On the left of the vertical zero axis, the cylinders rotate in opposite directions. Along this axis, only the inner cylinder rotates—this corresponds to the situation depicted in Fig. 7.36, and you can see that we cross the boundaries from steady flow to Taylor vortices (rolls) to wavy vortices to turbulent Taylor vortices as we ascend along this axis. (After: Andereck *et al.* 1986).

to calculate the patterns that the flows adopt. The Navier–Stokes equation for this situation can be written down, but to calculate the streamlines of the resulting flow the Russians needed a computer to solve the equation numerically, even when the flow takes place just in a two-dimensional flat plane. In this case, the flow commonly breaks up into a series of circulating cells arranged in a kaleidoscopic pattern (Fig. 7.39 and Plate 17). Notice, however, that a few streamlines sometimes trace a tortuous path throughout the whole of the system. The fluid mass does not actually 'get anywhere'; like a convecting fluid, it merely simmers with its own internal rhythms.

In many of these flows there exist important types of streamline called separatrices. Think of two simple flow streams that are heading straight for one another (Fig. 7.40). Where they meet, something clearly has to give. One possibility is that one flow bends to the right and one to the left; then the streams slip past one another in a shear flow. But it turns out that another option is for both flows to splay in two, with the two streams diverging to left and right. The streamlines in

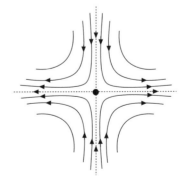

Fig. 7.40 Oppositely directed flows (*top* and *bottom*) splay into two diverging streams. The streamlines of each flow follow hyperbolic trajectories. A unique set of converging and diverging streamlines defines a separatrix (*dashed line*), meeting at a saddle point (*dot*). Here the direction of flow becomes indeterminate.

Fig. 7.39 A two-dimensional flow driven by a force that has eightfold symmetry throughout the plane breaks up into a complex pattern of circulating cells with eightfold symmetry. (Image: George Zaslavsky, New York University.)

each flow then follow trajectories in the shape of curves mathematically defined as hyperbolas. You might think that every streamline in the original flows has to bend either one way or the other, and this is largely true. But as we look closer and closer to the centre of each flow, we find opposed hyperbolic streamlines that approach one another ever more nearly until they almost 'kiss' before diverging. And right down the middle, dividing the streamlines that splay one way and the other, is a unique pair of streamlines that don't defect in this game of 'chicken': they meet each other head on. These define a separatrix (dashed line in Fig. 7.40), along which two streamlines converge at a single point, called a hyperbolic point or saddle point, in the centre of the converging and diverging trajectories. You can see in the figure that there is another separatrix in this flow along the centrelines of the diverging flows, where oppositely directed streamlines emerge from the saddle point. Separatrices typically separate different circulating cells in these complex flows.

The question is: what does a particle of the fluid do when it is carried along a separatrix to the saddle point? The answer is that the behaviour at this point is completely undetermined—we can't tell which way the fluid goes. The Navier–Stokes equation blows up at this point—it becomes 'singular', in physicists' language.

The flow in Plate 17 is particularly interesting, because it contains a central motif that has approximate tenfold symmetry, meaning that the pattern can be superimposed on itself by rotating it through a tenth of a circle (36°). You can see this simply by counting the number of obviously repeating elements (like the yellow triangles) around the circumference. This symmetry arises because the oscillating driving force has fivefold symmetry: at any point in the plane, the force drives the flow in five equivalent directions in space. (This fivefold symmetry in the force just happens to get doubled into tenfold symmetry in the flow, but that needn't necessarily be the case.) Zaslavsky has found flow patterns where this fivefold symmetry repeats again and again throughout a plane (Fig. 7.41a). Patterns like this have long fascinated scientists, because they know that a regularly repeating ('crystalline') two-dimensional pattern with true fivefold symmetry is impossible—just as it is impossible to fill a plane with a regular packing of pentagons (Fig. 2.2). You might be able to see that you can't simply superimpose the pattern in Fig. 7.41a on itself by shifting it in any direction in space—some points may match up, but not all. The pattern in fact has a kind of 'centre' (in the middle of the section shown here), and can't be superimposed by shifting this centre.

All the same, elements with fivefold symmetry repeat again and again in this pattern—you should be able to make out pentagonal arrangements of the circular features that recur throughout the plane. So although the pattern does not have genuine fivefold translational symmetry, it does have clear echoes of this symmetry in the details of the pattern. Although scientists have

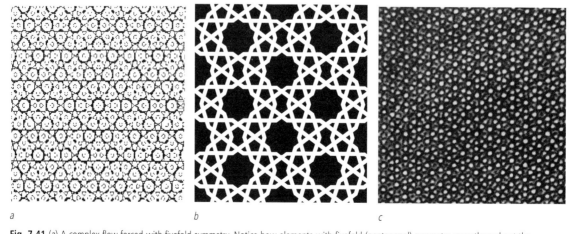

a b c

Fig. 7.41 (a) A complex flow forced with fivefold symmetry. Notice how elements with fivefold (pentagonal) symmetry recur throughout the pattern—but the overall pattern has no *translational* symmetry, as it cannot be superimposed on itself by simply translating the whole thing through space. (Image: George Zaslavsky.) (b) Fivefold symmetry was much used by Islamic decorative artists, as in this pattern found in the Alhambra palace in Granada, Spain. (c) Fivefold symmetry (and lack of true translational symmetry) is also evident in the atomic structure of quasicrystals, seen here under the electron microscope. (Photo: Kenji Hiraga, Tohoku University.)

become familiar with extended patterns containing fivefold symmetric elements only in the past few decades, they have decorated the architecture of the Islamic world for centuries (Fig. 7.41*b*). The geometric inventiveness of these structures became fully appreciated by scientists when in 1984 a new class of materials was discovered that also had a kind of fivefold 'quasi-symmetry'. These materials, called quasicrystals, look like crystals with 'forbidden' fivefold symmetries. (Some have eightfold or twelvefold symmetries instead, which are also 'forbidden' in true crystals.) They are in fact not perfect, periodic crystals but have complex stacking arrangements of atoms in which structural elements with these forbidden symmetries recur without perfect regularity (Fig. 7.41*c*). Zaslavsky has shown that the structures of some quasicrystals can be described by exactly the same mathematics that gives rise to his complex, quasisymmetric flow patterns.

All this complexity applies just to two-dimensional steady-state flows. When Zaslavsky and colleagues looked at the same class of problem in three dimensions, they found something else altogether. Streamlines in three dimensions turn out to have a particular property that gives the flow pattern the potential to be much more complicated: they can intersect and cross (Fig. 7.42). When this happens, the streamlines can get very entangled, and the resulting flow becomes chaotic—that is to say, turbulent. But what surprised the researchers is that in three dimensions the chaotic parts of the flow may be arranged regularly in space. The flows break up into cells in which the streamlines are well behaved, like the Bénard cells of convection patterns, arranged in a periodic pattern and separated from one another by a web of chaotic streamlines (Fig. 7.43), which Zaslavsky has called a 'stochastic web' (stochastic means governed by randomness). The web is bordered by separatrices, and within it particles follow wild trajectories that change direction at random. In other words, the flow in the stochastic web is turbulent. So these flows consist of

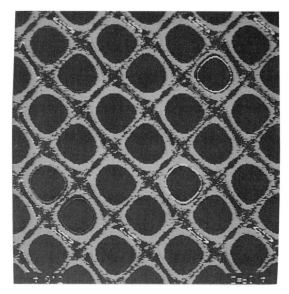

Fig. 7.43 The stochastic web. The flow is partly steady (within circulating cells shown as dark, roughly circular features here) and partly turbulent (inside the web). Where the flow is chaotic, the streamlines break up into an irregular 'dust' (dark specks on lighter background). The web is bounded by separatrices, and the regular cells are arranged periodically in the web. (Image: George Zaslavsky.)

a bizarre mixture of patterned regularity—the cells—threaded through with turbulence. The stochastic web provides a seed of turbulence that will grow to overwhelm the whole flow if the driving force (that is, the Reynolds number) becomes a little larger. Zaslavsky and colleagues found that the stochastic webs can have forbidden (for example, fivefold) quasisymmetries as well as the regular (fourfold) symmetry shown here.

Hidden order

These flows exhibit one of the most dramatic manifestations of the descent into the maelstrom of turbulence. But there are several other ways in which flows change from regular or structured to turbulent. In the case of convection, the transition may be abrupt, the flow patterned but laminar one moment and then suddenly chaotic when the driving force is cranked up a notch further. In the wakes of shear flows, on the other hand, turbulence comes and goes intermittently before taking hold fully. In Taylor–Couette flow, meanwhile, we saw that turbulence and regular patterns can coexist for a while in the form of turbulent Taylor vortices.

When turbulence finally sets in fully, however, we might be tempted to give up looking for any pattern in it

Fig. 7.42 In three dimensions, streamlines can cross and become entangled. The flows can then become unstable and chaotic.

Fig. 7.44 In turbulent flow, the motion of the fluid is chaotic. But even here we can recognize complex, coherent structures. (Image: Katepalli Sreenivasan, Yale University.)

at all. The trajectories of the fluid particles become extremely convoluted and constantly changing in time (Fig. 7.44), and the Navier–Stokes equations can be solved only numerically by laborious computer calculations, not by mathematical ingenuity. This chaotic picture is why, as indicated at the beginning of this chapter, turbulence has proved so resistant to theoretical descriptions. A turbulent fluid is in a state of continuous instability. Since our ability to predict the details of a flow decreases each time the flow changes as a result of an instability, this means that we generally lose all predictive capability for the details of a turbulent flow. (It does *not* mean, however, that the Navier–Stokes equations break down, but instead that these equations don't any longer have steady, time-invariant solutions.)

What this means is that, rather than trying to look at the detailed pattern of flow in terms of streamlines, we are forced to seek instead some average features of the flow. In other words, we can forget about individual trajectories of fluid particles and consider instead their statistical properties. What has emerged from this sort of approach is that even apparently random, structure-less systems like turbulent fluids may have characteristic *forms* if looked at statistically. We might find ourselves able to distinguish one kind of apparently chaotic process from another by deducing what their statistical forms are. This is an extremely important concept in theories of complex systems, and it will underlie some of the phenomena discussed in the remainder of this book: when a visually evident pattern vanishes into

chaos, we can nevertheless often still identify a kind of form that remains if we know how to look for it.

The idea of statistical form probably seems a little abstract at this point, but it might help if I point out that we have encountered at least one example of it already. In Chapter 5, I showed two branching patterns for mineral dendrites, which looked kind of similar but which, by eye, we'd be hard pressed to identify as the same or different. We saw that the concept of fractal dimension enables us to characterize these patterns with a precise numerical parameter, and thus to distinguish different classes of branching pattern. The fractal dimension is a measure of a pattern that is independent of the details. No two branched clusters grown by diffusion-limited aggregation will ever be identical, because their growth process involves a strong element of randomness, but nonetheless the *generic form*, as characterized by the fractal dimension, is identical for all.

Do turbulent flows have a generic form, which can be assigned some precise numerical parameters? This has been a subject of intense debate throughout the twentieth century, and it would be fair to say that the jury is still out. But the Englishman Lewis Fry Richardson took a critical step in addressing this question in the 1920s when he proposed that the universal properties of turbulence would become apparent only if we make a distinction between the mean and the fluctuating parts of the fluid velocity field. Most turbulent flows have mean velocities that are determined by the specifics of the situation—the turbulent wake of a shear flow, for example, or a turbulent jet of smoke have an overall average direction of travel. Richardson suggested that any generic behaviour would be superimposed on this case-specific behaviour, so that the latter must first be subtracted.

Richardson proposed that any structure contained in the fluctuating, chaotic part of the velocity field could be revealed by considering how the *differences* in velocity at two points within the flow vary as the points get farther apart. The flow as a whole may have a certain mean velocity in one direction, but it is in these point-to-point differences that the fabric of turbulence is to be discerned. If the flow is totally random over all distance scales, the velocity at one point will bear no relation to that at all other points—all differences in velocity will occur with equal probability as the points get further apart. If the flow has a structure, however, like a convection cell, the velocities at different points will tend to be *correlated*—knowing one allows us at least to guess at the other. For example, the velocities in adjacent edges of two convection roll cells are not independent: if the fluid at a point on one edge is going upwards, we can be sure that the fluid at a corresponding point on the edge of the other cell is also moving upwards at about the same speed, because adjacent rolls are always counter-rotating. The role of correlations in pattern formation is an important one that I shall come back to at the end of the book.

So if turbulence has inherent structures that distinguish it from utter randomness, there will be some correlation between velocities at different points. This is true on average even if such structures are short-lived, provided that they continually reform. Intuitively we should expect such correlations (if they exist) to decline with increasing distance, since it is reasonable to suppose that the behaviour of the fluid at one point in a turbulent flow takes ever less heed of the behaviour at another point the more distant it is. In a perfectly ordered array of Rayleigh–Bénard convection cells this is not the case—the correlations are very long-ranged. But perhaps surprisingly, experiments have shown not only that there *are* correlations in turbulence, but that these have a remarkably long reach, generally extending over almost the entire width of the flow. It is as though individuals in a jabbering crowd were able to converse with one another from opposite sides of a room.

These correlations make a description of turbulence much more subtle than a description of mere randomness, and they provide it with its elegant, baroque beauty. A fully random flow would show no features at all, whereas we can see in Fig. 7.44 that there are many swirling, vortex-like structures of many different sizes. It doesn't take a scientific training to appreciate that one of the fundamental structures of turbulence is the whirlpool-like vortex—Oriental artists had picked up on this long ago. These features are eddies, like those we saw earlier in non-turbulent shear flow. The difference is now that the eddies are formed over a very wide range of length scales (whereas in laminar shear flows only a certain size tends to be selected), they are transient, and they might appear anywhere in the flow.

Eddies carry much of the energy that is injected into a turbulent flow. Whereas in a laminar flow the energy is borne along in the direction of the fluid motion, in turbulent flow only a part of the motional (kinetic) energy of the fluid 'gets anywhere' (via the mean velocity of the flow)—the rest is just caught up in eddies until being finally dissipated in frictional heating owing to viscous drag. But dissipation of kinetic energy occurs ultimately at very small length scales, as the molecules in the fluid collide and take up the kinetic energy by undergoing more vigorous random motions. Somehow the energy that is fed into the flow at large scales, creating big eddies that we can see with the naked eye, has to find its way down to these small scales before being dissipated. What happens is that there is an *energy cascade*: big eddies transfer their energy to smaller eddies, which do likewise at ever smaller scales. Richardson appreciated this, and in 1922 coined a rhyme, inspired by Jonathon Swift's doggerel about fleas, to describe the process:

> Big whirls have little whirls
> that feed on their velocity,
> and little whirls have lesser whirls
> and so on to viscosity.

In the 1940s the Russian physicist Andrei Kolmogorov derived a law that put this energy cascade into precise form. He proposed that the energy contained in a turbulent fluid at a length scale d varies in proportion to the 5/3rd power of d—in other words, it increases with d at a rate proportional to slightly less than the square of d. This, like the definition of a fractal dimension, is an example of a scaling law: it shows that some property of the system (here energy, or mass in the case of a fractal cluster) is proportional to a variable (commonly a size scale) raised to some power (called the scaling exponent). Abstract and mathematical though they may seem, scaling laws are central to the science that underlies many natural patterns and forms.

Kolmogorov's law is in fact often found to be slightly off-target when investigated experimentally, since Kolmogorov made slightly too simplistic an assumption in deriving it. But more recent theories of turbulence have shown that things can be put right by including a few other variables in the scaling law. What has remained clear is that the basic idea of an energy cascade is correct. This means that eddies appear in turbulent fluids on all relevant length scales, so that (unlike the eddies that appear in Kármán vortex streets) you can't define any absolute length scale by looking at the sizes of the eddies in the system. In other words, the fluid has a kind of scale invariance, like the fractals that I described in Chapter 5. An explicit link between turbulence and fractals has been made by Benoit Mandelbrot, who has argued that a blob of coloured fluid develops a fractal shape as it disperses in a turbulent medium. Mandelbrot suggests that fractals provide the natural geometrical tool for describing the physical form of turbulence.

This idea gains support from the work of David Ruelle and Floris Takens, who showed in 1971 that the trajectories of fluid particles at the *onset* of turbulent flow can be described by mathematical objects called strange attractors, which, it later transpired, may have a fractal form in the 'phase space' of the variables that describe the flow. Whether the flow retains this cryptic form when the turbulence is fully fledged (the regime in which Kolmogorov-like scaling laws apply) is still not clear—but it looks unlikely. Fully fledged turbulence is often patchy, like a sluggish river, with regions of intense disorder superimposed on a more quiescent background. We have yet to discern the geometry appropriate to this situation.

Form from chaos

The statistical picture of turbulence has proved immensely valuable, but it does not always do justice to the reality. A statistical description works best if the system looks more or less the same, on average, at every point (which is not the same as saying that it is featureless). But turbulent fluids do not seem to fit this description very well. We can occasionally find relatively long-lived structures or even periodic patterns with well-defined length scales in fluids that should be fully turbulent. The turbulent Taylor rolls in Taylor–Couette flow are one such (Fig. 7.36c). Another example is the appearance of a coherent, somewhat regular eddy structure, much like a Karman vortex street, in a turbulent wake (Fig. 7.45). In general these latter structures do not last indefinitely and they are rather less regular than the patterns seen at pre-turbulent Reynolds numbers; but they are clearly quite distinct from random fluctuations. Where do they come from?

There may be no general answer to that, and certainly none has yet been identified. It seems likely that some of these so-called coherent structures in turbulent flows arise from the same kind of instabilities that create the regular patterns in pre-turbulent flows, but this time acting on the component of the flow that has an overall mean velocity—the part that Richardson subtracted to get at the underlying statistical nature of turbulence. Thus, for example, the instability that creates vortex streets in pre-turbulent flow might also generate roughly regular sequences of eddies in a turbulent jet that is, on average, travelling in the direction away from the nozzle through which it emanates.

Describing turbulence in terms of statistical scaling laws is therefore a little like describing a Seurat painting in terms of the probability of finding, say, a blue dot

Fig. 7.45 In this turbulent wake behind an obstacle, one can make out vortex-like features that have an approximate periodicity. (From: Tritton 1988.)

next to a yellow one. Such a description contains information about the picture, but it doesn't contain *all* the information—and sometimes not the most significant part! No concatenation of probability functions will allow us to anticipate that, on taking a step back, the dots combine at one place into a tree, at another into a parasol. Amongst all the chaos and unpredictability, we should not lose sight of the fact that there are strong ordering principles in turbulence.

A particularly robust and elegant coherent structure that emerges from a turbulent jet is the dipolar vortex (Fig. 7.46), a mushroom-like structure much like the plumes that rise through turbulent convecting fluids (Fig. 7.17). The initially disordered, turbulent head of such a jet will gradually organize itself into a two-lobed dipolar vortex when the flow is two-dimensional. We saw earlier that turbulent shear flows are generally three-dimensional; but GertJan van Heijst and Jan-Bert Flór from the University of Utrecht have created two-dimensional turbulent jets by using a stratified fluid, whose salinity and thus density increases with depth. Fluid motion in the up–down direction is suppressed in this system by the differences in density. To show just how robust these dipoles are, van Heijst and Flór fired two at each other from opposite directions. Instead of clashing and dissolving in a turbulent frenzy,

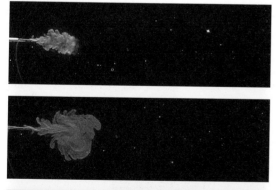

Fig. 7.46 A turbulent jet injected into a stratified fluid (whose density structure ensures that the flow remains largely two-dimensional) organizes itself into a coherent structure, the dipolar vortex. (Photos: GertJan van Heijst and Jan-Bert Flór, University of Utrecht.)

the vortices displayed a slippery resilience that puts me in mind of egg yolks. When they collided, the vortices simply paired up with their counterpart in the other dipole and, without mixing, set off as pairs in a new direction (Plate 18).

The giant's eye

One of the most celebrated and dramatic of coherent structures in a turbulent flow was first seen three centuries ago—and still persists today. Jupiter's Great Red Spot was observed in the seventeenth century by Robert Hooke in England and Giovanni Cassini in France. It is an oval feature, as tall as the Earth is wide and three times as long, in Jupiter's southern hemisphere (Plate 19). Jupiter's cloudy upper atmosphere is a mixture of hydrogen and helium with clouds of water, ammonia and other compounds that give rise to its spectacular colours. All this is stirred by the planet's rotation into a highly turbulent brew. Yet somehow amongst this planetary chaos coherent structures arise and persist for times ranging from months to centuries. The Great Red Spot is just the largest and oldest of these; other, smaller features with similar shapes come and go. Three white spots just to the south of the Great Red Spot appeared in 1938, and have remained ever since (one is visible in Plate 19). How do these structures arise, and how can they for so long defy the disruptive pull of turbulence?

Jupiter's atmosphere has a decidedly striking pattern even before we start to consider its spots. It is divided into a series of bands, marked out by the clouds, which recall the banded Taylor vortices of rotating Taylor–Couette flow. Each band is a 'zonal jet', a stream that flows around lines of latitude either in the same or the opposite direction to the planet's rotation. Unlike the Earth, where there are just two kinds of zonal jet—the eastward current of the trade winds in the tropics and the westward current of the jet stream at higher latitudes—both hemispheres on Jupiter have several zonal jets travelling to the east and west. The origin of these bands is still disputed, but they may be the product of small-scale eddies pulled and blended into latitudinal jets by the planet's rotation. Peter Olson and Jean-Baptiste Manneville of Johns Hopkins University in Baltimore have shown that a similar banded structure can arise from convection in a laboratory model of Jupiter's atmosphere. They used water to mimic the fluid atmosphere (since their densities are similar), trapped between two concentric spheres 25 and 30 cm across. The inner sphere was made of copper and chilled by filling it with cold antifreeze; the outer sphere was of clear plastic, so that the flow pattern could be seen. The effect of the planet's gravity was simulated by spinning the two spheres to create a centrifugal force. Now, in this experiment both of the forces driving the fluid flow—the temperature difference between the inner and outer spheres (like that between the inside and outside of the planet) and the centrifugal force—were directed in the *opposite* direction to those on Jupiter itself: gravity pulls inwards, of course, and the planet is warmer inside than out. But that didn't matter to the model, so long as the two forces acted in opposite directions *to each other*, as they do on the real planet. When the researchers added a fluorescent dye to the water so that the flow pattern became evident under ultraviolet light, they saw zonal bands appear around their model planet owing to convective circulation (Plate 20). This suggests that Jupiter's stripes may indeed be convection rolls.

The spot features in Jupiter's atmosphere are formed at the boundary of two zonal jets, where the flow of gases in opposite directions creates an intense shear flow. The Great Red Spot and the lesser white spots circulate like ball bearings between the flows above and below (Fig. 7.47). While the little ones come and go, the Great Red Spot remains. Is this just chance, or is a single

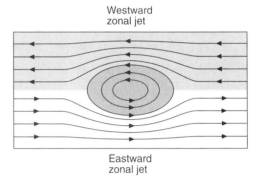

Fig. 7.47 Jupiter's Great Red Spot circulates between oppositely directed zonal jets that encircle the planet.

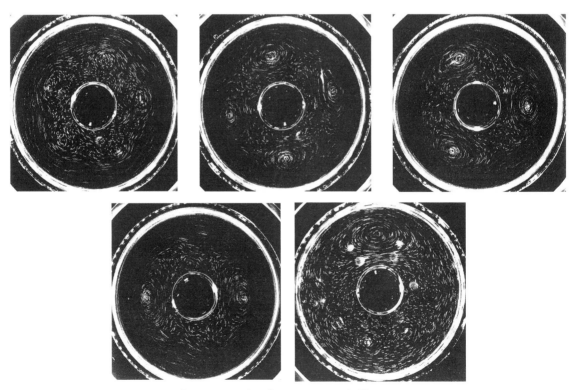

Fig. 7.48 In these experimental simulations of Jupiter's atmospheric flow in a rotating tank of fluid, the fluid is pumped so as to simulate the zonal jets and associated shear flows. Organized vortices arise spontaneously and persist in the flow. As the shear flow gets stronger, the number of vortices decreases from five to one. (Photos: Harry Swinney, University of Texas at Austin.)

big vortex a fundamental coherent feature of this kind of turbulent flow?

Both computer calculations and laboratory experiments suggest the latter. Philip Marcus from the University of California at Berkeley has carried out numerical calculations of the flow in a thin annulus of fluid—a washer-shaped disk with a hole in the middle, a kind of two-dimensional projection of one of Jupiter's hemispheres. The rotation itself sets up a shear flow: rings of fluid at successively larger radial distances from the centre flow past each other. Marcus found that when the shearing was high enough to cause full turbulence, small vortices would occasionally arise in the circulating fluid. If they rotated in sympathy with the shear flow, like the Great Red Spot, they would persist for some time; if they rotated against the shear, they would be pulled apart.

Then Marcus investigated what would happen in a flow containing pre-existing, large rotating vortices. A vortex rotating in the 'right' direction would remain, whereas one rotating the other way would be rapidly stretched and pulled apart. But even more strikingly, the persistent vortex would proceed to feed on smaller vortices with the same sense of rotation that arose subsequently in the turbulent flow, swallowing them up to sustain itself (Plate 21*a*). If, meanwhile, *two* large vortices with the right rotation were set up in the initial flow, they would rapidly merge into one, whose size and shape would then remain more or less steady (Plate 21*b*).

These calculations suggested that, once formed, a single large vortex is the most stable structure in this kind of flow. But how might it get there in the first place? Inspired by Marcus's calculations, Joel Sommeria, Steven Meyers and Harry Swinney from the University of Texas at Austin devised experiments to investigate this kind of flow in the flesh, as it were. They used a rotating annular tank into which they pumped water at various points in the tank's base equally spaced from the centre. Outlet ports located in the base of the tank allowed the fluid to escape again. The reason why the researchers used this pumping system rather than just filling a plain tank with water was that the interaction between the flow induced by pumping and extraction at

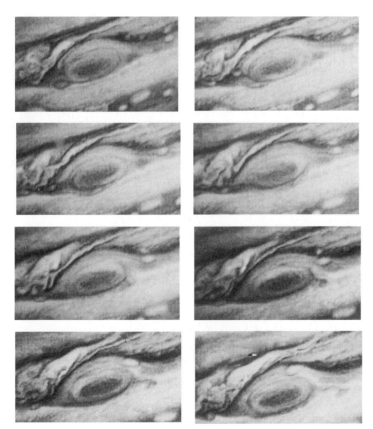

Fig. 7.49 The Great Red Spot consumes smaller vortices created in Jupiter's shear flow. In this sequence of images, taken over a period of about 2 weeks, a small spot can be seen to enter in the upper right corner and to be dragged into orbit around the Great Red Spot before disappearing into its vortex. (Photos: NASA.)

different radii and the flow induced by rotation of the tank set up counter-rotating zonal jets like Jupiter's, creating strong shearing forces.

The researchers found that stable vortices appeared in the tank at the boundaries of the zonal jets. The vortices sat at the corners of regular polygons, marking out a pentagon for five vortices, a square for four and a triangle for three. The number of vortices decreased as the shearing (which depended on the pumping rate) got stronger; eventually they were left with only a single large vortex (Fig. 7.48). Arising spontaneously from small random fluctuations in the turbulent flow, this vortex then remained stable and more or less isolated from the rest of the flow. Dye injected into it would remain there (Plate 22); dye injected outside would

remain excluded. But occasionally other small vortices, rotating in the same sense, would appear in the flow. These would last only for a short while before either merging with others or, ultimately, being swallowed up by the large vortex—just as Marcus had found. This very same process has been seen on Jupiter itself: as they passed the planet in the early 1980s, the Voyager 1 and 2 spacecraft repeatedly saw small white spots, approaching the Great Red Spot from the east, become trapped 'in orbit' around the Spot's edge before finally merging with it (Fig. 7.49).

We have good reason, therefore, to think that Jupiter's bleary eye is a robust and fundamental features of its turbulent skies. Even in chaos there may be more order than we would guess.

GRAINS

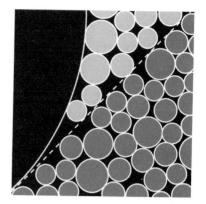

We forgave Bagnold everything for the way he wrote about dunes. 'The grooves and the corrugated sand resemble the hollow of the roof of a dog's mouth.' *That was the real Bagnold, a man who would put his inquiring hand into the jaws of a dog.*

Michael Ondaatje
The English Patient

On the 17th of October 1989 the residents of the Marina district of San Francisco Bay discovered the wisdom of the advice that Jesus offered to house builders—beware of building on sand. A shudder of the San Andreas fault centred in the mountains to the north-east of Santa Cruz brought many of the houses in this district tumbling to the ground. Miraculously there was no loss of life, but hundreds of millions of dollars' worth of damage was done by the magnitude-6.9 earthquake.

Unlike most other buildings in the Bay area, those in the Marina district stood on sand—more precisely, on sand-rich landfill sites. The ground certainly *felt* solid enough, and under normal circumstances it was. But when the earth shook, these wet, sandy soils turned to a slurry that flowed like treacle.

This property of a granular substance, naturally enough called liquefaction, is well known to seismologists and civil engineers. It is one of the most dramatic manifestations of the fact that a granular substance is a peculiar state of matter: composed of solid grains, yet able to show liquid-like behaviour. We all know that grainy materials like sugar and sand can flow, but at the same time they are clearly not true liquids, since they can resist shearing stresses and can support heavy objects.

The behaviour of granular media is of huge technological and industrial importance. All manner of substances are routinely handled in the form of granular powders, from cement to drugs to breakfast cereals, nails, nuts and bolts. Understanding the static and flow properties of these materials is crucial for their transport, storage and processing. Graininess is ubiquitous in the geological world, being central not only to the effects of earthquakes but to landslides, sediment transport, and the shape and evolution of sand dunes and beaches. Yet only in recent years have scientists begun to appreciate that, to explain how granular media behave, they must invent new physics. Engineers have long developed rules of thumb for handling these materials, but physicists want general principles that are broadly applicable and that account for observations at a fundamental level.

If there's one thing that has become clear, it is that granular media are seldom predictable. Shaking together different kinds of grains can either ensure good mixing or have the opposite effect of causing them to segregate according to their size. Sound waves can bend through a right angle as they travel through sand, while the stress below a sand pile has a minimum where the pile is highest. The pressure at the bottom of a tall column of sand does not depend on its height, and this is why a sand glass is a good timekeeper—the sand leaks away at a steady rate even though the column gets smaller. If water were like this, the pressure at the sea bed would be no greater than that a few metres below the surface.

One of the most striking outcomes of investigations into the fundamental nature of grainy materials is the realization that they represent rich ground for the appearance of patterns and form. Some of these patterns show many of the same features as those seen in other, completely different, systems; granular media can provide a convenient model system for studying complex phenomena as diverse as the fluctuations of stock markets and the formation of large-scale structure in the Universe.

Shaken, not stirred

I don't know about you, but I am not very good at using up the last of a packet of muesli. It is dreadfully wasteful, I know, but the fact is that by the time you get to the bottom, all the large pieces of fruit and nuts are gone and all that's left is a rather unappetizing residue of dry oat flakes. The big pieces always seem to stay on top, and the small ones settle to the bottom. This has become known to physicists as the 'Brazil nut effect'.

The sorting of grains of different sizes in a shaken granular medium is well known to engineers, but the reason for it is still disputed. You might think that shaking would simply mix up grains of different sizes, but clearly it is not so—usually, the larger grains instead rise mysteriously to the top. Even if the packet of muesli left the factory well-mixed (which is unlikely), the Brazil nuts and banana flakes are likely to have reached the top by the time the packets have made their way by rumbling juggernaut to the supermarket. The British engineer J.C. Williams was one of the first to study this effect systematically, in 1963. He saw a single large parti-

cle rise up through a bed of finer powder as it was vibrated up and down. Williams suggested that the large particle is *ratcheted* upwards: as all the particles jump up during a shake, the large one leaves a void beneath it, into which smaller grains fall (Fig. 8.1). So the small grains progressively prevent the large one from settling back to its original height after each shake.

In 1992 physicists Rémi Jullien, Paul Meakin and André Pavlovitch, conducted computer simulations of this shaking experiment to find out what makes the big grains rise. They looked at a column of spherical grains of different sizes, each of whose trajectories they could trace after each simulated shake; and they observed a ratchet process just like that proposed by Williams.

But this isn't the whole story, according to Sidney Nagel and colleagues from the University of Chicago. A year later they conducted experiments in which small glass beads were subjected to a series of vertical shakes in a glass cylinder. All of the beads were the same size except for one or a few larger ones, which gradually rose to the top. To follow the motion of individual beads, the Chicago team dyed some of them with ink.

What they found was a remarkably subtle kind of motion. They marked with ink an entire layer of small beads surrounding a large one placed on the central axis of the container and towards the bottom. The large bead rose up along the central axis, and the small dyed beads

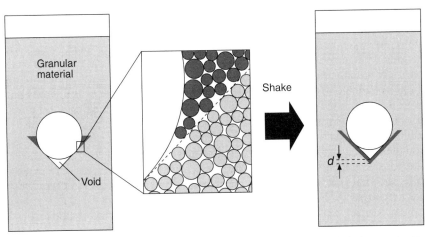

Fig. 8.1 Grains in a bed of granular material separate out according to size when it is shaken vertically, with the larger grains rising to the top. This may be due to a ratchet-like motion in which small grains fall into the space beneath larger grains as they rise during each shake. Each large grain tends to accumulate an empty space (void) below it. When the box is shaken vertically, the large white ball rises from the walls of the void, and the smaller grains in a ring (with a wedge-shaped cross-section) above it can slide down the walls of the void—here I have indicated these grains in dark grey to distinguish them from the other small (*light grey*) grains around them. When the ball settles again, it come to rests on the cone of dark grains, and so has risen a small distance *d*, roughly equal to the thickness of the dark layer. Note that the disparity in sizes is extreme in this picture, for clarity.

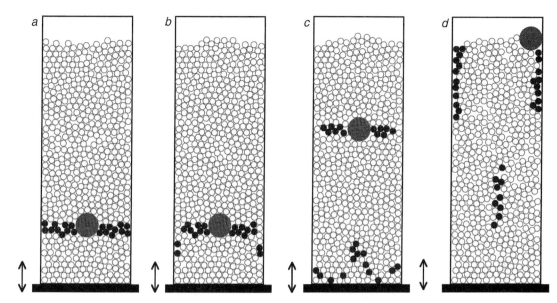

Fig. 8.2 Grains in a tall column undergo convection-like circulating motions: the grains in the centre rise upwards, and those at the edges crawl down to the bottom in a narrow band. The images shown here are reconstructions of an experiment in which some glass beads were dyed to reveal their motions. An initially flat layer near the bottom of the column (*a*) separates into down-going beads at the edges and rising beads at the centre (*b*). The latter move outwards at the top and then downwards at the walls (*d*); the former move upwards at the centre when they reach the bottom (*c, d*). A single large bead gets trapped at the top because it is too large to fit in the narrow down-welling band at the edges. So the convective motion causes size segregation. (Images: Sidney Nagel, University of Chicago.)

immediately around it also rose. But the dyed beads at the edges of the layer, in contact with the container's sides, began instead to make their way down to the bottom of the container (Fig. 8.2). As the central group of beads continued to rise, those that descended at the edges reached the bottom and then began to rise up again in the centre! Once the large bead and its surrounding small companions reached the top, the large bead stayed there but the smaller ones made their way to the sides of the container and began to travel downwards.

Does this pattern remind you of anything? The small, dyed beads are executing a circulatory motion just like that seen in a convection cell (Fig. 7.3)—rising at the centre and descending at the edges. The size segregation is merely a by-product of this convection-like motion: larger beads are pushed up on the rising column of the cell but, once at the top, are unable to follow the cycle further because the descending portion of the cell is confined to a very thin layer (about the thickness of the small beads) at the container's edge.

Convective flow in shaken granular media is a well-established phenomenon—Michael Faraday seems to have been the first to see it in 1831. But what drives the flow? In normal fluids, we saw earlier that convection is

a result of buoyancy due to density differences between layers of the fluid at different temperatures. But all of the particles in Nagel's granular medium have the same density, and they are all (with the exception of the large bead, whose presence isn't essential for convective flow) the same size. Nagel and colleagues figured that the important factor was the frictional force between the beads and the walls of the container, which hindered the upward jumps of the peripheral beads during each shake. In support of this idea, they found that more slippery walls reduced the circulatory motions, while rougher walls made them more pronounced. As the simulations of Jullien and colleagues involved no walls at all (their granular columns were free-standing), they would not have seen these convective effects.

For engineers, it seems likely that these explanations may be of only limited practical utility. Much of the controversy that still exists over the origins of convection and size segregation in shaken powders stems from the fact that so many experiments give different answers. For example, Colin Harwood of the HT Research Institute in Chicago found in 1975 that layers of powder sandwiched between two layers of another powder can display all manner of behaviours. Coarse powders usually move upwards to mix with the finer grains above; but if the size difference is small, the upward motion is small and some of the coarser material actually moves *downwards* slowly. But even if the sandwiched layer is *finer* than the powder above, it usually still moves upwards. If, meanwhile, the upper and lower grains have some degree of cohesion (wet sand is an extreme case), then there is only limited upwards movement of the sandwiched layer. All of which suggests that the only way to really tell what your cereal packet will do is to shake it and see.

Jumping beans

When Michael Faraday first shook the packet, he saw both circulatory (convective) motion of grains and the spontaneous appearance of heaps or 'bunkers' on the surface of the material. He suspected that the air that is present in the tiny spaces between the grains plays a part in causing these effects. When a layer of grains is shaken vigorously enough, the bottom of the layer jumps away from the floor, creating a cavity that is almost empty of air. The abrupt difference in air pressure between the gas

amongst the grains and this almost air-free cavity pushes some grains *underneath* the pile as it rises, and so creates unevenness in the layer, leading to heaping. Recently, researchers have performed experiments in which the pressure of the gas permeating the granular layer is changed systematically to investigate the effect on heaping. These experiments suggest that Faraday's mechanism does play a role, and so it seems that this, as well as friction between grains and walls, has to be considered in any complete explanation for convective motion.

But in the mid-1900s, Harry Swinney and colleagues Paul Umbanhowar and Francisco Melo at the University of Texas at Austin decided to see what happens to a shaken layer of grains under conditions where both friction and gas pressure play little or no part. They studied a very thin layer of tiny bronze spheres (about the same size as sand grains) in a shallow, sealed container that was pumped free of air and vibrated rapidly up and down. Only a tiny fraction of the grains in this system were in contact with the container walls (unlike the situation shown in Fig. 8.2, for instance), and so frictional effects could be expected to have virtually no influence on the behaviour of most of the layer. Would this mean

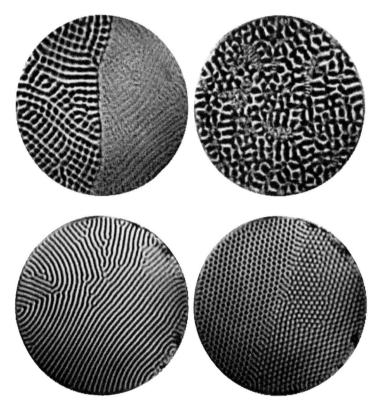

Fig. 8.3 When shaken vertically, a shallow layer of grains can develop complex wave patterns, including stripes, square and hexagonal patterns. (Photo: Harry Swinney, University of Texas at Austin.)

that there were no forces that could drive the grains into organized, structured motions?

Far from it! This set-up proved to be the most fertile breeding ground for grainy patterns so far known. The granular layer became organized into a series of dynamic ripples: stationary waves in which the grains are constantly rising and falling in step with each other. These wave patterns can be visualized by 'freezing' the little bronze balls at one point in their motion using a stroboscope: the light flashes on and illuminates the balls in step with their oscillatory rise and fall, and so always catches the pattern at the same point in its cycle. Swinney and colleagues saw patterns that are now familiar to us: stripes interspersed with dislocations, spirals, hexagonal and square cells, and more random, non-stationary cell-like patterns that appear to be turbulent (Fig. 8.3). All of these have analogues in the convection patterns discussed in Chapter 7 and the chemical Turing patterns of Chapter 4.

Transitions between one pattern and another can be induced by changing the shaking frequency and amplitude (that is, how far up and down the container is moved on each shake) (Fig. 8.4). The frequency of the wave patterns is a regular fraction of the frequency of shaking—either a half (so that the balls rise and fall once every two shakes) or, for larger amplitudes of shaking, a quarter. But sometimes different parts of the same basic pattern oscillate out of step with one another, so that one part is rising while the other is falling. Then the strobe light catches the balls out of

phase, illuminating peaks in one region and troughs in another (Fig. 8.5). Disorder (turbulent motion) sets in when the amplitude of shaking exceeds a certain threshold, more or less irrespective of the shaking frequency. This is because, once the balls are thrown up too high, they lose the capacity to organize all their motions in step.

By studying the pattern-forming process using a simple theoretical model in which the balls lose a little energy as they collide (making them more like rubber balls than billiard balls), the researchers found that this process is associated with a period-doubling bifurcation (p. 67). For low-vibration amplitudes, the entire granular layer rises and falls in step with the shaking. Above a critical amplitude there is a bifurcation in this flat layer, which leaves alternate stripes rising and falling out of step—this sets up either a stripe pattern at higher frequencies or a square pattern at lower frequencies. At a second critical amplitude a second bifurcation occurs— a period doubling, leading to a hexagonal pattern. At one point on the oscillatory cycle this pattern appears as an array of little spot-like peaks, whereas if the stroboscope is set up to capture the pattern half a cycle later one sees an array of hexagonal honeycomb cells. Thus the pattern repeats as a doubled-up oscillation, a sequence of peak–cell–peak–cell … and so on. You can make out both of these patterns (along with two intermediate configurations in out-of-step regions) in Fig. 8.5.

It takes remarkably few ingredients to make these patterns, as Troy Shinbrot of Northwestern University discovered. He chose to set aside the fact that the balls are

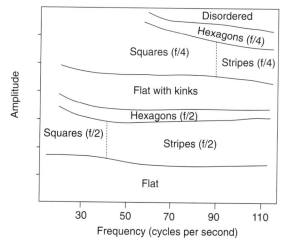

Fig. 8.4 Transitions between different wave patterns in a granular layer can be induced by altering the shaking frequency (*f*) or amplitude. (After: Umbanhowar *et al*. 1997.)

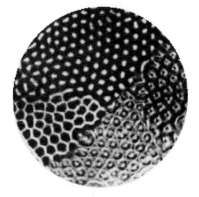

Fig. 8.5 When different domains of a pattern rise and fall out of step with one another, the stroboscope that captures a 'frozen' image of the patterns catches the domains at different points in their cycle, and so the patterns appear different even though they are in fact identical. (Photo: Harry Swinney, University of Texas at Austin.)

rising and falling under gravity as the tray is vibrated, and focused just on the horizontal component of their motions: the movements parallel to the tray's surfaces. Of course, in reality the balls move both vertically and horizontally as they collide and scatter, but Shinbrot figured that, as the patterns appear in the horizontal plane, perhaps the horizontal motion alone would be enough to account for them. In his model the balls are given an impulse in a randomly chosen direction on each vibration cycle, reflecting the randomizing influence of shaking. And again, each time balls collide they lose a little of their energy, which is dissipated as heat. There seems to be nothing in this prescription but a recipe for randomness, and yet after just a hundred shakes Shinbrot found an initially random scattering of grains organizing themselves into stripes, hexagons, squares and several other patterns besides (Fig. 8.6). Which pattern is selected depends on the strength of the randomizing effect of shaking and on the average distance that each ball travels before colliding with another. As well as reproducing the patterns observed

experimentally, Shinbrot found others that had not been seen before but which might, he suggested, become manifest if the right experimental conditions could be identified (such as Fig. 8.6d). This model is clearly a great simplification of the real system, but it seems to imply that almost any combination of randomization (which is analogous to diffusion) and energy dissipation (analogous to reaction) will suffice to make the patterns apparent.

Loners

On the whole I have so far spoken about patterns such as these in terms of 'global' instabilities of the whole system. In convection and Turing patterns, for instance, the implication has been that the whole structure emerges at once throughout the entire system. But there is another way in which we can describe these patterns: as ranks of individual elements that interact with each other to form ordered structures. Within this picture, each element maintains an optimal distance from its neighbours, and this ensures that the emerging pattern

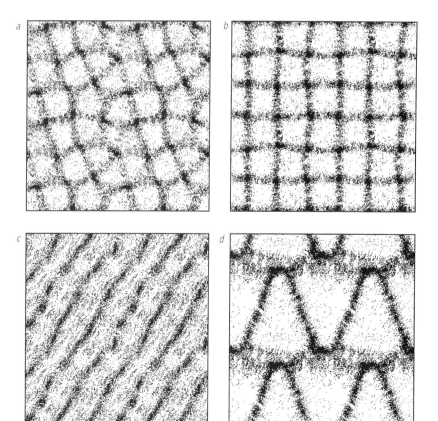

Fig. 8.6 Complex ordered and irregular patterns arise spontaneously in a simple model of a shaken thin layer of grains as a result of the interplay between randomization of the grain motions and collisions between grains. This model considers only the horizontal component of the grains' motions—it doesn't even need to include the effect of gravity! (Images: Troy Shinbrot, Northwestern University.)

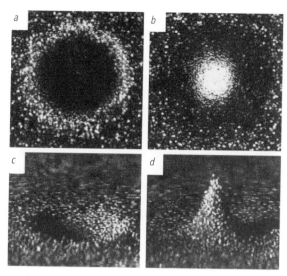

Fig. 8.7 Individual elements of the granular wave patterns—*oscillons*—can be isolated and studied. Each oscillon is a single peak, here seen from above (*a*, *b*) and from the side (*c*, *d*) at different points in the cycle. (Photo: Harry Swinney.)

ing the amplitude from above the threshold value at which the full patterns were produced. In this latter case, the oscillons can be regarded as parts of the pattern that get left behind as it fades out globally.

Swinney and colleagues discovered that they could conduct a curious kind of 'oscillon chemistry'. Oscillons can move around through the granular layers, and when they encounter one another, one of two things can happen. Each oscillon jumps up and down at half the shaking frequency, and because of this 'pinning' to the

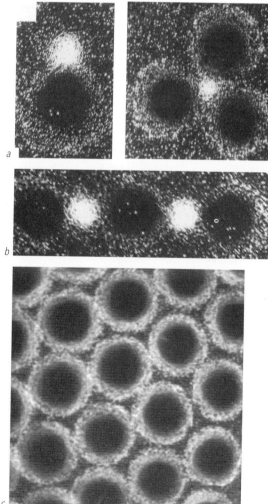

Fig. 8.8 Oscillons behave like particles that attract, if their oscillations are out of phase, or repel if they are in phase. Out-of-phase oscillons can form 'oscillon molecules' (*a*) or chains (*b*). A group of in-phase oscillons will maintain an optimal distance apart by adopting a hexagonal arrangement (*c*). (Photos: Harry Swinney.)

has a particular periodicity. You see much the same thing when people form a queue: they don't leave huge gaps (in case others might fill them!), nor do they stand with their noses brushing the head of the person in front. A more or less regular spacing is established.

You could regard this picture of a pattern-forming system as a kind of particle-based alternative to the wave-based picture of global instabilities with characteristic wavelengths. One of the most extraordinary discoveries from the shaking sand tray is that it has proved possible to capture and study the isolated particles of the patterns. Swinney and colleagues found that, for a certain range of layer depths and shaking frequencies and amplitudes, they could generate just a few lone oscillating peaks, or even just a single one, in the granular layer (Fig. 8.7).

They call these lone peaks *oscillons*—isolated 'packets' of oscillation. An oscillon is a peak of jumping balls one instant, and a crater-like depression the next. It looks rather like the splash made when something plops into a puddle of water—except that the splash doesn't die out in a series of spreading ripples, but keeps jumping back up as if captured in a time loop. These beasts appear when the shaking amplitude is just below that required for the appearance of the full pattern. But they don't form on their own from an initially flat layer—the researchers either had to perturb the layer in some way to trigger oscillon formation, or create them by reduc-

driving frequency, two oscillons must be either in step with each other or perfectly out of step, so that one rises to a peak when the other makes a crater. Two out-of-step oscillons attract each other, like particles of opposite electrical charge, enabling them to link up into 'molecules' (Fig. 8.8a). In some cases, whole strings of out-of-step oscillons could be observed, analogous to chain-like polymer molecules (Fig. 8.8b). The range of the attractive interaction is only small—about one and a half times the width of an oscillon—so they have to approach quite closely before sticking together.

Oscillons that are in step, meanwhile, repel each other, like particles with the same electrical charge. A party of in-step oscillons will form a hexagonal pattern (Fig. 8.8c), since this allows each oscillon to maintain an optimal distance from all of its neighbours. This is like a fragment of the global hexagonal pattern in Fig. 8.3.

Striped landslides

The fact that grainy materials can be part-time solids and part-time fluids leaves us in an uneasy relationship with them. The transition from one to the other can be catastrophic, as I illustrated at the start of the chapter. Mountaineers know this well enough: those sparkling white slopes can appear to be a rigid feature of the landscape … until some disturbance turns them into an avalanche. Grainy volcanic outpourings called pyroclastic flows, buoyed by hot gases, are amongst the fastest and most lethal of volcanic hazards. When grains are set in motion, we had better watch out.

But avalanches are, to a physicist, a delight. Grainy flows are full of surprises, not the least of which is their

Fig. 8.9 Two well-mixed grains of different sizes and shapes separate spontaneously into stripes when poured into a narrow rectangular cell. Notice also the segregation of grains, with one type at the left-hand top of the slope and the other at the right-hand foot. (Photo: Gene Stanley, Boston University.)

ability to create astonishing patterns. One such, in which stripes of one kind of grain alternate with stripes of the other, is shown in Fig. 8.9. This stripy (stratified) grain-segregation effect was discovered by Hernán Makse and Gene Stanley from Boston University and co-workers in 1995, and it is really rather simple to reproduce—I explain how in Appendix 7. All you need is two sheets of hard transparent plastic, like perspex or Plexiglas, held together on three sides to give a narrow, open-ended box. Into one top corner of the open end of the box you pour a mixture of two different grainy media with grains of different sizes *and shapes*, such as sugar and salt. It is best if the two types of grain are different colours, so that you can see the patterning easily. The grains should be initially well mixed by stirring (not shaking!). As you pour, a heap builds up in one corner, in which the grains are still well mixed. But at a certain point, the trickle that falls onto the topmost edge of this heap begins to trigger regular landslides down the sloping edge, during each of which the tumbling grains separate out into stripes of different colours. In addition, the grains become segregated at either extremity of the stripes, with the larger grains gathering at the base of the slope and the smaller grains down the topmost edge.

This is another example of a pattern-forming system that seems to deny intuition. It is as if time were running backwards: we might expect an initially segregated body of grains to mix as it flows, like ink dispersing in water; and yet here is just the reverse. And what is more, the flowing media don't just separate out—they separate into a pattern with a characteristic size scale (the width of the stripes)! You can imagine this effect taking place in all sorts of industrial processes, such as when mixtures of different cereal grains or sands are poured out of a hopper and onto a heap. It seems that no one has ever reported striped size segregation in a heap of this sort—but probably because no one has looked for it.

The stratification happens in a characteristic manner: each landslide generates a pair of stripes, which appear first at the bottom of the slope and run back up it in a kind of kink at the sloping face. The topmost stripe of the pair contains the larger grains. Makse and colleagues supposed that the basis of this sorting process is that the larger grains tumble down the slope more freely than the smaller grains—the latter are more easily trapped in small dips and irregularities of the slope on the way down. This same effect can be seen in rock slides, where the largest boulders crash to the bottom while the smaller ones get stuck further up the hillside.

In effect, the slope looks smoother to the large grains than to the small ones.

So the large grains reach the bottom first, which is why there is segregation of these grains at the foot of the heap. There they pile up to form a kink. Then as the subsequent grains tumble down to reach this kink, the small grains get stopped first, since the large ones are less easily trapped in the hollow of the kink's upper edge. So the small grains are deposited first, and the large ones come to rest on top, as the kink moves back up the slope.

The researchers set out to construct a simple theoretical model of this process. To do so, they had to establish criteria for when avalanches started and stopped. These criteria are well explored for piles of grains, and you can see them for yourself by tipping up bowls of granulated sugar and long-grain rice until they undergo avalanches. First, smooth the surfaces of the materials so that they are both horizontal. Then slowly tilt the bowls until a layer of grains shears off and runs out in an avalanche. There is a critical angle, called the angle of maximum stability (θ_m), at which sliding takes place. Moreover, when the avalanche is over, the slope of the grains in the bowl will have decreased to a value for which it is stable. This is called the angle of repose (θ_r), and the slope always relaxes to this same angle (Fig. 8.10). Both of these avalanche angles depend on the grain shape—you'll find that θ_m for rice is larger than that for sugar, whereas granulated sugar, caster

sugar and couscous (all with roughly spherical grains) all have a similar angle within the accuracy of this kitchen-table demonstration.

You can see the same thing by letting a steady trickle of sugar pass through a hole in a bag so that it forms a heap on the table top. The heap grows steeper and steeper until eventually there is a miniature landslide. Thereafter, you'll find that, however much more sugar you add, the slope of the pile stays more or less constant as it grows, with little landslides making sure that this is so. The angle of the steady slope is the angle of repose.

It was quite by chance that Hernán Makse had decided to conduct his initial experiments with sand

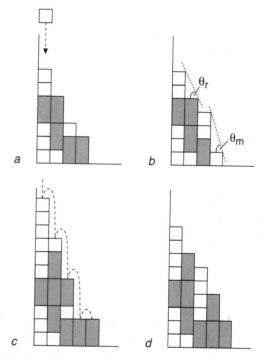

Fig. 8.11 The stratification that takes place when mixed grains are poured can be mimicked in a simple theoretical model in which the two grains have different shapes: square and rectangular. The model assumes that as they are poured, the grains stack up into columns, with all of the rectangular grains upright (a). Although this is a highly artificial assumption, it reproduces the effect of different grain shapes, which is the cause of the stratification. The angle of maximum stability θ_m is such that the difference in height between one column and the next cannot exceed three times the width of the square grains; and the angle of repose θ_r is equivalent to a height difference of two (b). If a new grain added to the top of the slope creates a slope greater than θ_m, it tumbles from column to column until it finds a stable position (c). But if the grain has to go all the way to the foot of the pile (as in c), this implies that the slope is equal to θ_m everywhere. The pile then undergoes a landslide to reduce the slope everywhere to θ_r or less (d). (After: Makse et al. 1997.)

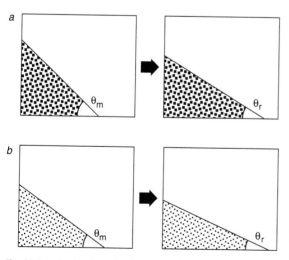

Fig. 8.10 A pile of grains will undergo an avalanche at a critical angle θ_m, the angle of maximum stability. The avalanche will cause the slope to 'relax' to a stable angle θ_r, the angle of repose. These angles will generally differ for grains of different shapes.

and sugar, which have slightly different shapes and therefore slightly different angles of maximum stability and repose—different sizes alone wouldn't have given the stratification. So in the model that he and his colleagues developed, they tried crudely to mimic this difference in shape. They considered two types of grain: small square ones and larger rectangular ones. These were assumed to drop onto the pile so as to stack in columns (Fig. 8.11). This model seems highly artificial—the experimental grains are clearly not squares and rectangles, nor do they stack up in regular vertical columns. But it's only a rough first shot, aimed at capturing the essentials of the process.

The heap was assigned characteristic angles of repose and maximum stability. When a grain drops onto the pile to create a local slope greater than θ_m, it tumbles down from column to column until it finds a position for which the slope is less than or equal to θ_m. But if a grain tumbles all the way to the bottom, which means that the slope everywhere is already equal to θ_m, then a landslide is considered to occur in the model: all the grains tumble, starting at the bottom, until the slope everywhere is reduced to the angle of repose θ_r.

Because the large grains are 'taller' and so more readily introduce a local slope greater than θ_m, they tumble more readily—just as in the experiments (remember that the large grains are less easily trapped on the slope). This accounts for the *segregation* of grains, with the larger ones at the bottom. The researchers found that all experiments showed this segregation when the grains were of different sizes.

Stratification—striped layers—requires something more, however. They found that this happened experimentally when the two types of grain not only have different sizes but also different angles of repose; that of the smaller grains being less steep than that of the larger grains. Because the particles in the model were not just of different sizes but also of different shapes, the model captures this feature of the experiments too. So when played out on the computer, it is able to produce piles that are both segregated *and* stratified (Fig. 8.12). The simple model, therefore, does a fair job—but it may neglect some important factors such as dynamical effects of grain collisions. These may explain, for example, why the pouring rate is also critical to obtaining good stratification.

Roll out the barrel

As bricklayers know well, an easy but generally effective way to mix two substances is to place them inside a rotating drum, like a cement mixer. But when the substances are powders, don't expect the obvious. This became clear to Julio Ottino and co-workers at

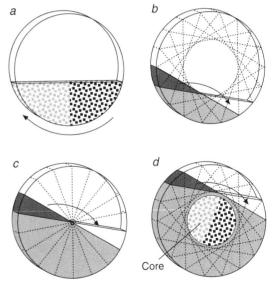

Fig. 8.13 Avalanches of grains in a rotating drum will mix different grains that are initially divided into two segments (*a*). As the drum turns, there is a succession of avalanches each time the slope exceeds the angle of maximum stability, transposing the dark wedges to the white wedges (*b*, *c*, *d*). If the drum is less than half-full (*b*), the wedges overlap, and the two types of grain eventually become fully mixed. If the drum is exactly half-full (*c*), the wedges do not overlap, so mixing takes place only within individual wedges. When it is more than half-full (*d*), there is a central core in which avalanches never take place, so this circular region never gets mixed.

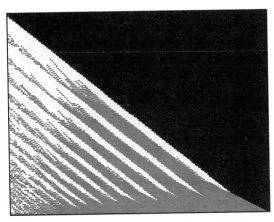

Fig. 8.12 The model outlined above generates the same kind of stratification *and* segregation as seen experimentally. (Image: Hernán Makse, Schlumberger–Doll Research, Ridgefield, Connecticut.)

Northeastern University in Illinois when they tried to mix two types of salt, identical except for being dyed different colours, in this way (Fig. 8.13a). If the drum rotates slowly enough, the layer of granular material remains stationary until the drum tips it past its angle of repose, whereupon the top layer slides in an avalanche (Fig. 8.13b). This abruptly transports a wedge of grains from the top to the bottom of the slope. The drum meanwhile continues to rotate until another wedge slides.

Each time a wedge slides, the grains within it get scrambled (because they are identical apart from colour). So if the grains are initially divided into two compartments separated by a vertical boundary (Fig. 8.13a), they become gradually intermixed by avalanches. But are grains also transported between wedges? They are if the drum is less than half-full, because then successive wedges intersect one another (Fig. 8.13b). But when it is exactly half-full the wedges no longer overlap (Fig. 8.13c), and mixing occurs only within individual wedges. If the drum is more than half-full, something strange occurs. There is a region around the outer part of the drum where avalanches and mixing take place, but in the central region is a core of material that never slides (Fig. 8.13d). The initially segregated grains in this core therefore *stay* segregated even after the drum has rotated many times. This, Ottino and colleagues observed, leaves a central pristine region of rotating, unmixed grains, while the region outside becomes gradually mixed (Fig. 8.14). In theory, you could spin this cement mixer for ever without disturbing the core.

Even if you start with a well-mixed concoction of different grains, it won't necessarily stay that way in a tumbling barrel. Engineers have known since 1939, when the effect was observed by Y. Oyama in Japan, that this process can cause grains to segregate out in a series of bands when the barrel is a long cylindrical tube (Fig. 8.15a). This happens if the grains have different angles of repose—for example, tiny glass balls (with θ_r of 30°) will separate from sand (θ_r of 36°). And in a rotating tube with a periodic change in width, so that it has a series of 'bellies' connected by a series of 'necks', grains will separate according to their size even if the angle of repose is the same for both (Fig. 8.15b). In the case shown here, small glass balls segregate into the bellies while large balls gather in the necks. Crucially, segregation in both uniform and bulging cylinders requires that the grains only partially fill the tube, so that there is a free surface across which grains can roll in a constant landslide.

Where I grew up on the Isle of Wight in the south of England, there is a place called Alum Bay that is famous for its multicoloured sands. Tourists are invited to fill glass cylinders—models of the Needles lighthouse—with stripes of these sands by carefully adding each colour in sequence. I very much doubt they would believe you if you were to suggest that they might get much the same result by mixing up all the sands and then rolling the tube!

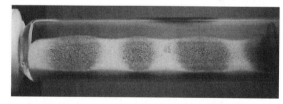

a

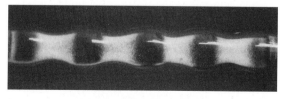

b

Fig. 8.15 (a) Grains of different shapes (and thus angles of repose) will segregate into bands when rotated inside a cylindrical tube. Here the dark bands are sand, and the light bands are glass balls. (b) In a tube with an undulating cross-section, a difference in size alone is enough to separate grains, which segregate into the necks and bellies. (Photo: Joel Stavans, Weizmann Institute of Science, Rehovot.)

Fig. 8.14 The unmixed core is clearly visible in experiments. (Photo: Julio Ottino, Northwestern University, Illinois.)

Joel Stavans from the Weizmann Institute of Science in Israel and co-workers proposed, when they investigated this phenomenon in 1994, that it might be put to good use as a way of separating different kinds of grains in a mixture. It certainly beats doing the job by hand. Stavans and colleagues suggested that the banding results from the complex interplay between two properties of the grains: their different angles of repose and the differences in their frictional interactions with the edge of the tube. They built these differences into a model of the tumbling process, and found it predicted that the well-mixed state of the grains was unstable: small, chance variations in the relative amounts of the two grains would be amplified such that the imbalance would grow. A tiny excess of sand in one region, for instance, would enhance itself until that region contained only sand and no glass balls.

But you might think that this would give rise to bands of random width, whereas the experiments showed bands whose widths were all more or less the same within a narrow range (Fig. 8.15a). In other words, a certain preferred length scale appears spontaneously in the pattern. The researchers pointed out that this is analogous to a phenomenon called spinodal decomposition, which takes place when a mixture of fluids is suddenly made immiscible (for example, by cooling the mixture to a temperature at which the two fluids separate). Spinodal decomposition takes place in quenched mixtures of molten metals: the two metals separate, as they freeze, into blobs of more or less uniform sizes. This too is driven by random fluctuations in concentrations of the two substances, which conspire to select a certain length scale.

For the tube with bulges (Fig. 8.15b), the two types of glass balls have the *same* angle of repose, yet segregation still occurs. This is because the variation in width along the tube imposes a change in slope on the free surface of the tumbling grains, and the large balls roll down the slope more readily than the small ones, which—as we saw earlier—are more easily trapped by bumps on the surface. Although it's not obvious without looking at the profile of the free surface, *downhill* carries the larger balls into the necks if the tube is more than half full but into the bellies if it is less than half full.

Thus, shaking, tumbling or even simple pouring of granular media can cause a mixture of different grains to mix, unmix or form striking patterns. At present there is no general theory that allows us to predict which of these will take place for a given system: again, you don't know until you try it.

Organized avalanches

So the slope of a granular heap, fed with fresh grains from above, remains at the angle of repose. This is a *dynamically* stable shape, because it is maintained in the face of a constant throughput of energy and matter (grains flow onto and off the heap). It is, in fact, a non-equilibrium 'dissipative' structure (see p. 255). In the past decade, it has become clear that in this everyday structure, familiar to millers and quarrymen everywhere, there is an unguessed complexity. The mass and profile of such piles are subject to continual fluctuations, as landslides remove grains from their slopes. Studies of the way in which sand piles maintain their shape while executing these fluctuations have led to a whole new field of research in non-equilibrium science and pattern formation.

In 1987, physicists Per Bak, Chao Tang and Kurt Wiesenfeld at Brookhaven National Laboratory on Long Island, New York, devised a model to describe the dynamics of a heaped pile of sand. They were led to do so not by the kind of practical concerns that would motivate an engineer, but because this system provided an easy-to-visualize model for studying a rather recondite question about the electronic behaviour of solids. In essence, their hypothetical model described a pile of sand grains with a well-defined angle of repose, to which new grains were gradually added. In the simplest version of the model, the sand pile was two-dimensional and was shored up against a vertical wall at one end. (This is like the experimental system of Makse and colleagues described above, with the plastic plates so close together that the pile is only one grain thick—and with the important provision that the friction between the walls and the grains is ignored.) Grains were added to this pile one by one at random points.

The pile builds up unevenly, so that its slope varies from place to place (Fig. 8.16a). But nowhere can the slope exceed the angle of maximum stability, the critical slope above which an avalanche takes place. If a single additional particle tips the slope locally over this critical value, a landslide is induced which washes down the 'hillside' and reduces the slope everywhere to a below-critical value (Fig. 8.16b).

But here is the curious thing: in their model, Bak and colleagues found that a single grain can induce a landslide of *any* magnitude. It might set only a few grains tumbling, or it might bring about a catastrophic sloughing of the entire pile. There is no way of telling which it will be.

dom? Not exactly. We can never be sure what effect any particular grain will have, but just as is the case for turbulence (p. 193), we can identify some robust 'form' to the behaviour of the sand pile by looking at the statistics of the problem. While landslides of all sizes are permitted, they are not all equally probable. Rather, little slides are more likely than big ones, and ones that send virtually the whole slope tumbling are rare indeed. The number of landslides decreases as the number of grains it involves increases, and the relationship between the two is a power law (also called a scaling law—see p. 193). Specifically, it is an inverse power law—rather like Newton's gravitational law, which says that the force of gravity exerted by a body falls off as the inverse of the square of the distance. The power law relating avalanche frequency to avalanche size in the model of Bak and colleagues falls off rather less sharply than this: the frequency (or probability, if you like) of an avalanche falls off as the inverse of its size (Fig. 8.17). Conversely, the size is proportional to the inverse of the frequency of occurrence, denoted f. This kind of inverse relationship between the size of an event and the probability that it will attain that size is commonly called a $1/f$ ('one over f') law.

It turns out that $1/f$ laws characterize the fluctuations observed in a great many diverse systems. An electrical current flowing through a resistor undergoes tiny

a

b

Fig. 8.16 The slope of a granular heap varies locally from place to place (*a*). In this pile of mustard seeds, small variations in slope can be seen superimposed on a constant average gradient. When the slope approaches the angle of maximum stability, the addition of a single seed can trigger an avalanche (*b*). This avalanche can involve any number of grains, from just a few to the entire slope. Notice that only grains within the first few layers move (here the grains in motion are blurred). (Photos: Sidney Nagel, University of Chicago).

This behaviour is reminiscent of that seen in chaotic systems, where the smallest perturbation can have effects quite out of proportion to its size—or it can remain just a small perturbation with a small effect. There is no characteristic *scale* to the system (in this case, no typical or favoured number of grains set tumbling when one more is added)—it is scale-invariant. We saw earlier that this characteristic is found in turbulent flows, and also in fractal objects such as DLA clusters and some fracture surfaces.

Does this mean that the landslides are totally unpredictable, that their magnitudes simply vary at ran-

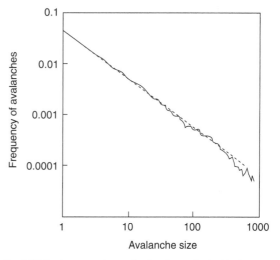

Fig. 8.17 The frequency of an avalanche of a certain size (that is, involving a certain number of grains) decreases in inverse proportion to its size, in a simple model of sand-pile avalanches. On a plot of the logarithm of frequency against the logarithm of size, this relationship defines an approximately straight line with a slope of around minus one (depicted by the dashed line). (After: Bak 1997.)

fluctuations, for instance, and the probability (or frequency of occurrence) of a fluctuation of a given magnitude depends on the magnitude according to a $1/f$ law—small fluctuations are more common than large ones. The Sun's luminosity fluctuates constantly owing to outbursts called solar flares, which are the result of magnetic instabilities in the hot plasma of the Sun's outer atmosphere. The magnitude of solar flares can be conveniently monitored by measuring the intensity of the X-ray emission that they generate, and this fluctuating X-ray emission obeys a $1/f$ power law over several orders of magnitude (factors of 10) in intensity. The emission from distant astrophysical objects called quasars shows the same kind of variability.

We will encounter other examples of $1/f$ behaviour later. In some of these cases, the relationship between the size and frequency of a fluctuation is not *exactly* a $1/f$ law: instead, the size varies in proportion to $1/f^\alpha$, where α is a constant that is greater than zero and less than two. This sort of relationship is, however, commonly included within the umbrella term of '$1/f$ behaviour'.

Now, the curious fact is that a $1/f$ law is *not* what one would predict if the fluctuations were purely random—that should instead generate a different scaling law. This is most tangibly (I should really say audibly) illustrated with reference to one of the most striking examples of $1/f$ behaviour, discovered by Richard Voss and John Clarke at the University of California at Berkeley in

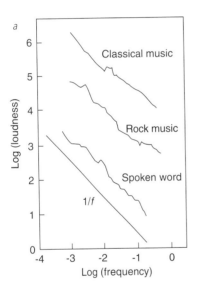

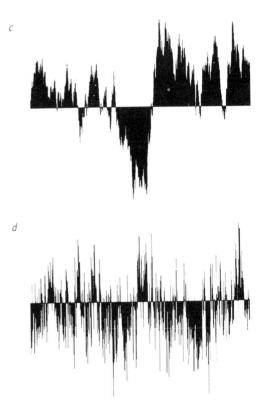

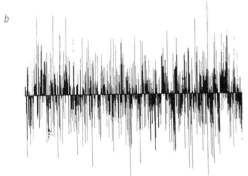

Fig. 8.18 The power spectra—loudness plotted against sound frequency—of a wide range of human-generated audio signals, from classical music to rock music to spoken word, exhibit $1/f$ scaling laws (*a*). 'White' noise is a featureless hiss with a flat power spectrum (loudness independent of frequency). Its time-varying signal (*b*) is fully random and unpredictable—and uninteresting. 'Brown' noise (*c*), with a $1/f^2$ power spectrum, is perceived as boring and rather monotonous. 'Pink' noise (*d*), lying between these two extremes with a $1/f$ power spectrum, has enough variation to be interesting, but not so much as to become indecipherable. (After: Voss & Clarke 1975.)

1975. They found that the power spectra of many pieces of classical music—roughly speaking, a plot of the loudness of the sound signal versus the sound frequency—display $1/f$ behaviour. What is more, by analysing the outputs from different radio stations over several hours, they found that the same relationship holds for rock music and for the spoken word (Fig. 8.18a). Looked at this way, you might as well be listening to the daily news as to a Bach concerto!

Of course, the two are very clearly *not* the same; but statistically, there is little distinction between them. Here again is that cautionary message for those who search for 'statistical form' in scaling laws: sometimes you risk losing the most crucial features of a system by throwing out the detailed specifics and focusing only on the statistics. On the other hand, who would otherwise have guessed at this 'hidden' kinship between the nine o'clock news and a baroque concerto?

The main point I wish to make is that both of these sound signals are clearly distinct from random noise. The latter is called white noise (Fig. 8.18b), and it is more or less what you get if you tune the radio between stations, or unplug the TV aerial: an unpleasant hiss. Audio signals that display $1/f$ behaviour, on the other hand, are examples of so-called 'pink noise' (Fig. 8.18d)—they contain an injection of low-frequency components in their power spectra, which white noise lacks. One can also create audio signals, called 'brown' noise, that have $1/f^2$ power spectra (Fig. 8.18c). These are perceived as rather dull by listeners. It seems that for some reason our ears find $1/f$ noise more pleasant than either the total unpredictability of white noise or the rather plodding monotony of brown noise—the level of variability is just sufficient to be deemed interesting.

Self-organized criticality

So $1/f$ fluctuations are unpredictable but are not due to some purely random process. Although common to many different physical systems, this behaviour has long been a mystery. When Per Bak and his colleagues saw $1/f$ behaviour in their model sand pile, they were consequently hugely excited. Here was a relatively simple model system, for which they knew all of the ingredients (because they had mixed them up themselves), that might offer some clues about the origin of the puzzling $1/f$ behaviour. This kind of scaling law is the consequence of abrupt avalanche-like events happening on all scales irrespective of the size of the perturbation that triggers them.

There is something very peculiar about the sand pile that displays this behaviour: it is constantly seeking the *least* stable state. We are used to the converse—water runs downhill, golf balls drop into holes, trees topple. The sand pile, however, is forever returning to the state in which it is on the brink of an avalanche. Each time an avalanche occurs, this precarious balancing act gives way; but then as further grains are added, the system creeps right back to the brink.

States like this, which are susceptible to fluctuations on all scales at the slightest provocation, have been known to physicists for a long time. They are called critical states, and are found in systems as diverse as magnets, liquids and theoretical models of the Big Bang. Every liquid achieves a critical state at a well-defined temperature and pressure, called the critical point. If you heat a liquid, it evaporates to a vapour once it reaches the boiling point: the state of the fluid changes

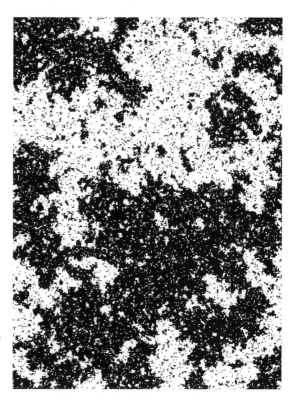

Fig. 8.19 At the critical point of a liquid and a gas, the distinction between the two breaks down. A critical fluid contains variations in density on all size scales, with domains of liquid-like fluid coexisting with domains of vapour-like fluid. Here I show the results of model calculations of the structure of a fluid at the critical point; the dark regions represent liquid-like (dense) domains and the white regions are vapour-like (rarefied). (Image: Alastair Bruce, University of Edinburgh.)

abruptly from a (dense) liquid to a (rarefied) gas. But above the critical temperature this abrupt change of state no longer happens; instead, the fluid passes smoothly and continuously from a dense liquid-like state to a diffuse gas-like state as its pressure is lowered. The critical point is the point at which there is no longer any sharp distinction between 'liquid' and 'gas', and no boiling point separating the two.

At the critical point of a fluid, its density undergoes fluctuations on all length scales: in some regions the fluid might have a liquid-like density, in others it is gas-like, and these regions are constantly changing over time and have no characteristic size or shape (Fig. 8.19). The fluid is poised right on the brink of separating out into liquid-like and gas-like regions, but cannot quite make up its mind to do so. It is extremely difficult to maintain a fluid at its critical point, however—a critical fluid is highly sensitive to the smallest perturbations, and will readily 'tip over' and separate into liquid-like and gas-like states. The susceptibility of the critical state to perturbations is, in fact, strictly infinite. It is precisely like trying to balance a needle by its tip: theoretically a balanced state exists, but it is unstable to even the slightest disturbance.

The theoretical sand piles of Per Bak and colleagues have this same critical character, being susceptible to fluctuations (avalanches) on all length scales through the action of the smallest perturbation (the addition of a single grain, say). But unlike the critical states of fluids, they seem to be robust, not infinitely unstable. Instead of constantly seeking to escape the critical state, the sand pile seeks constantly to return to it. Who would have guessed that a sand pile could be so perverse?

Bak called this phenomenon self-organized criticality (SOC), reflecting the fact that the critical state seems to organize itself into this most precarious of configurations. The natural assumption was that all the other physical systems that exhibited $1/f$ behaviour were also in self-organized critical states. Bak began to see signs of self-organized criticality just about everywhere he looked. In a theoretical model of forest fires, for instance, a forest can be split up into clusters of unburnt trees of all sizes, and newly initiated fires can propagate on all length scales, burning just a few trees in the immediate vicinity or spreading catastrophically over large areas. If the trees regrow slowly, the forest is maintained in a self-organized critical state by occasional fires.

It has been known for over four decades that earthquakes follow a power law, called the Gutenberg–

Richter law: earthquakes occur on all scales of magnitude (from a plate rattler to a city leveller) with the probability declining as the magnitude gets larger (Fig. 8.20). This smacks of self-organized criticality, and a simple mechanical model of earthquake faulting developed by Bak and co-workers shows power-law behaviour resembling the empirical Gutenberg–Richter law.

Power–law behaviour is also seen (or at least claimed) in volcanic activity, in the length of streams in river networks, and in the fossil record of fluctuations in the abundance of life on Earth through the geological past—to name just a few examples. There is clearly no link between the physical mechanisms that control these phenomena, but nonetheless it seems that they may show essentially the same statistical behaviour. There appears to be something universal about the probabilities that does not depend on the details of how the components interact.

When you think about it, this is not really so unusual—the same applies, for instance, to the statistical behaviour of purely random systems. I will obtain the same bell-shaped (so-called Gaussian) probability curve for a million executions of a random process with two possible outcomes, regardless of whether it involves

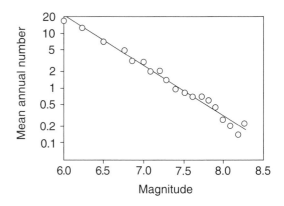

Fig. 8.20 The Gutenberg–Richter law for earthquakes provides an example of 1/f behaviour. It states that the number N of earthquakes of a given size (per year in a given geographical area) is related to the size (S) of the quake through an inverse power law: $N \propto 1/S^b$, with b being nearly equal to one for many regions throughout the world. So a log–log plot of number against size gives a straight line. Here I show the relationship for 'shallow' earthquakes, which occur at depths of 0–60 km in the Earth, as plotted by Gutenberg and Richter in 1949. (The magnitude (M) of an earthquake on the Richter scale is related to the logarithm of its size in terms of the energy released, so the linear magnitude scale here is a logarithmic scale of energetic 'size'.) This power–law relationship has been invoked as an indication that earthquakes are an example of a self-organized critical phenomenon.

tossing a coin, generating the numbers at random on a computer, rolling a dice with three white and three black faces or picking black and white balls from a bag. In each case, the physical mechanisms that determine whether black or white is selected are different.

Moreover, we saw something very similar in the growth of fractal forms in Chapters 5 and 6: they appear in physically diverse systems, because similar considerations (for example, that growth is more likely at tips than in valleys) apply to the statistics of the growth processes. Indeed, self-organized critical states have a scale invariance just like that of fractals, and the spatial distributions of their component elements (such as trees or streams, or the profile of a sand pile) can be truly fractal. This is potentially important, as there was previously no known *general* mechanism for generating fractal structures.

Per Bak believes that in self-organized criticality he has uncovered 'a comprehensive framework to describe the ubiquity of complexity in Nature'. There is no doubt that it is a fascinating new area of physics, and that many of the models developed to describe 'complex' systems in the real world do find their way into a self-organized critical state. It may even be that, given how we apparently perceive systems whose variability follows a $1/f$ law as complex and interesting (as opposed to monotonous or impossibly unpredictable), self-organized criticality has something to tell us about our aesthetic response to pattern and form. But what about the real world? Are natural systems (as opposed to simple models of them) also in this precarious state?

One of the difficulties in answering that question is that the statistics are often ambiguous. To be sure you are seeing a particular kind of scaling behaviour and not just something that looks a bit like it over a small range of size scales, you need a lot of data. And that's not always on hand. There may not have been enough mass extinctions since the beginning of the world, for example, to allow us ever to be sure that evolution operates in a self-organized critical state (as Per Bak has claimed). Another problem is that, whereas in a model you can usually be sure exactly what all the important parameters are, and can see the effect of changing each one independently, in reality complex systems may be susceptible to all manner of perturbing influences, some more obvious than others. Will a model of earthquake faulting that includes a more realistic description of the sliding process or of the geological structure of the Earth still show self-organized criticality?

In fact, it has been loudly and contentiously debated whether even real sand piles, the inspiration for Bak's original model, have self-organized critical states. You might imagine that this, at least, ought to be a simple experiment to perform: you just drop sand grain by grain onto a pile and observe how big an avalanche follows from each addition. But experiments conducted since Bak, Tang and Wiesenfeld first proposed their model have produced ambiguous results, partly because there is no unique way to measure the size of an avalanche. Sidney Nagel and colleagues at Chicago found in 1989 that real sand piles seem always to yield large avalanches, in which most of the top layer of sand slides away. But other experiments in the early 1990s seemed to generate power–law behaviour like that expected of self-organized criticality.

One of the problems is that real sand is not like model sand: the grains are not identical in size, shape or surface features, and these microscopic details determine how readily they slide over one another. In 1995 Jens Feder, Kim Christensen and co-workers from the University of Oslo in Norway attempted to settle the debate. They added a new twist to the tale: instead of studying sand piles, they looked at piles of rice. This was because rice grains do not roll or slip over one another as readily as sand grains do (just as rugby balls do not roll as well as footballs), and so they capture more accurately the assumed behaviour of grains in those computer models that show SOC (a rare example of an experiment being adapted to fit the model rather than vice versa). The grains tumble if they exceed the angle of repose, but moving grains are rapidly brought to rest when this is no longer so. That was more or less the situation modelled by Bak and colleagues, who didn't include the *inertial* aspects of the problem that resulted from moving and colliding grains.

Feder and colleagues looked at two-dimensional piles, in which the rice grains were confined to a narrow layer between two parallel glass plates (Fig. 8.21). By photographing and digitizing the profiles of the piles they could deduce the size of the landslides that took place.

Observing enough avalanches to provide trustworthy statistics was a slow and tedious process, and took about a year. But at the end of it all, the researchers concluded that the behaviour of these granular piles depended on the kind of rice that they used: specifically, on whether it was long-grain or short-grain. To physicists, these varieties differ not in terms of whether they are to be used for risotto or rice pudding but according to their so-called aspect ratio: the ratio of length to width. Long-grain rice has the higher aspect ratio, and Feder

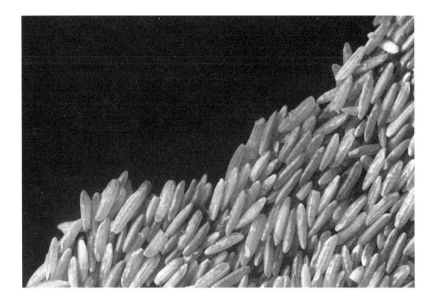

Fig. 8.21 A section of a rice pile confined between two glass plates. Notice how uneven the slope is on this fine scale. (Photo: Kim Christensen, University of Oslo.)

and colleagues found that it seems to show true self-organized critical behaviour, with a power–law relationship between the size of the avalanche (the researchers actually measured how much energy each one released) and its frequency of occurrence. But short-grain rice, which is more nearly spherical and so more like sand, showed different behaviour: instead of a simple power–law, the relationship between size and frequency was more complicated, having a mathematical form called a stretched exponential. Feder also showed that a stretched exponential can easily be mistaken for a power–law (and thus for self-organized critical behaviour) if the measurements are not taken over a wide enough range of avalanche sizes, possibly explaining why others had previously claimed to have seen SOC in sand piles.

The conclusion, then, was that piles of grains *can* show SOC, but that they will not necessarily do so—it depends on (amongst other things) the shape of the grains and how rapidly their energy is dissipated during tumbling. This both vindicates and modifies Bak's assertions: self-organized criticality seems to be a real phenomenon, not just a product of computer models, but it may not be universal or even particularly easy to observe or achieve. For the present time, sand piles appear to be an intriguing but limited metaphor for nature's complexity.

Shifting sands

All the same, I think that the most spectacular of granular patterns must surely be those that nature makes: the vast desert dunes that are the backdrop to our images of Arabian legend (Plate 23). These features are engraved by the wind, and their widths range from a few metres to several kilometres. Despite their immensity, sand dunes are not robust topographic features but are constantly shifting in a stately, writhing dance. As on the sea's wrinkled surface, it is the *pattern* that remains, not its individual components. Seen from above, linear dune fields (Fig. 8.22) resemble the fingerprint-like stripe phases seen in convection, Turing patterns and Langmuir films, with much the same kinds of dislocation defects where ripples terminate or bifurcate in two. Hans Meinhardt suggests that, at root, dune formation is akin to an activator–inhibitor system, in which short-range activation competes with long-range inhibition. Dunes are formed by deposition of wind-blown sand. As a dune gets bigger, it enhances its own prospects for growth, as it captures more sand from the air and provides more wind shelter for the grains on the leeward side. But in doing so, the dune removes the sand from the wind and so suppresses the formation of other dunes in the vicinity. The balance between these two processes establishes a constant mean distance between dunes which depends on wind speed, sand grain size (and thus their mobility in the wind) and so forth.

The essence of this idea is probably sound, but it's virtually certain that no single mechanism can explain the vast diversity of shapes and forms seen in the world's deserts. Indeed there are so many of these, with names that are often regionally specific due to their derivation

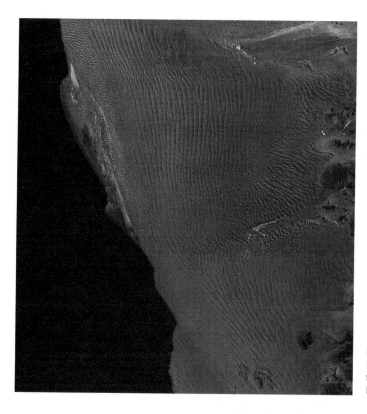

Fig. 8.22 A satellite image of linear dunes in the Namib Sand Sea in southwestern Africa. The width of the region shown here is about 160 km. (Photo: Nick Lancaster, Desert Research Institute, Nevada.)

from the local language, that even geomorphologists have trouble keeping track of them. For one thing, there seem to be several characteristic size scales of natural sand patterns. Dunes in the strict sense are repetitive features that recur with wavelengths typically of around ten to several hundred metres. Superimposed on them are much smaller sand ripples, which tend to be wavy, linear crests with wavelengths of between half a centimetre and many metres, and heights of about 0.5 to 25 cm. And dunes themselves are commonly superimposed on even larger features, often called draas after their name in North Africa. Draas have wavelengths of hundreds of metres to several kilometres.

The common feature of these structures, however, is that they are (on the whole) self-organized patterns, whose shapes and wavelengths arise from a subtle

conspiracy between sand and wind, rather than being imposed by any external agency.[*]

The challenge is to understand the rules of these pattern-forming processes.

Desert grooves

Sand ripples are the most common features of wind-blown sand. The process by which moving air sweeps sand into these regular little crests was elucidated in some detail in the 1940s by R.A. Bagnold, whose work on granular media has laid the foundations of much that is known today. In modern terminology, we'd say that Bagnold's explanation for the appearance of small-scale sand ripples on a flat surface bombarded with a steady flow of wind-blown grains is an example of a growth instability.

Think of a sandy plain from which a steady wind continually picks up surface grains and dumps them elsewhere. If the wind blows persistently in the same direction, the plain is gradually shifted *en masse* upwind. But if by chance a tiny bump appears on the surface where

---

[*] The class of dunes called coppice dunes, however, arise from the accumulation of sand by small patches of vegetation, while climbing dunes, echo dunes and falling dunes are initiated by large-scale topographic features such as hills.

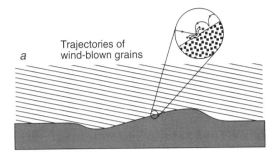

a   Trajectories of
      wind-blown grains

b

Excavation

New slope

Fig. 8.23 The formation of sand ripples involves a propagating instability. Wind-borne sand grains rain down on the desert surface at an oblique angle. Where the surface slopes, more grains impact the windward (stoss) side of the slope than the leeward side (*a*). Each grain scatters others from the surface as it strikes, and travels in the downwind direction for a few short hops (a process called saltation) before coming to rest. The accumulation of saltating grains at the slope crest means that the leeward foot of the slope receives fewer new grains than other regions, and so it begins to be excavated into a depression (*b*). This depression develops into a new, downwind stoss slope, and a new ripple is formed.

the deposition rate is momentarily increased, the windward side of the bump (called the stoss side) encounters a higher flux of impacting grains than the flat plain elsewhere. This is illustrated in Fig. 8.23*a*, where you can see that more lines (representing the trajectories of wind-borne grains) intersect the stoss face, per unit length, than a horizontal part of the surface. Conversely, fewer lines intersect the downwind (lee) side of the bump, where there is an 'impact shadow'. So once a bump is formed it begins to grow, making a flat sand bed unstable to fluctuations in its topography.

But there is more: the formation of one bump triggers the appearance of another downwind, so that a system

of ridges propagates across the plain. This is because the story does not end when the wind-blown grains hit the surface of the desert. They do not simply sit where they strike: the grains bounce. The wind carries these bouncing sand grains downwind in a series of hops, a process called *saltation*. Moreover, the initial impact of a wind-blown grain creates a little granular splash, throwing out other grains from the surface which can then also be carried along by saltation.

This process of impaction and saltation takes place all across the plain. It maintains a flat, horizontal surface if the rate at which sand is delivered by the wind is equal to the rate at which it is transported downwind by saltation. But when a ripple begins to form, these rates of grain delivery and removal are not everywhere balanced. Beyond the impact shadow at the foot of the lee slope, impacts followed by saltation lead to the downwind transport of sand (to the right in Fig. 8.23). But because there are relatively fewer impacts on the lee slope itself and in its impact shadow, this transport is not balanced by a flux of grains coming from the left. Therefore the foot of the slope becomes excavated, creating a new stoss slope to its right. At the top of this new slope, grains begin to accumulate by saltation, and another lee slope develops (Fig. 8.23*b*). And so the wavy disturbance propagates downwind as a series of ripples. These ripples have a characteristic wavelength: Bagnold proposed that this is determined by the typical distance that a saltating grain travels before coming to rest (which in turn depends on the grain size, wind speed and wind angle); but it now seems that the wavelength reflects a balance between rather more complex aspects of grain transport, and in reality there is typically a range of wavelengths in any ripple field.

Spencer Forrest and Peter Haff from Duke University in North Carolina have shown by computer-modelling that sand-ripple formation is a self-organized process. In their model, sand grains are fired at a flat sand bed at a certain angle and speed. The model is two-dimensional: the sand grains are confined to a single layer, like a vertical cross-section through a real desert. Each impacting grain ejects other grains from the surface according to a so-called splash function, which specifies the number of grains ejected and their velocities. The model is a cellular automaton (p. 57), since the behaviour of each particle is determined by well-defined rules that take into account what the particle's immediate neighbours are doing.

The researchers found that ripples quickly began to rise out of the flat surface (Fig. 8.24*a*). They finally

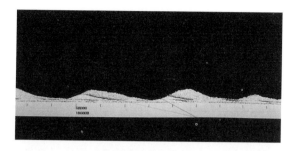

a

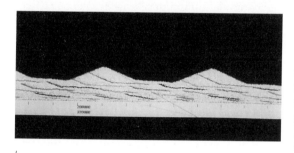

b

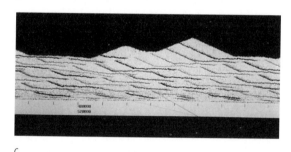

c

Fig. 8.24 Self-organized ripples appear in a cellular-automaton model of wind-blown sand deposition. The ripples develop on a flat surface as the growth instability amplifies small irregularities (*a*). These ripples move from left to right owing to saltation. Because of the difference in speed between smaller, faster ripples and larger, slower ones, they exchange mass until their size, speed and spacing is more or less uniform (*b*). 'Stained' grains injected at regular intervals reveal the patterns of layer deposition for different deposition rates (*b*, *c*). (Images: Peter Haff, Duke University, North Carolina. Reproduced from Forrest and Haff (1992). *Science* **255**, 1240.)

the individual grains. What happened was that the smaller ripples travelled faster than the larger ones, simply because they contained less material to be transported. But as they overtook larger ripples, small ones would acquire sand from the slower heaps in front until their sizes, and therefore their speeds, were more or less equalized. In this way, a roughly regular train of ripples was formed that moved in procession downwind (Fig. 8.24*b*).

In the simulations, more material was deposited than was removed by saltation downwind, and so the sand bed gradually increased in thickness. In the real world these depositional beds can be preserved for posterity as the gaps between the grains get filled in with a cement of minerals precipitated from permeating water. Such sedimentary rocks are known as aeolian (wind-borne) sandstones. By artificially colouring the wind-borne grains at periodic intervals in their computer model, Forrest and Haff were able to deposit 'stained' layers which acted as markers to show how the deposited material became distributed subsequently in the thickening bed. Depending on the rate of deposition, they found various patterns (Fig. 8.24*b*, *c*), which resembled those found in natural aeolian sandstones when some environmental factor allows material deposited at different times to be distinguished (for example, its composition and colour might change).

Stars and stripes

Many sand dunes share the same wavelike form as sand ripples, with linear, slightly wavy crests that lie perpendicular to the wind direction. These are called transverse dunes (Fig. 8.22). But not all dunes have this form. Some form crests *parallel* to the prevailing wind: these are longitudinal dunes. Others, called barchan dunes, are crescent-shaped, with their horns pointing downwind (Fig. 8.25*a*). Barchan dunes can merge into wavy crests called barchanoid ridges. And some dunes have several arms radiating in different directions: these are star dunes (Fig. 8.25*b*). How does the same basic grain-transport process (saltation) produce these different forms?

Many models have been proposed to account for the shapes of particular kinds of dune, and for their characteristic spacings. Some of these models invoke rather complex interactions between the evolving dune shape and the wind flow pattern. Bagnold, for instance, suggested that longitudinal dunes might be the result of helical wind vortices arising from the interaction between the wind and convective airflow as heat from

attained a triangular shape, and at the same time they began to migrate across the surface in the direction of the wind—just as they do in real deserts. Remember that this migration is not in any sense a result of the ripples being 'blown' by the wind; instead, the solemn procession is the indirect effect of individual grain impacts followed by saltation.

Forrest and Haff found that a characteristic ripple size emerged that was several hundred times the diameter of

a

b

Fig. 8.25 Large-scale sand transport can create dunes of several characteristic shapes, including the crescent-shaped barchan dunes (*a*) and the many-armed star dunes (*b*). (Photos: Nick Lancaster, Desert Research Institute, Nevada.)

the desert surface warms the air above. Another early pioneer of dune geomorphology, V. Cornish, suggested at the beginning of the century that star dunes form at the centre of convection cells above the desert floor. It's clear that a major influence on dune type is the nature of the wind field: whether it is steady or varying in direction, fast or slow. The amount of sand available for dune building is also important: transverse dunes may

be favoured if the sand supply is abundant, whereas longitudinal dunes form in a sparser environment. The fact that the dune itself changes the flow of air around it as it grows adds a further level of complication, as does the presence of vegetation.

But in spite of all this, geomorphologist Bradley Werner from the Scripps Institute of Oceanography in California has developed a cellular-automaton model of

a 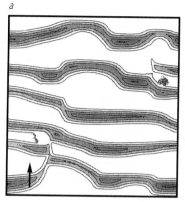 *b* 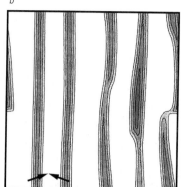 *c*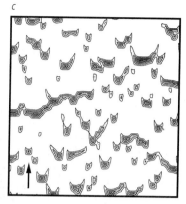

Fig 8.26 A cellular–automaton model of dune formation generates many of the major dune types, including transverse and longitudinal ripple dunes (*a*, *b*) and barchan dunes (*c*). Here I show the contours of the deposited material. The shapes depend on the wind direction and variability (indicated by arrows). (Images: from B.T. Werner (1995). *Geology* **23**, 1107.)

dune formation in which most of the major dune shapes arise as spontaneously self-organized forms out of the process of grain deposition under different wind conditions. In Werner's model, sand grains are scattered at random on a rough stony bed of rugged topography, and are assumed to be picked up at random in parcels by the prevailing wind. After it has been carried a fixed distance, each parcel has a chance of being redeposited. The probability of this is greater if the parcel encounters sand-covered ground at that point rather than stony ground, reflecting the known fact that saltating sand bounces off stony ground more readily than off sandy surfaces. If the parcel is not deposited, it is carried on for the same fixed distance before the possibility of deposition arises again. If at any point deposition brings the slope of the sandy ground in excess of the angle of repose, slabs of sand are allowed to slide downhill until the slope is again less than this angle.

With just these ingredients, Werner was able to reproduce all of the major dune types in his model—barchan, star and linear dunes (Fig. 8.26). He found that these characteristic patterns seem to represent *attractors* (see p. 54), towards which the sand deposit is drawn regardless of its initial configuration. For instance, when the wind was predominantly in a single direction, dunes formed with their crests lying perpendicular to the wind (transverse dunes; Fig. 8.26*a*), whereas if the wind direction was more variable, the dunes were oriented in the average direction of the wind (longitudinal dunes; Fig. 8.26*b*). Werner's model suggests that the stable attractor changes from the transverse to the longitudinal pattern as the wind becomes more variable.

While it is likely that specific, local influences affect dune sizes and shapes, Werner's idea has the appealing feature that the broad patterns that emerge are generic, not dependent on case-by-case details. Within this picture, star and barchan dunes are as inevitable a feature of nature's tapestry as the branches of a river or the stripes of a zebra.

Through the sieve

One intriguing feature of natural sand patterns (both small-scale ripples and large-scale dunes) is that the sand grains are segregated according to size into different regions of the pattern elements. For sand ripples, the coarsest grains appear preferentially at the crests, with a thin veneer also coating the stoss face. For large dunes the reverse is commonly the case: the finest grains collect at the crests, and the coarsest in the troughs. When there is net deposition, so that ripples are gradually laid down on top of each other, the result is a series of stratified layers in which a periodic sequence of coarse to fine grains recurs down through the sand bed. This characteristic sequence of coarse-to-fine grains distinguishes aeolian sandstones from fluvial sandstones, which are deposited as sandy material sediments out of water. In the latter case, the larger grains settle faster, so the stratigraphic sequence has fine grains on top and coarser grains below. For this reason, a coarse-to-fine sequence is said to be 'inverse-graded'.

How does this grain-size sorting occur in sand ripples? It recalls the stratified sorting seen in the landslide experiments described earlier; but the resemblance is coincidental, as the origin of the sorting is actually rather different. Robert Anderson and Kirby Bunas from the University of California at Santa Cruz have shown that segregation of grain sizes is a consequence of the saltation process. They used a cellular-automaton model rather like that of Forrest and Haff, except that it incorporated grains of two different sizes. The effect of grain impacts was again determined by a splash function, but this differed for the small and large particles: the latter ejected more secondary grains, for the same impact speed, than the former, since the collisions in that case were more energetic. The size and speed of the impacting grain, as well as the composition of the bed that it struck, also determined the relative mixture of small and large particles in the 'splash'. So the rules governing the impacts were in this case fairly complicated; but their net effect was that smaller grains tended to be ejected preferentially, and with higher speeds (which carried them further away). The general effect of impacts was therefore to make the surface of the sand bed coarser.

With that in mind, it is not hard to see why the researchers found that their model ripples had coarser material coating the face of the stoss slope (Fig. 8.27a),

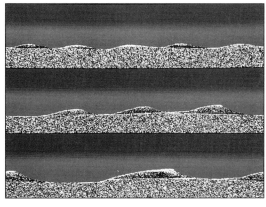

a

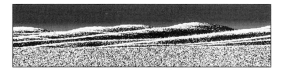

b

Fig. 8.27 Grains of different sizes are often segregated in sand ripples and dunes. Here a cellular-automaton model reproduces the tendency of sand ripples to accumulate coarse grains (*white*) on the stoss slopes and particularly at their crests (*a*). When there is net deposition, so that the deposited layer gradually thickens, the sand deposit becomes stratified (*b*). (Images: Robert Anderson, University of California at Santa Cruz.)

like the ripples found in the real world: the stoss slope receives more impacts, and so gets more coarsened, than other regions. The crests of the ripples were particularly enriched in coarse grains (as seen in nature), which the researchers explained as follows. The larger grains make smaller hops, because they are more massive and are therefore ejected from the impact splash with lower velocities. By means of these little jumps the large grains gradually make their way up the stoss slope and jump just over the crest, into a sheltered region just at the top of the lee slope within the impact shadow. Here they remain, protected from impacts, while further coarse material gradually climbs on top of them. The smaller grains, meanwhile, make bigger leaps and so are propelled further over the edge onto the lower parts of the lee slope.

Notice that in Fig. 8.27a the ripples are highly asymmetric, with a gently convex stoss slope and a steeper, concave lee slope. This shape is much closer to that of real sand ripples than is the triangular shape of Forrest and Haff's model, showing that the more sophisticated treatment of saltation and splashes captures more of the important physics of the process. Remember, furthermore, that these ripples are not static but are slowly moving from left to right in the figure. This means that the coarse material on the ripple crests is repeatedly buried and then exhumed again at the foot of the stoss slope as the ripples pass over it. The grains are forever climbing mountains.

Anderson and Bunas found that when their model was executed under conditions of net deposition, so that the sand bed gradually thickened, stratified beds were laid down in which coarse and fine layers alternated (Fig. 8.27b). This mimics the inverse grading of natural aeolian sandstone.

Do it yourself

I don't think that we have by any means exhausted the capacity of granular substances to generate spontaneous patterns, nor have I been able here to survey all of those that are currently known. What is particularly exciting about these systems is that not even the scientists studying them have yet acquired the kind of intuition that allows them to predict what they might see in a given experiment. You have to shake it and see! And I feel there is an attraction too in a kind of physics that returns to the spirit of the nineteenth-century pioneers like Michael Faraday, performing simple bench-top experiments with cheap, homemade equipment and a mind that is prepared to be astonished at the artistry of nature. There is much we can learn from playing in the sand …

COMMUNITIES

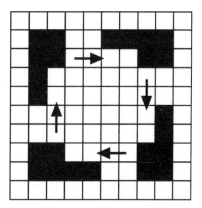

These are the fluid communities, some of long duration when circumstances favored ... some fleeting and vivid, encompassing in the time of their duration a moment only of the member's life; and in our world at least they are ramified and improvised, living and dying, growing and falling off almost as a life form itself.

J. Robert Oppenheimer
The Sciences and Man's Community

Who wants to be considered predictable? No, I thought not. We might be happy enough to accept that patterns can emerge from wind-blown sand, running water, strange brews of chemicals and even bacterial colonies; but human activities are surely something else altogether. OK, we all have our routines, there is an element of repetition in our lives (driven in large part by biology's exigencies of sleeping, eating and excreting)—but when it comes to *interactions*, to the ebb and flow of society, surely there is much too great a diversity in behavioural tendencies to establish patterns and forms of the sort with which this book is concerned? We might imagine that whatever organization in time and space our societies display is a consequence of careful planning and forethought, not an inevitable and spontaneous emergent property.

Well, of course I would not be setting up this straw man if I were not about to burn it, but I want to acknowledge at the outset that it becomes hard, within a discussion of societies and of behaviour, to keep a clear image of the definitions of pattern and form that I have developed so far. This is rich ground for confusion. On the one hand, human society is *riddled* with spatial patterns that are more or less complex, from the rice terraces of Asia (Fig. 1.5*b*) to the magical web-weaving of Islamic art (Fig. 7.41*b*) to the regular lattice of Manhattan's street plan. But we must try to bear in mind the self-made tapestry. These tapestries of agriculture, adornment and architecture are anything but self-made: someone has drawn the blueprints, someone has purposefully shifted soil and stone, someone has pulled the thread this way and that. Yes, these patterns are of obvious utility, they are widespread, and they are even beautiful—but not *inevitable*. We do not deduce the traceries of the Alhambra palace from a consideration of the laws of interaction between a Moorish artisan's chisel and stone: those mechanics permit just as readily of Barbara Hepworth's sculpture garden.

Some of the spatial forms and patterns that I shall touch upon in this chapter, on the other hand, are perhaps challenging to our self-esteem, because they evolve in spite of ourselves, unconsciously. We don't realize we are making them, until suddenly—there they are. And the final irony is that, when we look down upon some of these human-made designs, we may find that they are remarkably similar to the forms we have found elsewhere in this book, the products of inanimate matter. To that extent, life is nothing special. And I want to show too that natural ecosystems—networks of interactions far more diverse and complicated than those in a stream of water, a flame or a growing crystal—have their own intrinsic patterns too. These self-organizing tendencies of nature can be crystallized into simple models that define a kind of mathematics of the biosphere, a mathematics that allows room for choice, for reciprocity, for memories of past events. By identifying these patterns, there is much that we can learn about our place in the world and the way that we affect it.

Ecocycles

Life never used to be this complicated. Once there was no anxiety about whether your neighbour would have another all-night party, whether your car would start, whether your credit rating was falling. All you had to worry about was staying alive: eating, while not being eaten. Throughout much of the animal world, that's pretty much how it remains. You have your prey, and you have your predators. So long as you find time now and again to reproduce and spread your genes, you can devote the rest of the day to finding the former and avoiding the latter. Yet there are patterns and rhythms that emerge even from this most basic of lifestyles.

What controls the size of a community? Thomas Malthus thought he had the answer in the late eighteenth century, when he proposed that populations grow with geometrical rapidity (which is to say, as a power law) until the food runs out—and then things turn nasty. The historical record of China's population over the last 2000 years shows sawtooth cycles of gradual growth and abrupt, famine-induced decline, followed by periods of social turmoil, which look distinctly Malthusian. But although Malthus's idea of a cut-throat struggle for survival exerted a profound influence on Charles Darwin, in general it provides much too simplistic a view of population dynamics.

For what if one population feeds on another, while both are subject to comparable growth laws? Then for the predators (foxes, say), 'food' is no longer a field of grass of finite extent, but a resource (rabbits) that has the potential to keep pace with the expansion of the fox community itself. On the other hand, if the foxes gorge themselves too much, they deplete the prey to such an extent that there is not enough food to go round—and then the expansion of the fox community is itself held in check.

What about the rabbits? Well, they *do* face the intrinsic limitation of the amount of grass in the environment, although we can imagine that there is a steady rate of grass growth that can sustain a rabbit population of a certain size indefinitely. The more serious threat to the rabbit community is that of being eaten.

So here's the deal: the foxes will eat the rabbits and will, as a consequence, multiply in number. The rabbits, meanwhile, will reproduce to replace those who have fallen prey to the foxes (let's assume that the availability of grass is never a limiting factor). If the foxes grow too numerous, the rabbit population is depleted until there are too few to sustain the foxes, and they starve. So we might imagine that the fox population will grow only to such a size that it consumes rabbits more or less at the same rate as they are replaced by the fecund rabbit community. Then the two populations would coexist in a more or less steady state, neither growing nor declining (Fig. 9.1*a*).

But when Vito Volterra put these ideas into mathematical form in the 1920s, he found that this steady-state solution is unlikely. Volterra was analysing the dynamics of predators and prey in the fish populations of the Adriatic, and as I mentioned in Chapter 3, he appropriated Alfred Lotka's model of oscillating chemical reactions to do it. The changes in concentration of chemical species in Lotka's equations then become changes in numbers (or population density) of the predator and prey communities. In general these equations suggest that both populations (let's stay with the foxes and rabbits rather than with Volterra's fish) *oscillate* between greater and smaller numbers over time. The oscillations of the foxes and rabbits are out of step with one another, and they can in principle persist indefinitely (Fig. 9.1*b*). In other words, oscillating—not steady—populations appear to be the norm.

The idea of developing a mathematical theory for population dynamics was rather a bold one in Volterra's time; previously, one relied on intuitions like Malthus's. That's an understandable prejudice, since it's not hard to see that populations are subject to a lot of different factors which cannot easily be put in mathematical terms. Foxes and rabbits don't really go around hunting and hiding on a smooth grassy plain, independently of the rest of the ecosystem—food chains are extremely

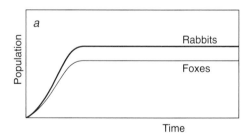

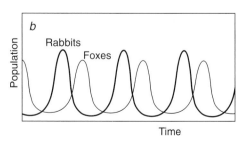

Fig. 9.1 The Lotka–Volterra equations for predator–prey interactions generate populations that reach a steady state (*a*) or, more commonly, that vary periodically in time (*b*). In the latter case the predator cycles lag behind those of the prey.

complex, and it can be very hard to predict the effect of a particular perturbation on a remote part of the chain. And there are external factors that surely have a bearing: the unpredictable fluctuations of the weather, the cycle of the seasons, the irregularities of terrain. But Volterra was a mathematician, not a biologist, and so had no qualms about making sweeping assumptions to simplify the dynamics of the ecosystem until they could take a precise mathematical form.

These dynamics can be expressed in three equations. Each equation contains quantities that appear in at least one of the other equations too, and so they are said to be coupled. Fortunately, we can put the equations rather easily into words. First:

$$\text{(rabbits and grass) lead to (more rabbits),} \quad (9.1)$$

which seems entirely reasonable. Notice, however, that this is an *autocatalytic* process, since the 'product' (rabbits, on the right-hand side) is also a 'reactant' (rabbits on the left). If left to its own devices, a rabbit population that obeys this rule will multiply exponentially. Of course, this kind of growth can't go on for ever, as Malthus understood; but we don't need to worry about the risk of overcrowding, because well before that possibility looms, the hungry foxes enter the picture. Then what happens is that:

$$\text{(rabbits and foxes) lead to (more foxes)} \quad (9.2)$$

and we can't argue with that. This is an autocatalytic step for the fox population, in the course of which rabbits are *consumed*: they are there on the left, but on the right we end up only with more foxes (some with rabbits in their bellies). Finally:

$$\text{(foxes) lead to (some dead foxes),} \quad (9.3)$$

which says that foxes die off at a steady rate.

Now as I say, Volterra didn't just conjure up this scheme of three 'reactions' to describe the interactions between predators and prey: he adapted it from Lotka, and it is called the Lotka–Volterra mechanism.

In general, the mathematical solutions for the way in which the populations change over time are oscillatory because the ecosystem is constantly overshooting. The foxes eat too many rabbits and find themselves without food. The fox population plummets as they starve, and this gives the rabbits some respite, allowing them to grow in number. But this works in the favour of the remaining foxes, who soon begin to multiply and deplete the rabbits again … and so the cycle repeats. There is a small range of growth rates that allow these oscillations to damp out and reach a steady state, but outside of this range the coupled equations possess a fundamental oscillatory instability with the character of a Hopf bifurcation (p. 67).

Although there have been attempts to apply the original Lotka–Volterra model to observations of real populations, these have revealed more of the model's shortcomings (due to the simplistic assumptions it makes) than its strengths. For one thing, the model assumes that predators are insatiable, and will keep eating no matter how much they've gorged themselves already. This isn't just unlikely, it is manifestly untrue. Most predators are better at capturing prey than the prey are at avoiding capture; and yet somehow predators typically contrive to establish an equilibrium with their prey whereby they content themselves with a sustainable number of catches.

Yet many populations of predators and their prey do indeed show oscillations in number with a more or less consistent periodicity. Cyclic fluctuations in the populations of herbivorous mammals, for instance, are common—small herbivores, like voles and lemmings, typically have roughly 4-year cycles; whereas larger ones, like muskrats, have 9–10-year cycles. Predators that prey on these creatures then tend to have population cycles of the same periodicity, and it's tempting to see these as the result of interactions of the Lotka–Volterra type. But reality is almost certainly not that simple. Take, for instance, the interactions between snowshoe hares and their predators the lynxes in eastern Canada. This is one of the few systems for which there are long records, because both animals have long been captured for their pelts by trappers for the Hudson Bay Company. Records of the number of fur catches have been kept since 1845, and if one assumes that the trappers always catch a fixed proportion of the population, the ups and downs of the catches should reflect those of the populations as a whole.

The numbers of both lynx and hare catches oscillate with something close to a 10-year cycle (Fig. 9.2), with the two oscillations slightly out of step—as predicted by the Lotka–Volterra scheme (Fig. 9.1). But if we look at the records closely, we can see that sometimes the predator cycles precede the prey's, implying that the hares are eating the lynxes! What is more, the life cycle of the lynx is such that its population grows considerably more slowly than that of the hares, and under these conditions Lotka–Volterra cycles are not expected because the predators cannot expand fast enough to overtake and control the prey population.

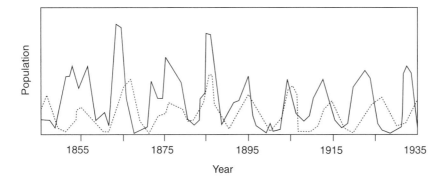

Fig. 9.2 Roughly periodic fluctuations in population size are evident in the records of lynx and snowshoe hare catches kept by the Hudson Bay Company since the mid-nineteenth century. Both the predator (lynx: *dashed line*) and the prey (hare: *solid line*) populations oscillate with a periodicity of about 10 years. Are these Lotka–Volterra cycles? Probably not.

Ups and downs

The Lotka–Volterra mechanism may be too simplistic, but modified versions that include more realistic assumptions will also generate oscillating populations. So for many years it was assumed that the fundamental character of simple predator–prey interactions is indeed oscillatory. But in the mid-1970s, as the discovery of chaos was blooming within fields as diverse as meteorology and economics, ecologists began to realize that even simple models of population dynamics display an unguessed richness and complexity. Variants of the Lotka–Volterra scheme of coupled equations exhibit the full range of instabilities that we saw earlier for oscillating reactions—if the rates are changed, such that the predator and prey populations become ever more acutely sensitive to fluctuations in each other's numbers, the solutions display period-doubling bifurcations that lead to increasingly complex periodic oscillations and ultimately to chaos—to fluctuations in population density without any apparent regularity at all.

What is more, you don't even need a predator to destabilize a population and tip it into oscillatory cycles: overcrowding alone will do the job. That even the simplest of population models can show dramatic and unpredictable ups and downs was demonstrated in the 1970s by Robert May, then at Princeton University, and George Oster at the University of California at Berkeley. They took a close look at a deceptively simple mathematical model of a population that breeds seasonally to produce generations that do not overlap. Many insect populations are of this sort. The mathematics of this model are outlined in Box 9.1, but you needn't trouble yourself with them if you are not mathematically inclined; it is enough to know that the size of each generation grows in proportion to that of the previous generation when the sizes are small, but is inhibited by

overcrowding when the size approaches some critical threshold. A relationship of this sort can be considered to apply not only to insect populations but also to economic phenomena, such as the dependence of the price of an item on its quantity, and to some situations in the social sciences. The crucial point is that the equation is *non-linear*—cause (the size of the preceding generation) and effect (the size of the ensuing one) are not related in direct (linear) proportion to one another.

The model shows markedly different behaviour for different sensitivities of each successive population to the size of the previous one. This sensitivity is determined by a single parameter a (see eqn 9.4)—the only adjustable parameter in the model. For values of a less than 1, the population simply dies out, because the reproductive success is not high enough. For a value of a between 1 and 3, the population settles down to a steady value—growth never becomes so great that it overwhelms the resources. But when a exceeds 3, things get complicated. The population first oscillates in size between greater and smaller values in successive generations. Then it displays a series of period-doubling bifurcations as a gets bigger, so that the cycles repeat every two oscillations, every four, every eight and so on (Fig. 9.3). Finally, when a is larger than 3.57, the fluctuations appear to be irregular—chaotic, in other words.

Ecologists had long known that real populations undergo irregular fluctuations in size that look random. But they had assumed that these were the result of the unpredictable influences of the environment—changes in the weather, in crop yields and so on. Such influences undoubtedly play a role in introducing randomness to population dynamics, but what May and Oster showed was that there can also be an intrinsic chaotic unpredictability in the size of populations, irrespective of external factors.

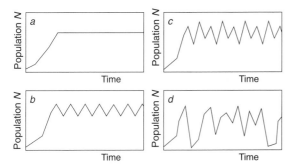

Fig. 9.3 A simple model of a population that grows exponentially until limited by overcrowding (eqn 9.6) generates complex behaviour. For different values of the 'sensitivity' parameter *a*, the model can produce a steady-state population (*a*), an oscillatory one (*b*), a period-doubled oscillation (*c*) or irregular, chaotic fluctuations (*d*).

One of the consequences of this behaviour is that the effects of perturbations to a population can be hard to predict. This, in fact, is what prompted Volterra to study the problem in the first place: he was attempting to understand why the relative proportions of predators and prey in fishermen's catches altered when the intensity of fishing activity altered (during wartime, for instance). The simplistic expectation would be that, while less fishing would deplete the fish stocks less rapidly and so lead to greater numbers in each catch, the

effect would be the same for predators and prey. But in a nonlinear system that need not be so, since the effects on future populations of a sudden decrease in both predators and prey (due to fishing, say) depend on the relative rates with which the two populations recover from this change, and will not generally be identical.

Claiming your patch

Volterra's analysis assumes that the predator and prey populations are always well mixed, so that any changes take place equally throughout the system. But real populations are usually patchy: at any instant, there is likely to be some clustering of creatures separated by more sparsely populated regions. This is not only because of chance fluctuations around the average population density, but because of geographical features, such as vegetation differences, that will influence the way in which the creatures distribute themselves over the terrain. As the BZ reaction indicates, small variations in density can have big consequences in non-linear 'reacting' systems, and the influence of a local disturbance can propagate to distant regions. In particular, the combination of oscillatory behaviour and inhomogeneous spatial distributions can lead to pulsed travelling waves that emanate from a source region and pass through the

---

Box 9.1: Complex behaviour from a simple population model

Let's assume that each insect in a population begets a certain average number (say, *a*) of offspring. Then the number in each generation *i* (N_i) determines the number in the next generation (N_{i+1}) via direct proportionality:

$$N_{i+1} = aN_i. \tag{9.4}$$

This relationship on its own gives rise to exponential growth of the population. But as Thomas Malthus appreciated, that kind of boom can't go on for ever—there is always a limit on resources. So we can assume that there is some maximum size of population that the environment can support, called *carrying capacity* and denoted here as *K*. The closer *N* gets to this maximum, the more the growth of the next generation will be curtailed. This can be represented by adding to equation 9.4 a factor $(1 - N_i/K)$, which is a measure of how close *N* is to *K*. The dependence of each generation on the previous one then becomes:

$$N_{i+1} = aN_i(1 - N_i/K) \tag{9.5}$$

which can be made simpler by introducing the scaling factor $X = N/K$, so that the equation is then:

$$X_{i+1} = aX_i(1 - X_i) \tag{9.6}$$

This is commonly called the logistic difference equation. How does it work? You can see that, once we scale the size of each generation *N* by the factor *K*, the scaled population size *X* must always remain below 1: if $X_i = 1$, then X_{i+1} goes to zero, so that the population collapses. Let's say that the population starts at a size $X_1 = 0.5$. Then the next generation, X_2, is of size $0.5a(1 - 0.5) = 0.25a$. So if $a = 2$, $X_2 = 0.5$; in other words, the next generation is the same size as the previous one. If $a = 3$, then $X_2 = 0.75$—the population grows. You might like to carry this iteration further for different values of *a*, to see what happens.

whole system. Once we take into account the fact that it takes a finite time for a perturbation to the populations in one region to propagate to a distant region—determined by the rates at which the animals move from place to place—our simplified Lotka–Volterra scheme becomes a reaction–diffusion mechanism, and it raises the possibility of spatial patterning in the number densities of predators and prey. Within the framework developed in Chapters 3 and 4, we can regard the rabbits as the *activator* species that multiply locally, and the foxes as *inhibitors* that act over long ranges: equation 9.1 is the activation step (the more rabbits there are in any given region, the more rapidly they multiply); but this population boom is inhibited over longer ranges by the foxes stalking through the land (eqn 9.2). By framing these models as activator–inhibitor schemes, we raise the possibility of finding stationary (Turing) patterns in the distributions of predators and prey, if their relative diffusion rates are conducive.

But although the possibility of travelling waves and spatial patterning in predator–prey models has been appreciated ever since the work of biologist Ronald Fisher in the 1930s, it is only in recent years that the idea has gained much acceptance. This is partly because there are now well-established computational approaches, such as the use of cellular automata, for elucidating the kinds of patterns that reaction–diffusion systems can generate; but it is no doubt also a consequence of the mutual support and impetus now afforded by studies of spontaneous pattern formation in many different systems. Population biologists can now cite the appearance of patterns in physiological systems like the heart (see Chapter 3) and in bacterial colonies (see Chapters 3 and 5) in support of the proposal that such mathematically defined structures can indeed arise even in living systems.

What these studies of spatial dynamics in ecosystems have revealed is that patchiness becomes of central importance when the interactions between communities are non-linear. This discovery has overturned much of the accepted wisdom about how populations distribute themselves in the wild. Field biologists who go out and measure the densities of populations in their natural habitat have long found that these can vary tremendously from place to place. But such variations were assumed to be the result of 'noise', of randomness in the environment, and were therefore regarded as a nuisance that simply obscures the underlying 'true' dispersal behaviour of populations. Talking about the particular predator–prey system of parasites and their hosts, the biologist Peter Kareiva from the University of Washington said in 1990 that:

> As recently as a decade ago, any field ecologist who recorded widely scattered rates of parasitism bearing no relationship to the density of hosts would probably have shelved the data as useless … but we now know that such 'disorder' can be a source of 'order' in species interactions.

What he meant by this is that apparently random patchiness in populations can in fact be *necessary* to allow predators and prey to coexist in a stable state. This patchiness can itself be an intrinsic aspect of the interaction, not a result of superimposed noise.

In the early 1990s Michael Hassell of Imperial College in London, working with Robert May and others, cast much light on this role of spatial variability in interactions between parasitoids and their hosts. Parasitoids are a particularly nasty kind of parasite: they are insects that lay their eggs in (or close to) the host's body, and the parasitoid larvae devour and kill the host once they hatch. So parasitoids are the predators, and their hosts are the prey.

Host–parasitoid interactions in nature can display periodic oscillations much like those seen in standard predator–prey systems (Fig. 9.4). There are some subtle differences from true predation, but nonetheless the

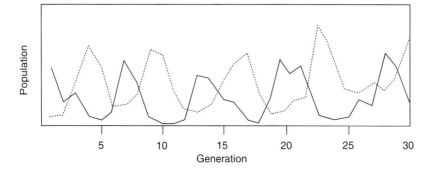

Fig. 9.4 Periodic variations are evident in successive generations of azuki bean weevils (*solid line*) and the wasp parasitoids (*dashed line*) that prey on them. This is a special case of predator–prey interaction called a host–parasitoid system. Whether these are true Lotka–Volterra cycles is again not clear.

interaction can be described by a mathematical model similar to the Lotka–Volterra scheme. But when the two populations are assumed to be distributed uniformly across a landscape, such a model generates oscillations of ever-increasing amplitude. This is an unstable outcome, which means that eventually the hosts and then the parasitoids that feed on them are driven to extinction. The implication is that, according to this model, the host and parasitoid populations can't coexist with each other in the long term.

But what if the parasitoids are distributed in an uneven manner? Hassell and colleagues showed that if the variability in the density of parasitoids searching for hosts was great enough, the two populations could both manage to persist indefinitely. For a sufficiently variable parasitoid distribution, there would always be regions in which the hosts would escape predation. Within this picture, the spatial patterning (patchiness) seen in nature is not just noise scattered over the underlying population dynamics, but an essential stabilizing factor.

In 1991 Hassell, May and Hugh Comins took this idea a stage further by investigating the *kinds* of spatial patterning that host–parasitoid interactions could support. They modelled the ecosystem using a cellular automaton: the environment consisted of a square grid of cells, each of which contained a certain number of parasitoids and hosts. The ecosystem evolved in a series of time steps, and at each step two processes took place. First, the number of parasitoids and hosts in each cell changed according to the mathematical equations that describe the breeding of both populations and the killing of hosts by parasitoids. Second, both hosts and parasitoids could move to neighbouring cells: at each step, some fixed fraction left each cell and became distributed evenly amongst its eight neighbours. So there is both reaction (multiplication of both species, and killing of hosts) and diffusion (local cell-to-cell movement) in this model.

When started from a patchy distribution of hosts and parasitoids, the cellular model produces a variety of spatial patterns, which depend on the rates of diffusion—the fraction of creatures that depart from each cell during each time step. For a certain range of diffusion rates, dynamic spiral waves appear in the population densities (Fig. 9.5a)—something that other predator–prey models had not previously shown. For other values of the diffusion rates, disordered, chaotic and constantly shifting patterns are seen (Fig. 9.5b). And if the parasitoids disperse much more quickly than the hosts, the populations can 'freeze' into a crystal-like

lattice of small patches spaced at roughly regular intervals and containing high densities of hosts, surrounded by large regions heavily populated with parasitoids (Fig. 9.5c). These are comparable to Turing structures generated by long-ranged inhibition. The point about all of these patterns is that they represent stable states—even though the chaotic patterns change constantly, we never find the population collapses that take place in this same model if either the populations are uniform or if cell-to-cell migration (diffusion) is not included. So again the spatial patterning here has the non-trivial ecological consequence of allowing otherwise unstable predator communities and their prey to survive in the same environment through a game of predatory hide-and-seek. This behaviour seems to be borne out by laboratory studies of predatory mites and their prey conducted by biologist Carl Huffaker of the University of California in 1958. He found that by imposing patchiness on the mite populations and maintaining it by restricting the mites' freedom to move around (something that involved a maze of Vaseline barriers placed amongst the food), the predators and prey could coexist for almost seven times as long as they could if the mites were unobstructed (so that the patchiness could be smoothed out).

There is at least one important lesson in these discoveries for our attempts to manage wildlife habitats: space matters. Some ecosystems need space to spread over, so that they can organize themselves into patchy communities that coexist where uniform ones cannot. The more we carve up the environment into isolated parcels by building roads and other barriers to the dispersal of species, the more we inhibit the opportunity of populations to use spatial patterning as a means of survival.

Self-organized community structures are in fact not at all uncommon in a wide range of ecosystems. Many animal and plant species gather together in patches, a distribution that is called *contagious*. Fish and plankton, for example, form schools, while plants such as alpine shrubs colonize a uniform ground in clumps of roughly the same size. On the other hand, plants that need a lot of root space, or animals that are strongly territorial, such as song birds, will distribute themselves so as to stay an optimal distance from their neighbours, a situation said to be *negatively contagious*. In the former case it is as if each individual in the community is attractive to others, and in the latter case as if they are repulsive. We can see these same kinds of quasi-regular distributions in non-living systems of particles or domains

Fig. 9.5 A cellular–automaton model of host-parasitoid interactions produces complex spatial patterns when patchy initial distributions are imposed. These include: (a) spiral waves, (b) chaotic patterns, and (c) stationary, almost regularly positioned islands of prey amongst a sea of predators. (Images: Michael Hassell, Imperial College, London.)

that attract and repel one another, and I showed an example in Fig. 2.21. For some species, there is an ideal balance of attraction and repulsion: there is safety in numbers (the predator is less likely to pick *you* if you're one of a crowd) but too great a number incurs a risk of overcrowding and depletion of local resources. In striking a balance between these two factors, creatures may find themselves aggregating into small clusters that are more or less periodically spaced. This is seen, for example, in the pattern of nesting colonies of the bluegill sunfish, which form self-organized regular clusters of about 150 individuals on lake beds.

A lot of noise?

That anything so organized as a spiral wave could arise spontaneously in natural populations is too much for many ecologists to accept, and models like the cellular automaton of Hassell and colleagues have been dismissed by some as little more than a means of generating pretty patterns on a computer. One criticism that has been raised is that the models are purely deterministic—nothing is left to chance. In real ecosystems there are randomizing elements, such as variations in landscape features and vegetation, which, say the critics, would wash out any elegant patterns like spirals. But Hassell and colleagues, as well as Graeme Ruxton and Pejman Rohani at Cambridge University, have shown that spiral patterns can persist even when a strong degree of random noise is injected into these models (Fig. 9.6).

Given, however, that population dynamics are non-linear, and that non-linear systems can be highly sensitive to small perturbations, it is hard to make any general predictions about the effect of the random environmental noise that is inevitably present in any ecosystem. Noise may inject a degree of fuzziness that washes out the finer points of ideal model equations—for example, it can reduce to unobservability some of the later period doublings in the oscillatory regime of the logistic equation. But in other cases noise seems to do more than smear out the ups and downs of a population—it can radically alter the whole picture. For example, Kevin Higgins of the University of California at Davis and colleagues have shown that the large fluctuations in population size of the Dungeness crab off the North American west coast, which show some indication of a 10-year periodicity, can be reproduced by a mathematical model of the ecosystem dynamics only when small but significant random environmental perturbations are injected into the equations. Without the noise, the same model predicts a stable population size, which is quite different to what is observed.

Similarly, Spanish physicists Ricard Solé and José Vilar have suggested that noise is essential to account for the *spatial* patterns in some predator–prey communities. They have looked at the patchy distribution of plankton in the sea. Here the predators are the zooplankton, microscopic animals whose prey are the phytoplankton, the sea's tiny plant life. The zooplankton can swim around and so have a faster diffusion rate than their prey, which are simply carried passively by the ocean currents. Both predator and prey are distributed in a patchy fashion, but the phytoplankton patches are bigger than the zooplankton patches (Fig. 9.7a). Deterministic activator–inhibitor models of this ecosystem, based on Lotka–Volterra-type equations, give rise to blotchy Turing-type distributions of the two communities—but with precisely the opposite characteristics, the predator patches being larger. Solé and Vilar showed that by adding a random, noisy element to the distribution of predators in the activator–inhibitor scheme they could obtain just the kind of patchiness found in reality (Fig. 9.7b, c). You might think that this is no surprise—that noise is bound to make the predators more patchy. But the fact is that noise is normally a *smoothing-out* influence, since it disrupts spatial patterning by tending to impose an average blandness. Here it apparently has a very different effect.

Although they may be robust against noise, spiral waves appear only for rather specific conditions in the cellular model of Hassell and colleagues. If the parasitoids disperse to neighbouring cells more rapidly than the hosts (which is likely in practice), then disordered, chaotic patterns are favoured instead. So it might be very hard to find cases of spiral patterning in nature; but a search for chaotic patterns could be more fruitful. What does such a search reveal?

Well, these ideas of spontaneous patterning are still new enough in population biology that no one has yet made a serious attempt to verify them in field studies. Although we know that natural populations do tend to be distributed highly unevenly, it will be no mean feat to distinguish chaotic 'intrinsic' patterns from the variability imposed by external noise. Were the patterns seen in Carl Huffaker's experiments chaotic, for instance? No one has yet looked.

Strategic planning

The interaction between a predator and its prey—between the fox and the rabbit—lacks much sophistication, at least in the model described above. If the fox

a

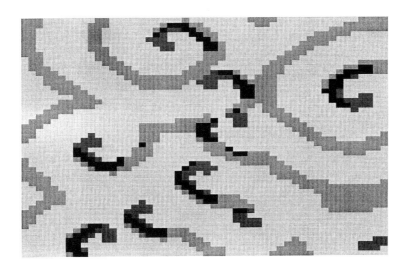

b

Fig. 9.6 Spiral waves that appear in host–parasitoid models (*a*) can persist even when there is an appreciable element of randomness (noise) in the system (*b*). (Images: Pejman Rohani, University of Cambridge.)

finds the rabbit, it's lunchtime. If the rabbit is found by the fox, it's curtains. Of course, in reality the rabbits will not sit around passively while a fox strolls up to eat them; but the fact that they'll try to escape needn't influence our model in any major way. We can simply assume that on average, foxes encountering rabbits will catch them with a certain fixed probability. This prob-

ability merely affects the rate in equation 9.2 of the Lotka–Volterra scheme.

What we *don't* expect is that the rabbit will persuade the fox of the unethical nature of carnivory. Nor will the rabbit suddenly access a furious boldness and eat the fox instead, or cajole its colleagues to beat off the fox through weight of numbers.

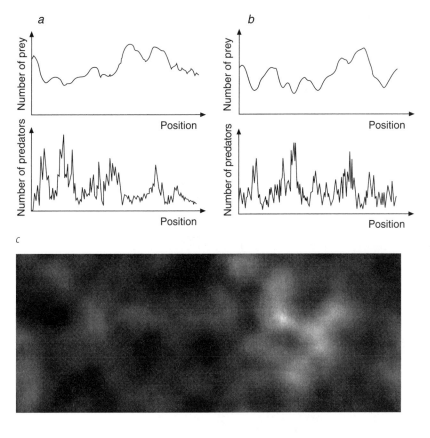

Fig. 9.7 (*a*) A transect through the ocean shows that both phytoplankton and their predators (zooplankton) are patchily distributed, with the latter more patchy than the former. (*b*) A reaction–diffusion-type predator–prey model can capture this kind of patchiness only when it incorporates noise, reflecting the randomizing influences in the environment. (*c*) A snapshot of the two-dimensional distribution of prey (*white*) in this model. (Image (*c*): Ricard Solé, Universidad Politecnica de Catalunya, Barcelona.)

Yet in human interactions, almost anything is possible. When we interact with our fellows, we are faced with choices. We don't, on the whole, eat one another, but we certainly do exploit each other for our personal gain. Yet we are also capable of showing much philanthropy: we might agree to cooperate to the benefit of us both, or we might even accept some personal loss for the good of others.

We may like to think that, by exercising these choices, and in particular by being prepared to make personal sacrifices for the greater good (or even just for the good of our neighbour), we are demonstrating our ability to rise above the brutish world of the wild, where the fittest survive and the meek are shown no mercy. But other animals show cooperative and selfless behaviour too. Vampire bats, unfairly caricaturized as the most demonic of beasts, will sometimes share their bounty of blood with their fellows. Sticklebacks will act collectively to investigate potentially dangerous intruders, some even putting themselves at greater risk than the others

by volunteering to reconnoitre ahead of the pack. Chimpanzees will painstakingly groom one another.

Tempting though it is to anthropomorphize this behaviour, there is very little likelihood that ethical choices are involved. Animals act in the way that will serve their interests best. Many kinds of cooperative or apparently selfless behaviour are surely programmed into the animal's genetic make-up, because in the end these seemingly generous acts are beneficial to the chances of the individual's genes being propagated. This is *not* necessarily the same as benefiting the creature's chance of engendering offspring, because sometimes it is enough to know that your kin will benefit from your actions even if you don't. Your kin carry much of the same genetic make-up, and you might improve the chances of passing on your genes to future generations by making personal sacrifices for the good of your family. So genes that encourage altruism towards one's kin can be favoured by natural selection. This is part of the 'selfish gene' idea that Richard Dawkins has done so much to popularize: creatures are subject to genetically determined impulses, which act to enhance the genes' replicative success without regard for the whole organism.

To what extent our own actions are determined by this kind of genetic 'kin selection' is rich ground for controversy, but for the present purposes I want to consider a different kind of motivation for complex behaviour in interactive populations. Both we and (as we'll see later) some other animals often help or cooperate with others with whom we have no apparent kinship at all. I personally believe that humans generally do so for reasons that are not susceptible to rational analysis; but because this kind of behaviour is also seen in the animal world more widely, there is good reason to suspect that it *can* have a basis in 'rational' self-interest. Creatures that are habitually altruistic without receiving some recompense will not survive for long.

So why should selfish individuals be nice? One answer has been supplied by the discipline of game theory, which considers interactions that present the opportunity for individual choice. Game theory is the almost single-handed invention of the Hungarian physicist John von Neumann, who in the 1920s began an attempt to develop a mathematics that describes parlour games such as poker. One of von Neumann's earliest conclusions was alarming for those who enjoy games: chess, he proclaimed, is not a true game. Mathematically speaking, he said, a game must involve some element of uncertainty about the opponent's intentions. In poker, the 'game' element comes not from the question of who has the best hand, but from the business of who can persuade the other players that *he or she* has the best hand. The trouble with chess is that there is always a best next move. You'd have to have a brain many times more powerful than the best supercomputer to work it out, by playing through all the possible future sequences—but the best choice exists, and a purely rational player would always take it. Then, two such players will always know what the other will do next, and the game might as well be played by automatons. But automatons are no good at bluffing.

Von Neumann laid the foundations of game theory in 1944 with the publication of the book *Theory of Games and Economic Behaviour* with his colleague Oskar Morgenstern. Their work has gone on to influence economists, politicians and sociologists, but it has also made itself felt in the realm of behavioural biology.

What have games got to do with animal behaviour? In short, the games that von Neumann's theory considers are those where the players (generally just two) compete to maximize their gain. The games have a pay-off for winners and a penalty for losers. It's a cut-throat affair every bit as ruthless as Darwin's theory can so easily appear.

But what game theory shows (which Darwinism *per se* does not) is that maximal ruthlessness and self-interest does not always give the best return. There are some games in which it pays to cooperate, even if both players are determined to triumph over the other. Some of these are games that we can find being played out in nature.

The most notorious of them, called the Prisoner's Dilemma, developed from studies of perplexing games by Merrill Flood and Melvin Dresher at the Rand Corporation, a US governmental think-tank, in the early 1950s. The game is perplexing because it presents a paradox: the 'best' way for rational, selfish opponents to play doesn't give either of them the best result. Although Flood and Dresher initially conceived of it in somewhat different terms, the standard scenario for the Prisoner's Dilemma now runs as follows. Two people have been detained on suspicion of having committed a crime. The evidence is not watertight, but is sufficient to guarantee a 1-year conviction for both. However, if one of the prisoners were to testify to the guilt of the other, that would provide sufficient evidence to send the other down for 5 years. This would be a much more satisfactory outcome for the police, and so they offer to let each prisoner go if he testifies against the other. But if *both* prisoners shop the other (thus implying their own

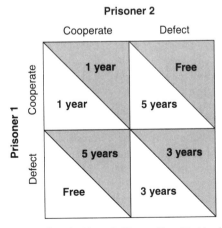

Fig. 9.8 Pay-offs in the Prisoner's Dilemma. The white triangles show the sentences for prisoner 1, and the grey triangles those for prisoner 2. What choice should each make?

innocence), that would suggest that one or both are lying, and the punishment for this deceit is set at 3 years' imprisonment for them both (Fig. 9.8).

What should they do? Well, the first prisoner is naturally tempted to shop his colleague, which offers the reward of freedom—he has no qualms about incriminating the other. This choice is termed *defection* within the jargon of the Prisoner's Dilemma. But he knows that this same temptation will be presented to his colleague. If his colleague is going to defect—well then, he'd be crazy not to, since otherwise he runs the risk of getting 5 years while his colleague walks away free. So he should defect, right?

This is *surely* the best thing to do. Whatever his colleague does, he'd be better off defecting. If his colleague doesn't defect (if instead he *cooperates* with his companion by not professing the other's guilt), then defection by the first prisoner will buy him his freedom, and his colleague goes down. But if his colleague does defect, he had better make sure he defects too, since that'll get him 3 years imprisonment instead of the 5 years he'd get by cooperating. So yes, he should defect.

The second prisoner, being equally rational, figures the same way too. So they both defect, and get 3 years each. But wait—if they'd both cooperated, they could *both* be better off, getting just a 1-year sentence. So why didn't they both cooperate instead?

The infuriating thing about the Prisoner's Dilemma is that rationality seems to frustrate the mutual good. The reason is that the game includes temptation: the players can see very plainly that there is an outcome that works for the mutual good (both cooperate), but they are tempted by an outcome that offers each individual an even better deal: freedom, instead of a 1-year sentence. They don't even have to succumb to this temptation themselves; just the fear that the other player will succumb is enough to drive each one of them to defection, since otherwise they could end up becoming an exploited sucker.

So the best way for a rational player to play is to always defect, since this way he is better off whatever his opponent does. That's a depressing conclusion: it implies that we're always best advised to try to exploit those with whom we interact. It doesn't make for the best of all possible worlds, but it would be a lot worse for *us* if we went round cooperating all the time. When in the 1950s and 60s the USA and the Soviet Union stood facing each other off with the threat of nuclear attack, the Prisoner's Dilemma was much talked about by military policy makers. It seemed to offer an unwelcome

message: it's best to stockpile nuclear arms. OK, so that consumes huge amounts of money and runs the risk of global destruction, but if you don't and the others do, you stand to be exploited. No wonder the rhetoric of 'better dead than Red' temporarily gripped the West.

But the story is not necessarily so disheartening. In real life, we don't just interact once with our fellow beings—we are likely to do dealings with them again and again. We replay the Prisoner's Dilemma many times, if you will. This makes a critical difference to how we should play.

But surely the most rational way to play a game can't alter simply as a result of playing it repeatedly? Well, it can. The whole basis of the dilemma revolves around temptation and trust. If we play just once, our choice is clear—defect, because you'd be a sucker to trust the other person to cooperate. But if we keep replaying this same pattern, sooner or later you would expect rational opponents to stop, scratch their heads, and say 'wait a minute—wouldn't we be getting more long-term advantage if we both decided to cooperate?'

It's clearer to illustrate this by inverting the payoffs so that each player stands to gain something, rather than lose or at best break even, by playing. In other words, points are awarded on each round, and the idea is to maximize these over many iterations of the game. The payoffs in Fig. 9.9 present the same options as the sentences in Fig. 9.8: you do better to both cooperate than to both defect, but you do best of all if only you defect. Distrustful defectors might go for several rounds before

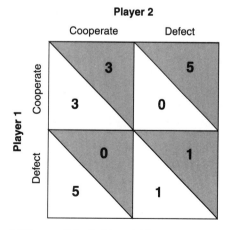

Fig. 9.9 The pay-offs for the Prisoner's Dilemma are more easily kept in account by making them rewards rather than punishments. Here the points awarded entail the same range of choices and outcomes as the penalties indicated in Fig. 9.8.

realizing that, if only they'd cooperate, they could each be earning three points per round instead of one.

So is it always better to cooperate in the iterated Prisoner's Dilemma? That would make for a nice society to live in; but unfortunately, it would also be one that is utterly defenceless against exploiters. If you were to decide to cooperate in all interactions, no matter what the outcome, then you'd be a positive magnet for anyone looking to gain by defecting. They'd gleefully take away the highest payoff from each interaction, while you get nothing.

What, then, *is* the best way to behave in a society based on repeated Prisoner's Dilemma interactions? If there is a single answer to that question, no one has yet found it. What we do know is that some strategies are better than others—but no single strategy seems to be best in all circumstances. This conclusion has emerged from Prisoner's Dilemma tournaments orchestrated by Robert Axelrod, a political scientist at the University of Michigan. In 1980 Axelrod announced a competition in which he invited all comers to submit strategies for playing the iterated Prisoner's Dilemma, and then he pitched all of the contestants against one another on a computer. Each strategy played against every other, and the winner was the one that emerged with the highest points.

In the first tournament, Axelrod received 14 entries. Some strategies were fairly simple; others were highly complex, involving the calculation of probabilities and

so forth. But the simplest of all was the clear winner. It was submitted by Anatol Rapaport from the University of Toronto, and he called it Tit-for-Tat. Its principle was: cooperate in the first round, and then do whatever your opponent did in the previous round.

The reason Tit-for-Tat is so effective is that it can be cooperative (for the mutual good) without being easy to exploit. If it is pitched against a completely cooperative opponent, both cooperate for the duration of the bout (Fig. 9.10). If pitched against a habitual defector, it loses the first round and then matches defection for defection thereafter. So it loses to its opponent in that bout (because of the first round), but it goes on to do much better than a defecting strategy when playing against nicer but not naive strategies—like itself. The strange thing is that Tit-for-Tat *never* does better than its opponent in any individual bout, but it does better than them all overall.

Or does it? One drawback with Tit-for-Tat is that it can get locked into cycles of recrimination. Suppose it plays another Tit-for-Tat strategy that is modified to try its luck just once. The two are happily cooperating, and then suddenly the opponent slips in a defection. So the next round, the pure Tit-for-Tat defects too. The opponent, which is geared to return to Tit-for-Tat after its single opportunistic defection, will respond with a defection in the *next* round, and the vicious circle repeats for the rest of the game—even though both strategies would happily return to cooperation if they 'could' (Fig. 9.11).

In reality, one could imagine these cycles of recrimination being initiated by accident. Sometimes, even with the best of intentions, we make mistakes; or we misinterpret the other person's actions. Robert Axelrod's tournaments were conducted with digital accuracy, but an iterated Prisoner's Dilemma would be a better model for real-world behaviour if it injected a small element of randomness into each strategy, so that occasionally they depart from their set rules and choose

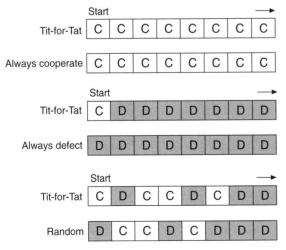

Fig. 9.10 The Tit-for-Tat strategy of the iterated (repeated) Prisoner's Dilemma begins by cooperating and then makes whatever choice its opponent made in the last round. This very simple strategy performs surprisingly well when faced with a variety of opponents.

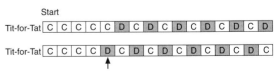

Fig. 9.11 The problem with Tit-for-Tat is that it is unforgiving. If a Tit-for-Tat-like opponent has a single aberration of exploitative behaviour (shown by the *arrow*), both players will become locked into a cycle of alternating exploitation from then on—even though both would return to mutual cooperation if they 'could'.

to cooperate or defect at random. When this happens, Tit-for-Tat's flaw becomes evident—it is too unforgiving. A single mistake between two Tit-for-Tat players dooms them both to alternating defection for the rest of the bout. Under these circumstances, a more forgiving strategy called Generous Tit-for-Tat can fare better: this cooperates while the other player cooperates, but also has a certain probability of cooperating after the other player defects—it allows the benefit of the doubt (sometimes) that the other player defected only by accident, and so breaks the recriminatory cycle.

Axelrod's initial tournaments were conducted as a round-robin: each strategy was pitched against every other in a one-to-one showdown. But a better model for an interacting community would be one in which many different strategies are constantly interacting with one another. And rather than just tallying up the points won by each strategy, we can express their Darwinian 'fitness' in reproductive terms: after each match, the strategies replicate to an extent determined by how many points each has won. When Axelrod conducted tournaments like this, he found that Tit-for-Tat was still the 'fittest' strategy, even in the face of a small (1%) error rate in the responses.

But game theorists Martin Nowak from the University of Oxford and Karl Sigmund from the University of Vienna have shown that Tit-for-Tat's intolerance of mistakes represents a weakness that, while not debilitating in Axelrod's tournaments, may show up in a larger evolutionary arena. Because of this unforgivingness, Tit-for-Tat does not fare so well amongst its own kind: in a predominantly Tit-for-Tat population, mistakes are highly costly, and an unconditionally cooperative strategy may do at least as well, because it is not exploited by Tit-for-Tat but not penalized by mistakes. Nowak and Sigmund investigated strategies that base their next move on the outcome of the previous round, so that instead of just asking 'What did my opponent do last?', as Tit-for-Tat does, they ask 'How did I fare in that last encounter?'. And their strategies also differed from those in Axelrod's competition in that their behaviour was probabilistic rather than deterministic. This means that, rather than saying 'If the previous round produced result X, then I'll cooperate', they might say 'If it produced result X, I'll cooperate with 90% probability'. This way of introducing chance differs subtly from that which allows for random errors, and reflects the possibility of a creature forgetting what happened to it in the previous exchange.

Nowak and Sigmund found that Tit-for-Tat was not the most successful strategy in these tournaments. Instead, it acted as a kind of catalyst that encouraged the population to become more cooperative but then faded away in favour of more forgiving strategies like Generous Tit-for-Tat. A population cannot afford to be overly generous while there are exploitative defecting strategies around, because these latter will just prey on the nicer ones. But Tit-for-Tat deals harshly with exploiters, paying them back measure for measure. It is kinder to cooperative ones, however, allowing them to benefit from their good behaviour. So Tit-for-Tat players act as a kind of police force, staying around long enough to deal with exploiters but disappearing when the community is cooperative enough not to need their protection. In the language of the business, Tit-for-Tat is not an evolutionarily stable strategy.

Nowak and Sigmund expected some variant of a Generous Tit-for-Tat strategy to be the fittest in these games. But in 1993 they discovered that another strategy, called Pavlov by mathematicians David and Vivian Kraines, can become dominant under these rules of play. Pavlov is not so nice: it bases its next move on the outcome of the last round, sticking to the last move it made if it gained a good pay-off but switching if the outcome was poor. To be more precise: if defection brought five points (because the opponent cooperated), it'll defect again; if defection brought one point (because the opponent defected too), it'll switch to cooperation, giving the opponent the chance to cooperate too. If cooperation brought it the sucker's pay-off of zero, it will defect on the next move; and if cooperation brought three points (because it was mutual), it'll cooperate next time around too. These option are shown in Fig. 9.12.

Last move	Opponent's move	Outcome	Pavlov's next move
C	C	3	C
C	D	0	D
D	C	5	D
D	D	1	C

Fig. 9.12 Pavlov is a strategy that could best be described as opportunistic. It will make its choices depending on how well it fared in the last round: if the outcome was 'good' (*white*), it plays the same way in the next round, but if it was 'bad' (*grey*), it switches. Pavlov appears to be nice if that is to its benefit, but it will exploit cooperators ruthlessly if given the chance.

Pavlov is not a new invention: Anatol Rapaport, Tit-for-Tat's creator, knew of it in 1965, and dismissed it as a 'simpleton' strategy. This is because, if pitched against strategies that always defect, Pavlov does rather poorly: it switches to cooperation every other round, and so gets repeatedly exploited. But in a mixed population, Pavlov is canny. It cooperates when it pays to do so (against the Tit-for-Tat police, for example), but unlike Tit-for-Tat it does not run the risk of being overwhelmed by nice strategies, such as Generous Tit-for-Tat, because it has no qualms about exploiting them with constant defection, if it is clear that this will bring no recrimination. The problem with highly cooperative populations is that, while they fare well amongst themselves, they are constantly at risk of being attacked and overtaken by defectors (which can arise by random mutations). Pavlov, however, is an exploiter that can masquerade as a cooperator when it pays to do so. And Nowak and Sigmund found that, if Pavlov has just a small element of randomness in its responses, it can even resist attack by habitual defectors.

Do real creatures show these strategies? In Axelrod's tournaments one could submit strategies that were as complicated as you like (and some were *highly* complicated); but animals (including us) do not base their interactions on the calculation of detailed probabilities or on the precise recollection of many past events—they tend to adopt very simple strategies. In this sense, Tit-for-Tat and Pavlov are plausible candidates for behavioural tendencies, since they base their choices on a simple consultation of what happened last time.

There is some evidence for Tit-for-Tat strategies amongst birds, bats, fish and monkeys. It is always important in these studies to distinguish between cooperative and sharing behaviour amongst kin, and that amongst creatures who are not closely related: as I indicated earlier, there are good reasons for the former behaviour to be genetically programmed irrespective of whether the 'altruistic' creature itself benefits from the exchange. Gerald Wilkinson of the University of California at San Diego showed in 1984 that vampire bats may share the blood that they have foraged not only amongst kin but also amongst non-kin members of the community. Significantly, he found that individual bats that behaved more selfishly could be identified and excluded from sharing by the others—just the kind of behaviour that Tit-for-Tat strategies reserve for defectors. Michael Lombardo of Rutgers University in New Brunswick saw Tit-for-Tat behaviour amongst tree swallows: he made it appear that some non-breeding

birds that were helping parents to tend their young had killed some of the nestlings. The parents responded with hostility to the 'framed' birds, but returned to a more cooperative interaction when it appeared that the framed individuals were willing to continue cooperating at the nest. (If this experiment seems a trifle unjust to the framed suspects, you might be reassured to know that they were only stuffed models.) And in a remarkable study by Manfred Milinski of the Ruhr University in Germany, stickleback fish displayed Tit-for-Tat tendencies as they investigated a predator (a pike). Using a series of mirrors, Milinski persuaded individual sticklebacks that they were accompanied in their forays by companions who would either cooperate (stay with them) or defect (swim away). The sticklebacks tended to cooperate with a cooperative 'virtual' partner, continuing to approach the predator while their partner did so; but they would defect—refusing to approach closely—if the virtual partner appeared to do likewise.

The magic carpet

So far I've talked only about well-mixed populations, in which everyone encounters everyone else. But the world is not like that, of course—and we saw earlier that for simple Lotka–Volterra-style relationships between predators and prey, spatial variability can give rise to complex patterns. What about evolutionary Prisoner's Dilemma games—do they have characteristic patterns too, when played out over space? We can already see from the discussion above that there is the potential for regional differences in populations to arise and be sustained. Cooperative strategies do well together, but do terribly amongst defecting strategies; amongst the latter, only fellow defectors can survive. So we can see the possibility of segregation between cooperators and defectors. But these divisions need not be rigid or invariant: a single defector placed amongst a cooperative colony can undermine it, while Tit-for-Tats can convert a defecting population to a cooperative one.

A naive expectation, therefore, might be to see some crude segregation of cooperators and defectors in Prisoner's-Dilemma-Land. But Martin Nowak and Robert May got something of a shock when, in 1992, they set out to study how, in the simplest of scenarios, these two types of creature dispersed across a two-dimensional checkerboard landscape. What they found were astonishing, kaleidoscopic patterns that put them in mind of Persian carpets (Plate 24). With only the simplest of rules, the strategic landscape becomes painted in complex and richly varied ways.

Nowak and May abandoned all the strategic nuances of Tit-for-Tat, Pavlov and their cousins, and chose to work with just two kinds of player: those who always cooperated and those who always defected. No player had any memory of the previous encounter; they just acted out their cooperations or defections monotonously. And everything was deterministic—there were no errors, no probabilistic changes of strategy. The rules were simple. Each square of the checkerboard grid contains a player, and each player interacts with the eight all around (or fewer for sites on the edges of the board).[*]

The payoffs from each of these interactions are counted up according to the usual rules for the Prisoner's Dilemma, and for the next round, the square is inherited by whichever of the nine (the square's original occupant and its eight neighbours) had the highest score. This simulates the reproductive advantage of the fittest competitor in that group (Fig. 9.13).

We can see that defectors have an advantage over cooperators: defectors can hold their own amongst their own kind, but they also do well (much better, in fact) when on their own amongst cooperators. Lone cooperators, on the other hand, are immediately snuffed out by defectors. So one possibility is that defectors will just take over the entire board, presenting the depress-

ing sight of an inexorable spread of selfishness. This will happen if the reward for defecting against a cooperator, designated d, is large enough ($d = 5$ in Fig. 9.9, for example). But if this reward is not too great, cooperators can gain a foothold, because mutual cooperation is more profitable than mutual defection. A cluster of cooperators can then support each other, while the defectors at the cluster's edges undermine their attempts to exploit the cooperators by their frustrated attempts to exploit each other too. Under these conditions, cooperators do better and better the more they spread, while defectors do worse and worse.

Nowak and May found that their communities could settle into states in which the patterns, while constantly shifting, would maintain a distinctive appearance. The relative proportions of cooperators and defectors in these 'dynamic steady states' reach an essentially constant value, which depends on the size of the reward parameter d. Figure 9.14 shows a pattern that results from relatively low rewards (values of d between 1.75 and 1.8). Here black squares are cooperators (C), grey squares are defectors (D), and white squares are those that have switched from C to D in the last round. We see that under these conditions, defectors don't do so

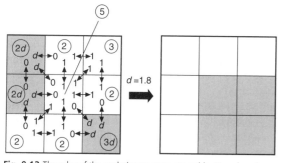

Fig. 9.13 The rules of the evolutionary game staged by Nowak and May. Each square is occupied by a contestant that competes by unconditional cooperation or defection against all its neighbours. The points for each of these interactions (either 1, 0, or a reward d for defection in the face of cooperation) are added up, and the square is colonized by a player of the same type as the one that scored highest amongst each player and all those it encountered. In the example shown here the players at the edges of the board have fewer neighbours and so fewer interactions. (Note that each square also competes against itself, to make the computation easier; but I haven't included this self-interaction here for simplicity.) White squares are cooperators, and grey squares are defectors.

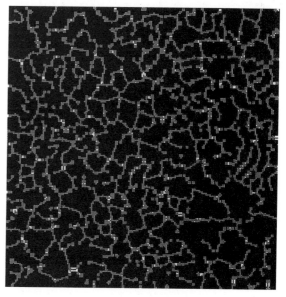

Fig. 9.14 Patterns of cooperative and defecting communities. *Black squares* denote cooperators and *grey squares* defectors. *White squares* show those sites that have changed from cooperator to defector in the last round—that is, sites where boundaries are shifting. This pattern occurs under payoff rules that favour cooperators. (Image: Martin Nowak, Oxford University.)

[*] The players also interact with themselves, since this makes the calculations easier. But much the same behaviour is seen when this rather artificial self-interaction is excluded.

well—they can just about maintain a tenuous web through the background of cooperators. Also notice that the pattern is pretty static—only a few squares change the nature of their occupants on each round.

But for a value of d greater than 1.8, something interesting happens. Then, the payoff is big enough for a two-by-two cluster of D squares to grow in a 'sea' of C, accumulating more D's around its periphery (particularly at the corners) on each round. This sounds like bad news for the C's, except that, so long as d remains below 2, the same applies to C's: a two-by-two cluster of C can support itself well enough to grow within a sea of D. So we are faced with the interesting situation where a 'critical cluster' of D can invade a C community and vice versa. The patterns in this case become much more dynamic, with blobs of C and D continually expanding, colliding and breaking up (Fig. 9.15a). Under these conditions, there is always a lower proportion of C's than D's in the dynamic steady state: specifically, the landscape contains about 32% of C (Fig. 9.15b).

For values of d between 1.8 and 2, the most startling results are obtained when one starts with a sea of C and places a single D invader at its centre. The invader can expand because it exploits all the C's around it; but within this range of d, the C's retain the capacity to fight back. The result is the symmetrical, intricate battle depicted in Plate 24, in which the deployment of troops is constantly changing. Nowak and May claimed that this conflict will eventually generate 'every lace doily, rose window or Persian carpet you can imagine'. The patterns are, in fact, fractal—features appear on all possible size scales between the limits of the grid size and the board size.

Life is just a game

For all the infinite variety of patterns here, one can pick out a menagerie of characteristic forms that tend to recur again and again—rather like the coherent structures that occasionally arise out of turbulence. These forms seem to have a life of their own—they possess certain properties, and carry out specific roles within the community. For example, one grouping of cells appears to glide across the landscape (Fig. 9.16a)—the cells themselves don't really move, of course, but the shape of these gliders is faithfully transmitted from place to place. Regions of D are often invaded by

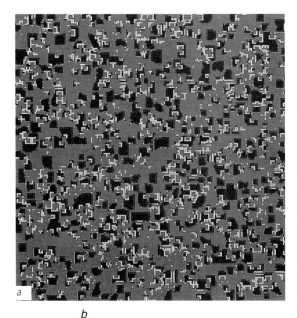

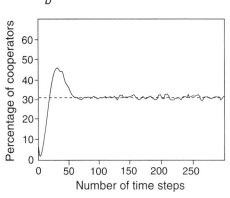

Fig. 9.15 When the payoffs for defection are slightly greater, the pattern becomes much more dynamic, with communities of both types constantly expanding and overwhelming one another (a). The grey scale here is the same as in Fig. 9.14. While the community structures change, the average proportion of cooperators and defectors remains more or less the same (b). (Image: Martin Nowak, Oxford University.)

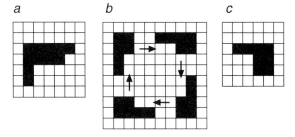

Fig. 9.16 Characteristic cooperator structures that survive and propagate in a D community include: (a) Gliders, (b) Rotators and (c) Growers. The last of these expands into a set of 'jaws' that eats its way into the surrounding defectors.

configurations of C that expand from 'growers' to eat up the D's like a set of jaws (Fig. 9.16*c*).

Nowak and May had seen such 'virtual creatures' before. Their game of cooperators and defectors is yet another cellular automaton, since the behaviour of each cell depends on that of its neighbours. Cellular automata, as we saw in Chapter 3, were the brainchild of John von Neumann and Stanislaw Ulam in the 1930s. Von Neumann was interested in the idea of automata—robotic entities—that could interact according to simple rules. His dream was to create automata that could reproduce, and which could give birth to other automata that were more complex and sophisticated than themselves. In this way, he speculated, automata might evolve into thinking machines. Ulam helped von Neumann to develop this idea into a simple, tractable model. Instead of actually trying to build mechanical devices, they envisaged a periodic array of cells that could hold *information* by existing in one of several different states. In its simplest form, each cell holds a binary bit—a 1 or a 0. The state of each cell, however, is determined by the states of those around it, according to simple, deterministic rules. In this way, information can be transmitted from place to place, as each cell readjusts its state to reflect those of its neighbours. Von Neumann hoped that it might be possible to write into these cellular automata a pattern of information that would be capable of duplicating itself elsewhere on the checkerboard lattice.

There are innumerable ways in which each cell can influence its neighbours, and the spatial Prisoner's Dilemma model of May and Nowak represents just one of the possibilities. Robust, propagating cell clusters with distinct shapes and behavioural characteristics, like those in their scheme, are also a feature of one of the most famous of all cellular automata games, called the Game of Life. This was devised in the late 1960s by Cambridge mathematician John Horton Conway. It is a grand name to call a game, of course: at that time no one was used to thinking of these checkerboard experiments as metaphors for living systems, so to call the game 'Life' introduced a provocative new perspective, even though arguably much more biologically realistic cellular automata have since been proposed.

Conway's game resembles that of the cooperators and defectors insofar as it considers two types of cell which 'compete' for dominance of the landscape. (This might seem the most obvious first choice for a cellular automaton, but von Neumann initially considered 29 cell states!) The two states are considered to represent cells that are either living or dead. The state of each cell in each round is determined by that of its eight neighbours in the previous round, according to the following rules:

1. A 'living' cell will stay alive if it has two or three living neighbours. If there are fewer or more than this, it dies.

2. A 'dead' cell will stay dead unless it has exactly three live neighbours, in which case it too comes alive.

We can justify these rules in biological terms, although the precise numbers are somewhat arbitrary. Living cells surrounded by too many other living cells die of overcrowding—they starve. Living cells surrounded by too few others, meanwhile, die of 'exposure'—you could say that they don't encounter enough others to reproduce. But groups of a certain size that surround a 'dead' cell can colonize it (make it come alive) by reproducing. OK, it takes a lot for granted about the way life works; but in Conway's Game of Life simplicity is a virtue, because it makes it relatively easy to explore the possible range of behaviour.

And that range is extraordinary. From these very simple rules spring forms and patterns that you'd never be able to predict from an analysis of the rules. The only way to appreciate the Game of Life is to play it. As an increasing number of enthusiasts did so in the 1970s, they discovered a diverse zoo of robust cellular groupings, with colourful names such as the Snake, Ship, Beehive Loaf and Pulsar (Fig. 9.17). There were also Gliders, like those found in the game of cooperators and defectors. The ways in which these denizens of the two-dimensional checkerboard world interact are suggestive of the encounters between different species—some ignore each other, some prey on each other, others

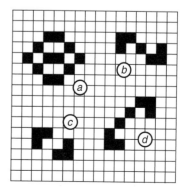

Fig. 9.17 Denizens of the Game of Life. (*a*) Honeycomb, (*b*) Long Snake, (*c*) Aircraft Carrier, (*d*) Sinking Ship.

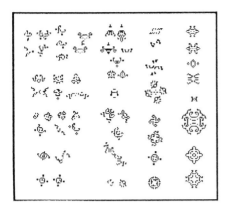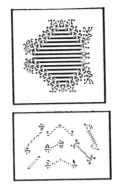

Fig. 9.18 Snapshots of 'live' (black) cells on the checkerboard of the Game of Life. These are just a few of an immense gallery of patterns that have been observed.

replicate. Meanwhile, elaborate, shifting spatial patterns of living and dead cells unfold across the landscape (Fig. 9.18). The proportions of live and dead cells fluctuate as the game progresses, and significantly, these fluctuations appear to follow a $1/f$ rule (see p. 211): mostly the variations are small, but on rare occasions one cell type or the other can sweep across the grid in profusion. In other words, this system seems to operate in a self-organized critical state. Within the Game of Life there is a whole universe of artificial life; it is a paradigm of the science of complexity, and has been described in some detail in several recent books.

There are endless variations on the rules governing the Game of Life cellular automata—you could alter, for example, the thresholds above which cells survive or die. Stephen Wolfram at the Institute for Advanced Studies in Princeton has managed to bring some structure to this multiplicity of universes by identifying four general classes of behaviour in cellular automata, of which the constantly shifting, complex patterns of the Game of Life exemplifies the most interesting. But I have little more to say about these games. They are a fascinating new facet of computer science, but it isn't at all clear that their relevance extends beyond this. Artificial-life enthusiasts are captivated by the gliders, ships and so forth for their own sake, not because they withstand any clear analogy with ecosystems in the real world. There is a great deal of social organization in the animal kingdom, but rarely does it take the form of a little cluster of individuals that progress across the landscape in a rigid, geometric arrangement. We might be reminded of the flying formations of birds, or the collective swarms of bees; but artificial life has so far had little to say to behavioural biologists about such traits.

Urban sprawl

The Game of Life might put you in mind of the models of bacterial communities described at the end of Chapter 5—there too, we saw model representations of living organisms that, confined to a cellular grid, reproduced and ran the risk of starvation through over-population. (They were also able to move from place to place, however.) Biologically motivated rules of this sort were sufficient to generate the branched community structure, which looked for all the world like a metal electrodeposit grown by diffusion-limited aggregation.

As bacteria are rather simple creatures, we can probably accept this resemblance to the patterns of the inanimate world with equanimity. How humbling, then, to find that a very similar picture emerges when we look at the shapes of our own communities—cities!

Take a look, for instance, at a map of the employment density of London (Fig. 9.19a). It's a fragmented, irregular cluster of little units, and from this perspective we can't discern any sign of the regularity that urban planners might try to impose on the city's structure. Instead, this structure is highly reminiscent of that which one sees when microscopic polymer spheres stick together in solution (Fig. 9.19b), a phenomenon that mimics the flocculation of silt in a river. In this case, the central dense cluster steadily accumulates new particles at random points on its ramified periphery in a process that is essentially a variant of diffusion-limited aggregation (Chapter 5). Alternatively we might compare the city's shape to that of a bubble formed as air is injected under pressure into a liquid-saturated porous rock (Fig. 9.19c). Is this resemblance superficial, or are there really any similarities between these growth processes?

That question was addressed in the early 1990s by Michael Batty from the State University of New York,

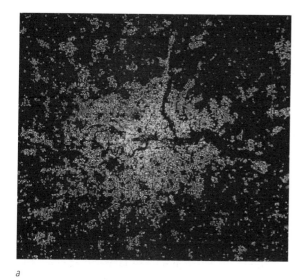

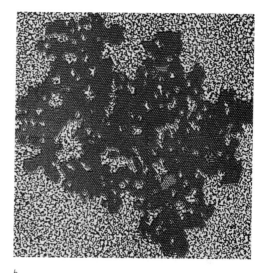

a

b

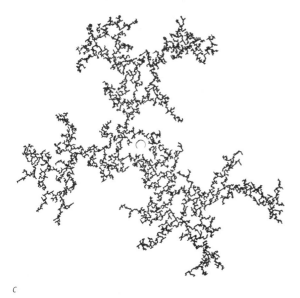

c

Fig. 9.19 The shape of the city of London, as represented by this map of employment density, is an irregular cluster of small units (*a*). Rather similar shapes can be seen in the aggregation of polymer spheres in solution (*b*), and the slow percolation of air into very viscous oil (*c*). (Images: (*a*) Michael Batty, University College, London; (*b*) Arne Skjeltorp, Institute for Energy Technology, Kjeller; (*c*) Roland Lenormand, Institut Français du Petrole, Rueil-Malmaison.)

who, with Paul Longley from Bristol University, has shown that models based on diffusion-limited aggregation (DLA) can indeed reproduce much of the characteristic form and growth behaviour of major cities.

The models proposed by Batty and Longley represent a break from traditional studies of urban structure. It is scarcely surprising that, since the major preoccupation of urban planners is with the *design* of cities, they have generally attempted to analyse city forms in terms of the effects of their efforts. That is to say, theories of urban planning have tended to focus on cities in whose form the guiding hand of human design is clearly discernible.

The trouble is, hardly any cities are like this. In spite of the efforts of planners to impose a simplistic order, most large cities present an apparently disordered, irregular scatter of developed space, in which residential areas, business districts and green areas are mixed haphazardly. By focusing on regions where planning has created some regularity (like Manhattan's grid-iron street plan), urban theorists have often ignored the fact that overall, a city grows organically, not through the dictates of planners.

This is seldom what is intended. As geometry became a dominant aspect of ancient Greek thought, its influence extended beyond architecture into the way in which the buildings themselves were arranged in settlements. The grid street plan has a history older than Greek culture, being evident also in Babylon and the cities of Assyria. It found its apotheosis, perhaps, in the towns built by Imperial Rome, since it provided a scheme by which these settlements—often starting as military encampments—could be erected quickly. The grid-iron pattern has been used extensively in North American towns and cities (Fig. 9.20a).

Also strongly featured throughout history is the radial or circular city plan, which became particularly favoured during the Renaissance (Fig. 9.20b). In the twentieth century these two forms have been augmented by other, more exploratory forms, such as the curvilinear plan, and an expansion into the third dimension in the form of tower blocks. Yet for all the bold schemes of planners, most cities have always ended up with an irregular appearance like that of modern London—an indication that they have grown 'too fast to plan'. Cities began to grow with particular alacrity during the Middle Ages in Europe, and their expansion really took off in both the New and Old Worlds in the nineteenth century. But even the 'classical' cities of Athens and Rome ended up with an organic appearance, although what we can now see of their ancient remains distorts this picture because the parts that remain tend to be those on which most forethought and money were lavished.

So it appears that cities are, and have probably always been, non-equilibrium structures. This is a notion that might seem unpalatable to planners, an indication that in the broad scheme of things they have little control. Yet the idea of an 'organic city' is an old one, though it has gained more acceptance in the past century. Cities have been compared to living organisms, with a heart (the central business district), a vascular system (transportation networks), lungs (green spaces) and so forth. The concept has been received with ambivalence, however. *Should* cities be allowed to grow this way, or should we try to impose some structure on it all? Does irregular growth mean that cities will get out of hand, that we will see an accretion of slums and a decline in public services; or do cities grow as they 'need' to, so that the imposition of a rigid geometry constrains social and functional structures and ends up creating more problems than it solves?

No doubt these are questions that will continue to be debated, and it's very unlikely that there is a universal answer independent of the social and economic context in which growth occurs. But what bothered Batty and Longley is that there weren't even any good models to *describe* how cities grow. The value of a good predictive model for urban growth would be tremendous—it would, for instance, allow planners to make accurate predictions about the likely future requirements of an area in terms of transportation, water, gas and electricity supplies, and so forth. In their 1994 book *Fractal Cities*, Batty and Longley state:

> There is a need for a geometry that grapples directly with the notion that most cities display organic or natural growth, that form cannot be properly described, let alone explained, using Euclidean geometry.

And Batty believed, as his title proclaimed, that this new geometry was to be found in the concept of fractals developed by Benoit Mandelbrot—the 'geometry of nature'.

One clue to the fact that cities have fractal characteristics lies in the observation that they obey *scaling laws*. You might recall from earlier chapters that these laws relate some property of the system (say, population density p) to some other variable (say, distance from the city centre x) through a power law: $p \propto x^n$. Urban planners have long known that urbanized areas do obey scaling laws: for example, they describe the relationship between the size (in population, say) of a settlement (a city, town or village) and the number of such settlements in a given area. There are many more small villages than there are towns, and still fewer cities, in any geographical area. This sounds obvious, but the scaling law quantifies that obviousness by assigning it some scaling exponent n, and so gives urban modellers something concrete to test their models against.

But for decades, urban theorists have been stumped by the known scaling laws describing the shapes and growth processes of cities. They could measure them, but they couldn't then figure out how these particular laws arose from the underlying economic and demographic processes that determine the evolution of an urban area. A physical model that captures the growth and scaling behaviour of cityscapes would not just provide an empirical predictive tool; it might suggest to planners what the underlying rules are that determine a city's form.

Fractal geometry provides a means to characterize both the structure of a city and the way that this changes over time. In the early 1990s, Batty and others used the methods of fractal analysis to deduce the fractal dimensions of cities from maps like that in Fig. 9.19a. They found that these span a range of values, typically between about 1.4 and 1.9: for example, for London in 1962 the fractal dimension was 1.77, for Berlin in 1945 it was 1.69 and for Pittsburgh in 1990 it was 1.78. The

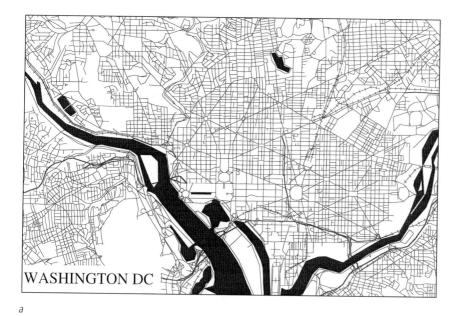

a

b

Fig. 9.20 Geometric cities: (*a*) the grid-iron plan is a common feature of North American cities, such as Washington DC; (*b*) the radial design of the Renaissance city of Palma Nuova in Italy. (Images: (*a*) Michael Batty, University College, London; (*b*) from Batty & Longley 1994, after Morris 1979.)

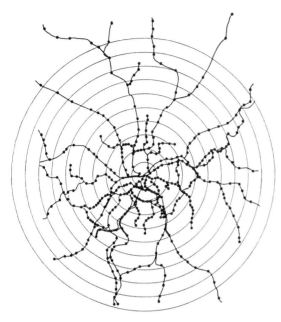

Fig. 9.21 The Paris metro is a branched network with a fractal form. (Image: M. Daoud, CEN Saclay.)

closer the fractal dimension is to 2, the more dense the city is—the more it resembles a blob that spreads without gaps over the landscape. In general, this dimension increases slowly over time, reflecting the fact that more and more of the 'free' space between centres of development tends to get filled in. In addition, two teams of French researchers have found that the transportation networks of Lyon, Paris and Stuttgart are branched fractals with dimensions ranging from only just over 1 (a very sparse network) to almost 1.9. The Paris metro and suburban rail network, for instance, has a fractal dimension of 1.47 (Fig. 9.21).

What do these numbers mean? Well, not much by themselves—the challenge is to understand how they come about, to look for a model of a growth process that reproduces the observations. Now, Batty and Longley realized that the mean fractal dimension of the cities that they and others had analysed—about 1.7—is rather close to the fractal dimension of DLA clusters, 1.71 (see p. 115). So as a 'baseline' model for urban growth, they decided to use the DLA model developed in the 1980s. In DLA, particles execute random walks until they strike the perimeter of a growing cluster, whereupon they stick where they strike. Batty and Longley suggested that maybe something similar happens as cities grow: new development units (such as business or residential neighbourhoods) are gradually added to the city with a probability that is greater at the city's perimeter (since there is more space there for the development). Of course, this highly simplistic model ignores a great deal that is important for urban development—not least, all efforts of planners to impose some order on it! All the same, the researchers found that some of the scaling laws observed in real cities, such as those describing the dependence of variables like the number or density of 'development units' on the distance from the city centre, are similar to those that pertain to the number or density of particles in a DLA cluster.

But for all that they might have similar fractal dimensions and scaling laws, DLA clusters (Fig. 5.7) and fractal cities (Fig. 9.19a) don't *look* very alike. This is a crucial consideration for disordered patterns: while, as I mentioned in Chapter 5, it is not necessarily enough to judge similarities by appearances alone, we can't neglect them either. The denser, more compact forms of cities can be more closely approximated by relaxing the 'stick-where-you-hit' rule for DLA. If the particles are allowed the chance to make a few hops around the cluster's periphery before finally becoming immobile, the result is a form like that shown in Fig. 9.19b. In effect, a model of this sort assumes that the particles are less sticky, so that they don't necessarily become attached where they first strike the cluster but only have a certain probability of sticking each time they strike. By allowing for improved packing, the model brings the resulting non-equilibrium cluster a little closer to the equilibrium form of a close-packed crystal.

But the real challenge is in finding a physical model of city growth that makes sense in terms of the processes involved—we know, for example, that urban development does not really bounce from site to site before finally coming to rest. Batty and Longley explored one such variation on the DLA model: the dielectric breakdown model (DBM) introduced in Chapter 6. You may recall that this model generates fractal electrical discharge patterns, like lightning, which can have forms and fractal dimensions very closely related to those of DLA. But the DBM is a more physically realistic model for urban growth, because it creates clusters that 'push their way out' from a central point, rather than ones that grow by accumulating wandering particles at their edges. The former is a better approximation to the way that cities expand: by a kind of pressure on new development to spread outwards and colonize the surrounding land.

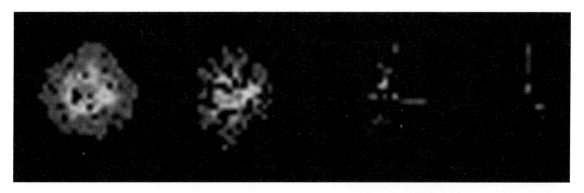

Fig. 9.22 The dielectric breakdown model (DBM) provides an attractive model for city growth because the density of its branches can be adjusted by changing a model parameter, denoted η. When η is large, the branching pattern is highly linear, with a fractal dimension close to 1 (*right*). As η decreases, the pattern becomes denser, and the fractal dimension approaches 2 (*left*). From left to right, $\eta = 0, 0.5, 2, 4$. (Images: Michael Batty, University College, London.)

But what really attracted Batty and Longley to the DBM was the fact that the form of the clusters it generates can be tuned from highly tenuous, almost linear shapes to dense, more circular ones. The DBM discharge advances from its perimeter at random, but with a probability that is higher where the electric field around the perimeter is highest (that is, at the tips). The discharge patterns can be tuned by varying how strongly the higher-field positions are favoured as the locations of further growth. As this bias gets smaller, the pattern becomes more circular and dense (the fractal dimension approaches 2), whereas if the bias is made very large then growth is completely dominated by the very first tips, and the clusters are highly linear (the fractal dimension approaches 1) (Fig. 9.22). Batty and Longley suggested that this bias could be loosely associated with the degree to which planning influences the city's growth: highly linear cities are not very 'natural', requiring a strong degree of planning, whereas cities that are entirely unplanned should resemble more closely the amorphous mass on the left of Fig. 9.22. The strength of the bias is determined by a parameter denoted η (pronounced 'eta') in the model. When η is equal to 1, the model generates DLA-like clusters; when η is zero, the cluster is dense.

Using this approach, Batty and Longley attempted to simulate the growth of the city of Cardiff (Fig. 9.23). They conducted a simulation of DBM growth constrained by the local geography: by the presence of the coastline to the southeast and the rivers Taff (to the west of the city centre) and Rhymney (to the east). The cluster was seeded from a point between these rivers. Its probability of growth became zero (sensibly enough) beyond the coastline; and the rivers too acted as impenetrable barriers to growth except at two points, where the cluster could 'squeeze' across bridges. In the model, these bridges were located where real ones exist in Cardiff. The results of the model, for different values of the probability bias parameter η (Fig. 9.23*b–e*), show that somewhat realistic approximations to the city shape can be generated when η is a little less than 1.

But physicists Hernàn Makse, Gene Stanley and Shlomo Havlin from Boston University were not persuaded that these models of fractal cities mirror the reality. For one thing, the models predict that cities form a single large fractal cluster, which is densest in the centre (around the central business district) and gets rapidly more tenuous, growing almost exclusively from the tips of its outer periphery. That doesn't sound quite right. Local areas of development commonly spring up around the verges of a city, creating little satellite clusters of population. As the city grows, these local developments get swallowed up by the sprawl. You can see several clusters of this sort beyond the edges of the main cluster in the structure of London (Fig. 9.19*a*).

The Boston physicists decided that a new growth model was needed to capture this sort of structure. They realized that in urban areas, localized development can be *correlated*. This means that, rather than new units appearing entirely at random, the development in any area is sensitive to what is happening in the immediate vicinity. In short, development attracts further development. Once two small clusters of population appear in close proximity, there is a good chance that development will spring up between them—shops to serve the new inhabitants, or local businesses keen to gain a

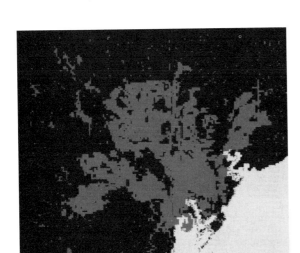

a

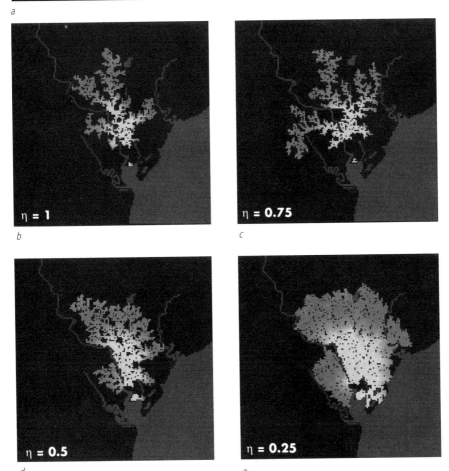

Fig. 9.23 The DBM has been used to model the growth of the city of Cardiff, which is constrained by a coastline and by two rivers. (*a*) The structure of the real city today, with the city shown as dark grey and the sea as light grey. (*b-e*) Model simulations for different values of the parameter η in the model. The best match occurs for a value of η around 0.75. Note that bridges are included to allow the city to spill over the rivers in two places. In these simulated images, earlier growth is shown as lighter. (Images: Michael Batty.)

$\eta = 1$

b

$\eta = 0.75$

c

$\eta = 0.5$

d

$\eta = 0.25$

e

foothold in an 'up-and-coming' area. To simulate this sort of local correlation, Makse and colleagues borrowed from physics the so-called correlated percolation model, developed to mimic the kind of fluid permeation process discussed in Chapter 6. Within this model, new particles (representing units of population) are added to a growing cluster at random, as in DLA; but in addition, growth in one region enhances the prospects of growth nearby, with a probability that falls off quite sharply with distance. So the addition of new particles is not fully random—it depends on what happens in the immediate neighbourhood.

This model can generate a jumbled scattering of clusters of different sizes. But real cities are firmly rooted to a core, which is usually the central business district—it has been long known that the population density tends on average to fall off exponentially with distance from this core. So to provide their simulated cities with a root, Makse and colleagues imposed the condition that the probability of adding a new unit declined exponentially from a central point. This rule on its own generates a compact, roughly circular cluster in which the population (that is, the density of particles) falls off fairly smoothly with increasing distance from the centre (Fig. 9.24a). The short-ranged correlation between units is added on top of this basic pattern, and as the degree of correlation becomes stronger, the cluster breaks up into fine-scale sub-clusters and tendrils which resemble the structures that spread from a real city (Fig. 9.24b, c).

You can probably see straight away that this shape looks more realistic than Batty's DLA- or DBM-based

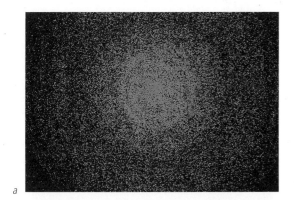

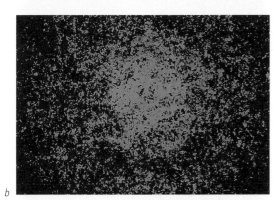

Fig. 9.24 Simulations of urban growth using the correlated percolation model. For increasing degrees of correlation between the growth units, the shape changes from more or less circular (*a*, for no correlation) to increasingly fragmented and clumpy (*b*, *c*). (Images: Hernán Makse, Schlumberger-Doll Research, Ridgefield, Connecticut.)

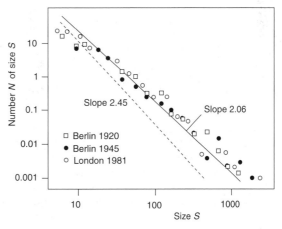

Fig. 9.25 The performance of the correlated percolation model can be tested by looking at the scaling law that it predicts for the number of towns *N* around the city that have a size *S*. The real data from Berlin in 1920 and 1945, and from London in 1981, all show a power law with an exponent of about 2.06 (which shows up on the log-log plot here as a straight line with a slope of 2.06). The strongly correlated model (*solid line*) gives the same scaling law, whereas in the absence of correlations the slope is steeper, corresponding to an exponent of 2.45. (After: Makse *et al*. 1995.)

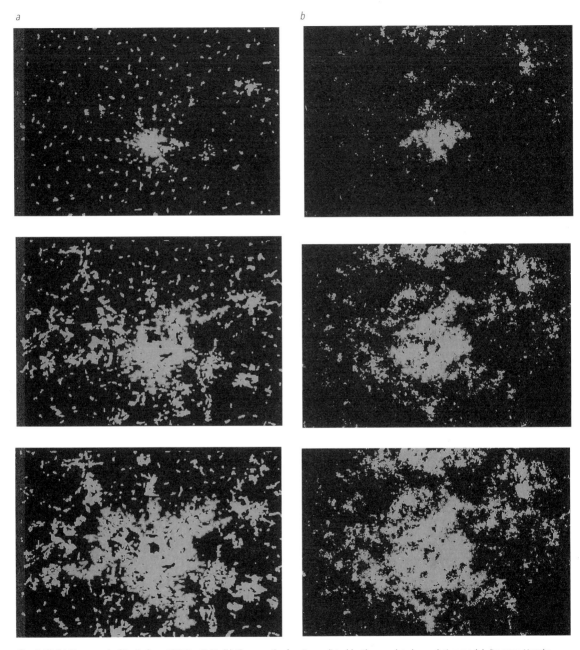

Fig. 9.26 (*a*) The growth of Berlin from 1875 to 1945. (*b*) The growth of a city predicted by the correlated percolation model. (Images: Hernán Makse.)

forms, at least for a city like London. But Makse and colleagues wanted more quantitative tests. First they looked at the scaling law relating the number of towns surrounding a large city (N) to the size of those towns (S). The models predict a scaling law in which N

increases in proportion to S^n, with the numerical value of the exponent n depending on how strongly correlated the growth is. For the completely uncorrelated model (Fig. 9.24*a*), n has the value 2.45; for strongly correlated growth, n is equal to 2.06. As we saw in the last chapter, log–log plots of these scaling relationships give straight lines with slopes equal to the value of n. The difference in slope seems rather subtle; but when the researchers plotted the real data for N and S for three different urban environments—Berlin as it was in 1920, Berlin in 1945 and London in 1981—they found that the data seemed to fall on or close to the less steep line, corresponding to the strongly correlated model (Fig. 9.25). In other words, these cities do indeed seem to show correlated growth.

As a further test, the researchers showed that their model generated a pretty good picture of how city shapes evolve over time: Fig. 9.26*a* shows how Berlin and its environs have developed from 1875 to 1945, while Fig. 9.26*b* shows the kind of growth predicted by the strongly correlated model. Not only do these look similar to the eye, but they show similar scaling relationships in terms of how the population density falls off with increasing distance from the centre of the city.

What does any of this tell us about urban growth? By showing that at least some aspects of city shapes can be reproduced by physical models, such as diffusion-limited aggregation or the dielectric breakdown model, Batty and Longley demonstrated that randomness—uncoordinated local decisions about development, akin to the uncoordinated aggregation of particles in DLA—is by itself enough to generate the characteristic clumpy, messy sprawl of cities. But in addition, they showed that this randomness is modulated by (in Batty's words) a 'deeper order'. In other words, different cities share scaling laws in common because they follow a certain class of growth process—one that includes other, apparently far simpler phenomena such as silt flocculation or electrical sparking. Says Batty, 'The time is now ripe for the new approach to cities and urban form for which we have been waiting for more than a generation'.

He points out that, while his simpler DLA and dielectric breakdown models work fairly well in describing the shapes of cities that sprung up during the early industrial era—which tended to be single clusters organized around the central business district—the model of Makse and co-workers is better suited to capturing the growth and form of post-industrial cities. In these, new communications technologies have made work in the centre of the city less prevalent, and the structures therefore tend to be less centralized. In addition, the centres of many large cities have become less desirable as places of work or residence—they have fallen prey to 'inner-city' decay, while affluent population centres evolve around the city edges. The concept of an 'edge city' is providing a new paradigm for urban life in the USA.

The success of these physical models, based on random growth influenced by local interactions, in describing the forms of cities has a salutary message for planners too. It raises the question of whether centralized planning for an 'organism' as complex as a city has any chance of succeeding. In the 1960s, planners sought to influence the way in which London sprawled into the surrounding countryside with a Green Belt policy that would restrict urbanization. Yet there is no sign that these policies have had any effect on the city's growth, which has gone right on expanding to the tune of the same scaling laws. It will take more than this, it seems, to undermine the inexorable physics of cities.

PRINCIPLES

Nature uses only the longest threads to weave her patterns, so each small piece of her fabric reveals the organization of the entire tapestry.

Richard Feynman
The Character of Physical Law

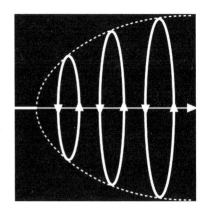

I hope you have not been holding your breath in anticipation that in this final chapter I will disclose a grand, unifying picture of pattern formation. I had better start out by saying that I shall not be presenting one, since I don't know that such a thing exists. Some physicists have in recent years encouraged an unfortunate aspiration towards grand unified pictures, but I fear we must accept that the world that we encounter, the world of real stuff that we see and touch, is far too messy for that.

At the same time, it is far more beautiful. Perhaps it is a matter of taste, but I feel that there is much more wonder in a world that weaves its own tapestry using countless elegant and subtle variations, combinations and modifications of a handful of common processes than one in which the details become irrelevant, in which a few recondite equations are supposed to provide us with all we need. To my mind, the astonishing thing about, let's say, butterfly wing patterns, is not just that we can implicate a few basic processes that lie at the heart of them all (reaction–diffusion mechanisms? morphogenetic fields?) but that small changes in the details, in the specific initial or *boundary* conditions, produce such fantastic variety. By the same token, the patterns of a river network and of a retinal nerve are both the same and utterly different. It is not enough to call them both fractal, or even to calculate a fractal dimension. To explain a river network fully, we must take into account the complicated realities of sediment transport, of changing meteorological conditions, of the specific vagaries of the underlying bedrock geology—things that have nothing to do with nerve cells.

Physicist Rolf Landauer has expressed very succinctly this need to resist over-enthusiastic attempts at universalization:

> A complex system is exactly that; there are many things going on simultaneously. If you search carefully, you can find your favorite toy: fractals, chaos, self-organized criticality, phase transition analogies [which I'm coming to], Lotka–Volterra predator–prey oscillations, etc., in some corner, in a relatively well-developed and isolated way. But do not expect any single simple insight to explain it all.

I guess it is a shame to begin a summary chapter with a caution against too much summarizing, but I think it is best that I do so. For the ideas that form the backbone of our understanding of spontaneous pattern formation seem so powerful, so all-encompassing, that they are all too often paraded as the keys to a theory of everything. Even D'Arcy Thompson would not have wanted us to believe that.

And yet what is extraordinary and thrilling is that so many pattern-forming systems have so much in common, to the extent, that by understanding one, we can predict a great deal about the others. This realization has made a delicious mockery of the traditional, rigid divisions between scientific disciplines, so that physicist, economist, ecologist, chemical engineer and geologist can all talk to one another—*and in the same language.* When this happens, something very exciting is going on in science.

And yet we've seen that many of the ideas behind pattern formation are not new. Oscillating chemical reactions were known in 1901; convection cells showed up around 1900; and Kepler perceived an underlying sixfold symmetry in the snowflake's make-up in the seventeenth century. But D'Arcy Thompson was unable to persuade most of his peers of the importance of form and pattern in the 1920s, and only in the past two decades or so has pattern formation emerged as anything like an identifiable field of study in its own right. Why so?

Surely one of the most pivotal reasons is the explosion in computer power. Many of the theoretical ideas about patterning are tough to test experimentally, since there are so many factors to bring simultaneously under control; but as we've seen in every previous chapter, computers allow researchers to perform 'ideal' experiments in which everything can be repeated exactly and complicating factors included or excluded at will. Many theoretical models are simple in principle but utterly impossible to test by manual number-crunching—the calculations would take astronomical times if done by hand. But although this increase in computer power has provided scientists with perhaps the most important technological tool currently at their disposal, I think it also serves to underline the phenomenal achievements of early researchers into complex systems, like Lord Rayleigh, Geoffrey Taylor and Andrei Kolmogorov, who had to rely on their exquisite intuition alone to deduce the essential physics of pattern-forming processes.

I believe that there is another factor, little emphasized but equally important, in the recent development of ideas in pattern formation. This is the maturation in the past two decades of a field of theoretical physics that provides much of the framework for understanding the features that accompany spontaneous patterning, such as abrupt, global changes of state and scaling laws. The field is the study of phase transitions and critical phenomena, and it is the bedrock of all of physics today. I shall say more about this discipline in what follows.

So let me now try to pull together some of the threads that have run, more or less perceptibly, through all of the previous chapters. They do not collectively constitute a 'theory of patterns' (much the same could be said of the mixed bag of concepts that are popularly touted as 'chaos theory'). Rather, these ideas are like stepping stones that lead us through the turbulent ebb and flow of pattern and form in the physical and natural world.

Competing forces

Spontaneous patterns represent a compromise. The ordered bicontinuous phases of block co-polymers and surfactants (Chapter 2) are an elegant solution to the conflicting demands of minimal surface area, minimal curvature and optimal molecular packing. The dynamic spiral waves and static Turing structures of non-linear chemical reactions result from a delicate balance between reaction and diffusion, between short-ranged activation (autocatalysis) and long-ranged inhibition. The bulbous pseudopodia of viscous fingering are the manifestation of competition between the Saffman–Taylor instability, which promotes branching, and surface tension, which limits its size scale. When anisotropy is thrown in to the balance, we get the Christmas-tree arms of dendrites, so long as noise does not overwhelm them. Vortex streets appear in fluid shear flows when the wavy instability wins out over viscosity.

Competition lies at the heart of beauty and complexity in pattern formation. If the competition is too one-sided, all form disappears, and one gets either unstructured, shifting randomness or featureless homogeneity—bland, in either event. Patterns live on the edge, in a fertile borderland between these extremes, where small changes can have large effects. This is, I suppose, what we are to infer from the clichéd phrase 'the edge of chaos', beloved of complexity enthusiasts. Pattern appears when competing forces banish uniformity but cannot quite induce chaos. It sounds like a dangerous place to be, but it is where we have always lived.

Symmetry breaking

At the beginning of the book I explained that spontaneous patterns generally arise from a state of higher symmetry, so that the patterning process *breaks* symmetry. In this way we saw (I hope) the important distinction between symmetry and pattern: high symmetry does not by any means imply the richest patterns, and indeed those that appear to us to be most striking often have rather low symmetry (such as Plate 17), or none at all (such as Fig. 5.6). On the other hand, symmetry tends to break a bit at a time, so that the first patterns that appear as a system is driven away from equilibrium are often highly symmetrical, such as the honeycomb array of Bénard convection cells. It is time in this final chapter to look at *why* a system might break its symmetry. In this respect, I will say for now only the following: that symmetry breaking is not like laying a tiled floor, or indeed like making a real honeycomb. A hexagonal cellular arrangement like that of Rayleigh– Bénard convection is not a matter of imposing cells of an arbitrary shape, one by one, on an inert medium. Rather, the medium becomes everywhere at once imbued with a 'hexagon-forming tendency', once the critical Rayleigh number is exceeded. It then takes only the minutest fluctuation to trigger an expression of this 'hexagon-ness' globally. The hexagonal array seems to rise out of the floor, you might say.

Non-equilibrium

Nearly (but not entirely) all of the pattern-forming systems that I've discussed in this book are out of equilibrium. That is to say, they are not in their thermo-dynamically most stable state. Once scientists considered such systems to be unapproachable, perhaps even unseemly. Thermodynamics, the science of change that developed initially as an engineering discipline in the nineteenth century, was intended to describe the equilibrium state of systems. It told us about the direction of change, and about what amount of useful work we could expect to extract from such a change; but what actually took place *during* a change was something that classical thermodynamics could barely touch. It was a pretty good tool for chemical and mechanical engineers who wanted to figure out whether they were getting the best from their machines; but it ran up against the difficult fact that some processes *never* seem to reach equilibrium. A river does not simply empty itself into the sea in one glorious, ephemeral rush—the water is cycled back into the sky and redeposited in the highlands for another journey. And so will it always be, while the Sun still shines.

Systems out of equilibrium were scrupulously avoided by the early pioneers of thermodynamics, such as the German Rudolf Clausius and the American Josiah Willard Gibbs. As a result, thermodynamics—for all its practical value—presented a rather artificial view of the world, in which everything happens in a series of jumps between stable states that have no intrinsic time variation. Not much like the world we know!

Out of this somewhat restrictive picture, however, emerged the idea of a directionality to the process of change. It is a familiar enough observation that nearly all processes seem to have a preferred direction—they go one way but not the reverse. Heat flows from hot to cold, an ink droplet disperses in water. These processes are said to be *irreversible*. That they have a directionality in time appeals to our intuition, but it becomes something of a puzzle when we look closely at the micro-

scopic physics behind such processes. In the equations that describe the motion of an individual ink particle in water, there is no 'arrow of time': you could play a film of the particle's progress backwards and not notice the difference, nor appear to break any physical laws. It is only when you look at the behaviour of the whole ensemble of particles that you'd notice anything odd when time is reversed: the droplet coalesces from a uniform solution of ink.

Irreversibility is connected with the second law of thermodynamics, which states that in a system isolated from its surroundings (so that it can't exchange energy or matter), the direction of change is always towards greater *entropy*. This is a probabilistic law, one that emerges when a system has a large choice of accessible states that it can adopt. We encountered the second law in Chapter 3, where I explained that it is regarded as an infallible tenet of nature and was therefore raised as an objection to Boris Belousov's oscillating chemical reaction in the 1950s. And as entropy is in some sense a measure of disorder, the second law seems to pose a big problem for the spontaneous appearance of pattern.

But the thermodynamics of non-equilibrium systems is concerned not with some end point in which entropy has increased relative to the initial state; rather, it considers the process of *becoming*, of how change occurs. As irreversible processes are ones that have a time direction specified by the requirement that they end up with more entropy than before, such processes *produce* entropy. When a system reaches equilibrium, entropy production ceases.

In the 1930s the Norwegian-born scientist Lars Onsager began to delve into the factors that govern entropy production while equilibrium has not been attained. He considered the case of only small deviations from an equilibrium state, under which circumstances one can assume very simple (linear) mathematical relationships between, say, the rate of entropy production, the rates of the irreversible processes taking place, and the forces driving them (such as differences in temperature, which, as we saw in Chapter 7, drive convection). In this linear regime, we can assume, for example, that the rate of change of some parameter of the system varies in direct proportion to the forces that drive the change. Onsager's great achievement in this arena was to show that under these conditions there are universal laws relating the various forces and rates which do not depend on the specific details of the system being considered—whether it is, for instance, a chemical process like the BZ reaction or a convecting fluid. For this work, Onsager was awarded the Nobel prize for chemistry in 1963.

In 1945 the Russian-born chemist Ilya Prigogine in Brussels attempted to extend this picture by suggesting that in the near-equilibrium (linear) regime, non-equilibrium systems tend to adopt a behaviour that minimizes the rate of entropy production. That is to say, if prevented from reaching equilibrium (where the rate of entropy production is zero), the system will (he said) instead settle into a state in which entropy is at least produced at the lowest possible rate. For a long time, Prigogine's principle of minimal entropy production was deemed to provide a criterion for determining the most stable state out of (but close to) equilibrium.

And what about patterns? Nothing in these early attempts to prescribe non-equilibrium behaviour gave any hint that, away from equilibrium, one might suddenly stumble into states that have long-ranged order, like Bénard's convection cells or Turing's spots. States like this become manifest rather far from equilibrium, where the linear equations formulated by Onsager and others no longer apply and where Prigogine made no claims for his rule of minimum entropy production. Where do they come from?

During the 1950s and 1960s, Prigogine and his colleague Paul Glansdorff attempted to extend the treatment of non-equilibrium thermodynamics to the more interesting non-linear regime, where it was clear from experiments like Bénard's and models like Turing's that complex structures and patterns can appear. By making a series of approximations, they suggested that further from equilibrium, the state of minimal entropy production reaches some crisis point at which it breaks down and becomes transformed to another state. Technically speaking, there is a *bifurcation*—literally, a branching in two—at which the evolution of the steady state splits into two branches, presenting a choice of new states that the system can adopt (Fig. 10.1).

What are these new states? The theory of Prigogine and Glansdorff couldn't say much about that; but it seemed reasonable to suppose that they might correspond to the self-organized structures and patterns that were known to appear far from equilibrium.

These ideas seemed at least to provide a promising start. But in 1975 Rolf Landauer showed that the criterion of minimal entropy production cannot hold in general for deciding which steady state a system will adopt slightly away from equilibrium. Even in this linear regime, he concluded, there can be no such simple rules for determining the most 'favourable' non-equilibrium state. Non-equilibrium thermodynamics is not, it seems, conquered so easily.

Dissipative structures

What is it that distinguishes the regular or ordered non-equilibrium states we have seen in the earlier chapters from superficially similar ordered states in equilibrium systems, such as crystals?

Regularity is not uncommon. A swinging pendulum, a bouncing ball, the 'grocer's stall' atomic packing of a crystal, the yearly passage of the Earth around the Sun—all are periodic in space or time. But the regular hexagonal lattice of Rayleigh–Bénard convection differs from the hexagonal lattice of a crystal like copper metal. The latter is an equilibrium structure whose periodicity is determined by some characteristic dimension of the components—the sizes of the atoms. The former, meanwhile, is maintained away from equilibrium by a throughflow of energy, which it dissipates in the process (thereby generating entropy). Stop the input of energy (that is, let the top-to-bottom temperature gradient equalize), and the pattern goes away. Likewise, the oscillations of the BZ reaction persist only when the reaction is fed with fresh reagents and the products are removed in a continuously stirred tank reactor. Structures that are supported away from equilibrium by the generation of entropy are called *dissipative structures*, a term first alluded to by Landauer in 1961.

In contrast to most equilibrium structures, the spatial scale of the pattern features in a dissipative structure bears no relation to the size of its constituents (the size of convection cells is much, much larger than the size of the circulating molecules), and this scale is robust in the face of perturbations. A transient perturbation may disrupt the structure temporarily, but the disturbance will

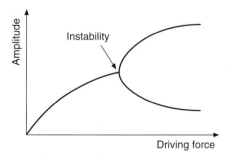

Fig. 10.1 A bifurcation occurs when a stable state develops an instability that offers the system a choice of two new states. A pitchfork bifurcation of the type shown here commonly occurs as a system is driven further from equilibrium.

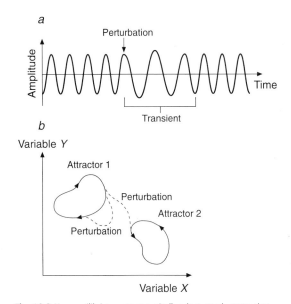

a

Amplitude

Perturbation

Time

Transient

b

Variable *Y*

Attractor 1

Perturbation

Attractor 2

Perturbation

Variable *X*

Fig. 10.2 Non-equilibrium systems typically adopt steady states that correspond to dissipative structures, which act as *attractors* towards which the system returns after a disturbance. An oscillatory dissipative structure, for instance, will return to its original period of oscillation after a perturbation (*a*). Dissipative structures often have several attractors in the 'phase space' of their variables, and large enough perturbations might allow the system to be trapped by a different attractor, producing a different steady-state behaviour (*b*).

pass and eventually the structure will regain the same period as before (Fig. 10.2*a*). Thus the characteristics of a dissipative structure are not at the mercy of perturbations, but are set by the intrinsic interactions in the system. These structures are said to possess an *attractor* in the set of variables that describe the system (the so-called phase space), because the system will always be drawn back to this particular set of variables (provided that it is not knocked so far that it falls into the basin of *another* attractor—see Fig. 10.2*b*). An example of such an attractor is the limit cycle of the oscillating BZ reaction (Fig. 3.2). The converse of a dissipative structure is a *conservative* structure, which possesses no attractors and so can be altered arbitrarily. An example is the orbit of a planet around the Sun: if the radius of the orbit is altered (say, by a catastrophic collision), it stays that way rather than returning to its former value.

Even chaotic non-equilibrium states are dissipative structures of a kind, since they too have corresponding attractors in phase space. The difference from ordered states is that the attractors are fractal—the trajectories spin a web with an infinite hierarchy of structure, so

that the behaviour of the system never repeats itself exactly. All the same, these chaotic or 'strange' attractors have a characteristic form (Fig. 10.3), a 'hidden' pattern that constrains the extent to which the behaviour of the system can meander through and explore the phase space of its variables.

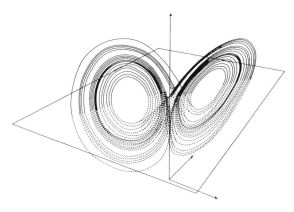

a

b

Fig. 10.3 The 'hidden' forms of chaotic systems are revealed in portraits of their strange attractors in phase space. Here I show two such attractors. (*a*) The famous Lorenz attractor which meteorologist Edward Lorenz discovered in the 1960s in a model of atmospheric circulation. This is commonly regarded as one of the key discoveries in the study of chaos. (*b*) The attractor generated in Taylor–Couette flow as it becomes turbulent. (Image: Tom Mullin, University of Manchester.)

Instabilities, thresholds and bifurcations

Most of the patterns that I've talked about appear suddenly. One moment there is nothing. Then you turn the dial up a notch, and everything is different: stripes appear, or dunes, or pulsations. This seems to be the nature of most symmetry-breaking processes: they happen all at once. They bear a close analogy with so-called phase transitions, which are the leitmotif of the discipline of statistical physics, the result of attempts to put thermodynamics on a fundamental, microscopic basis.

Phase transitions are generally abrupt jumps from one *equilibrium* state of matter to another—from ice to water, water to steam, magnet to non-magnet, insulator to superconductor. Like spontaneously formed patterns, phase transitions are global transformations. When you cool water through its freezing point, you don't get part of it turning to ice and the rest remaining liquid. (In practice you're probably used to seeing this coexisting mixture quite a lot, for example in a layer of ice on top of a pond. But this is because the water may not all be below freezing point, or because the process of freezing takes time so that part of the water might drop below freezing point before it has a chance to freeze. Both are non-equilibrium situations.) Below 0°C, all of a pond is imbued with ice-forming potential—given enough time, all will freeze solid. Moreover, it is all or nothing. At a fraction of a degree above freezing point, the water is *all* liquid once equilibrium is reached. And a fraction of a degree below, it is all ice.

In other words, there is a threshold in some control parameter, like temperature, that, once crossed, leaves the entire system globally unstable to some change in state. Just the same is true for, say, Turing patterns, which appear in some activator–inhibitor systems at a certain temperature threshold (p. 83) and may change from one pattern to another at a different threshold. Similarly, there is a temperature threshold (more properly, a Rayleigh-number threshold) for convection patterns.

In addition, the change in state during an equilibrium phase transition can involve a change in symmetry. Crystalline ice has an ordered structure (in fact ice has many ordered structures, as well as some disordered ones, but let's not worry about that). Liquid water is disorderly—the molecules are free to move about almost at random. You could be forgiven for thinking that symmetry is therefore broken during melting, when the periodic structure of the ice is lost; but in fact it is the other way around. Symmetry is broken during freezing because, whereas the liquid state is isotropic (all directions in space are equivalent), the crystal structure of ice picks out certain directions as 'special'.

So equilibrium phase transitions, like the abrupt transitions that characterize much of pattern formation, are spontaneous, global instabilities that set in when a threshold is crossed; and they may involve symmetry breaking.

A particularly important and common class of global instabilities is the bifurcation (Fig. 10.1). We have encountered examples of mathematical bifurcations, like the Hopf bifurcation that takes place when a predator–prey system or an autocatalytic chemical reaction starts to undergo oscillations (pp. 67 and 224). I showed in Chapter 9 how a population kept in check by overcrowding can become oscillatory (Fig. 9.3). We can think of these oscillations as alternate switches between two populations, one more numerous than the other: these are called 'fixed points', and they represent solutions to the logistic equation (p. 227) in the oscillatory regime. As the sensitivity parameter a in these equations is increased from a value giving a steady state (Fig. 9.3a) to one that gives oscillations (Fig. 9.3b), a bifurcation like that in Fig. 10.1 takes place as the equations acquire two stable fixed-point solutions instead of one. For obvious reasons, this is called a pitchfork bifurcation. In the case of a Hopf bifurcation (the 'onset of a wobble', remember), the choice of which branch to follow is not made once and for all; rather, the system oscillates between the two branches. It might therefore be more accurate to depict the Hopf bifurcation in the manner shown in Fig. 10.4.

We have also seen examples of bifurcations in real (as opposed to mathematical) space: in the tip-splitting of

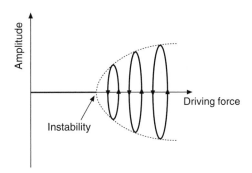

Fig. 10.4 A Hopf bifurcation occurs at the onset of certain oscillatory instabilities. The system undergoes an oscillation between the two branches of a pitchfork bifurcation.

bubbles during viscous fingering, or the forking of a tree's limb. Figuratively these two types of bifurcation are analogous; but we should not take the analogy too far. For one thing, in the case of 'real' branches, the pattern-forming system follows *both* of the two arms of the fork, whereas in mathematical bifurcations the two arms represent *choices*, of which any part of the system can at any instant take only one or the other.

Pitchfork bifurcations are familiar to physicists who study equilibrium phase transitions. They represent, for example, the behaviour of a magnetic material like iron as it is heated or cooled through its magnetic transition. Iron is an example of a ferromagnet, which means that in its magnetized state all of the iron atoms act like little bar magnets with their north and south poles aligned. If you heat a magnetized piece of iron above 770°C (its so-called Curie point), this alignment is lost—the jiggling effect of heat destroys the alignment, and the directions of the atomic magnetic poles are randomized. The magnetic fields due to each atom then cancel each other out on average, and the piece of iron as a whole is no longer magnetized. This abrupt change at the Curie point, from a magnet to a non-magnet on heating, or vice versa on cooling, is an example of a phase transition.

You might think that this phase transition involves a single choice—either the iron is magnetic or not. But in fact there are two choices as the metal is cooled from a non-magnetized state through the Curie point: the atomic magnetic poles can all point either in one direction or the other (Fig. 10.5a). Which direction do they choose? There is no way to answer that—both directions are entirely equivalent. What happens is that the choice is made at random, and is at the mercy of the smallest fluctuations, which can tip the balance one way or the other. (You may wonder how (or if) this random choice can be made in the same direction throughout the entire system. I'll come back to this.) The situation is very much like that of a ball perched on top of a perfectly symmetrical hill (Fig. 10.5b): it is unstable at the top, and has to roll down one side or the other, but which way it goes is unpredictable and subject to the whims of the most minuscule disturbance.

The same arbitrary choice of two equivalent alternatives is faced by a heated fluid about to develop convection rolls. Adjacent rolls turn over in opposite directions; but what determines whether a particular roll goes clockwise or anticlockwise? Again, the choice is down to chance.

To develop this analogy further, I must clarify an important characteristic of both equilibrium phase

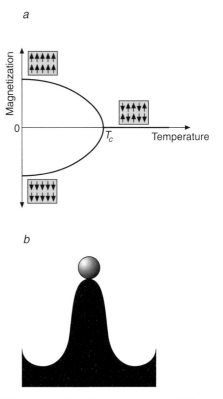

Fig. 10.5 (a) A magnetic material like iron cooled through its Curie temperature (T_c) undergoes a bifurcation at which its magnetization changes from zero (due to randomly oriented atomic magnetic poles) to a non-zero value (due to aligned poles). The alignment can point in one of two equivalent directions. (b) This kind of bifurcation is analogous to the situation of a ball perched on top of a hill, which must roll down one side or the other. The choice is arbitrary.

transitions and non-equilibrium bifurcations which bears on their relationship to symmetry breaking. I illustrated the idea of phase transitions initially with the freezing and melting of water, because this is a familiar example; but I spoke subsequently about the magnetization of iron, which is considerably less so. I'm not simply trying to test your perseverance here; the fact is that the two are different classes of phase transition, and only the latter is an appropriate analogy for what I wanted to say about symmetry-breaking bifurcations in non-equilibrium pattern formation. Freezing and melting are (at least in the familiar forms that we know them) examples of *first-order* phase transitions, in which some characterizing parameter of the system (say, the density of water), changes in a discontinuous, step-like manner as the control parameter (say, temperature) is altered. When water freezes, it so happens that there is a change in symmetry; but first-order phase transitions do not *necessarily* involve symmetry breaking. Moreover, exactly *at* (but not above or below) the transition temperature, the two states (here liquid and solid) can coexist with each other at equilibrium. And one commonly finds that first-order phase transitions display hysteresis, meaning that the transition occurs at a somewhat different temperature if approached from above or below, since in the vicinity of the transition the system can get 'trapped' in a state that is not the most stable one.

The spontaneous magnetization of iron at the Curie point, on the other hand, is an example of a *second-order* phase transition. Here the the magnetization changes abruptly but *continuously* as the system goes through the transition—there is no sudden jump from one value to another (see Fig. 10.5a). Second-order and other continuous phase transitions *always* involve symmetry breaking. Furthermore, there can be no co-existence of the states between which the system switches, even exactly at the transition point: it is all or nothing. And there is no hysteresis. Now, many of the pattern-forming bifurcations that I've discussed are analogous to continuous phase transitions—they are called *supercritical* bifurcations, and they lead to symmetry breaking. That's why I had to resort to the less familiar magnetic system for an appropriate analogy in equilibrium phase transitions. The onset of convection is like this, as is the switch between hexagonal and striped Turing patterns shown in Plate 4. But a few pattern-forming processes are *subcritical* bifurcations, analogous to first-order phase transitions—like the switch between spiral and target patterns in convecting fluids (p. 173), which, as Fig. 7.14 shows, can coexist. These distinctions may seem rather esoteric; but I'll show shortly that the details of what happens at a continuous phase transition provide important clues to how patterns arise away from equilibrium.

Model behaviour

These analogies with phase transitions are useful, but they don't provide any kind of rigorous mathematical description of pattern-forming bifurcations. Physicists like analogies, but they prefer rigour. Attempts to develop a more concrete description of what happens at a symmetry-breaking bifurcation in non-equilibrium systems began in earnest in 1916 when Lord Rayleigh analysed the case of convection studied experimentally by Henri Bénard. Geoffrey Taylor conducted a similar analysis for the case of Taylor–Couette flow between

rotating cylinders in 1923. As I explained in Chapter 7, a thorough treatment of any problem in fluid flow must start with the Navier–Stokes equation, which relates the changes in fluid velocity at all points to the forces that act on the fluid. The approach taken by both Rayleigh and Taylor was to look for the solution to the Navier–Stokes equation close to equilibrium for the particular conditions in their respective systems, and then to examine the stability of this 'base state' in the face of perturbations as the system is taken increasingly further from equilibrium. For the case of convection, the base state is one in which no flow at all occurs—if the temperature gradient between the top and bottom plates is small enough, the heat flow can be accommodated purely by heat *conduction* through the fluid, which is just like conduction through a solid material. For Taylor–Couette flow, the base state is one in which the fluid velocity varies in a particular fashion with distance from the axis of the cylinders, but does not vary in the vertical direction.

Rayleigh and Taylor conducted what is known as a linear stability analysis of these base states as the forcing parameter—the Rayleigh number and the rotation rate of the inner cylinder, respectively—is varied. That is to say, they imposed an infinitesimally small wavy disturbance on the base state, with a particular wavelength, and calculated how the disturbance evolved in time. Let's consider Rayleigh's case; Taylor's follows similar lines. Below the critical Rayleigh number Ra_c, perturbations of all wavelengths die away over time, and the system returns to the base state. But exactly at Ra_c, something stirs: the perturbation with a wavelength corresponding to the critical wave vector 3.12 neither decays nor grows, although perturbations with all other wavelengths still decay. Infinitesimally above Ra_c the wavy perturbation with wave vector 3.12 grows to a finite amplitude: a pattern with this wavelength develops. The instability corresponds to a pitchfork bifurcation (Fig. 10.6). I should mention that above Ra_c the base state is still a possible solution to the Navier–Stokes equation—but it is an unstable solution, since the slightest disturbance will trigger the appearance of a pattern at one of the allowed wave vectors. It is like the ball balanced on top of the hill, or a needle on its tip.

Now, the real point I want to make about this kind of analysis is that it can be generalized to a wide range of systems that undergo spontaneous patterning via a pitchfork bifurcation—not just those in fluids that obey the Navier–Stokes equation. A similar kind of analysis can be applied, for example, to the kind of equa-

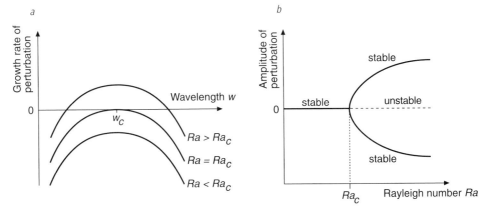

Fig. 10.6 (a) The growth rate of wavy perturbations to the featureless base state of Rayleigh–Bénard convection, as revealed by a linear stability analysis. When the growth rate is negative, the perturbations decay over time. At the critical Rayleigh number Ra_c, the growth rate is zero for a perturbation of the critical wavelength w_c—this perturbation neither grows nor decays. Above Ra_c the growth rate is positive for a range of wavelengths. (b) The onset of the instability is a pitchfork bifurcation. Note that the unpatterned base state remains a solution to the appropriate Navier–Stokes equation for this system even above Ra_c (dashed line); but it is no longer a stable solution.

tions that describe reaction-diffusion systems. Some researchers have attempted to construct simplified 'model equations' that capture the generic features of pattern-forming systems such as these without encountering the complexity (and thus the intractability) of, say, the full Navier–Stokes equation. One of the most popular of these model equations was proposed by Jack Swift and Pierre Hohenberg in 1977, and it has proved to be an excellent 'toy' for exploring the behaviour of this class of pattern-forming systems around the onset of the instability of the unpatterned base state.

Perhaps the most important message to have emerged from these studies of instabilities is that we do not necessarily have a complete understanding of a system once we know the equations that govern it; what we really want to know are the *particular solutions* to those equations. The latter need not be obvious from the former. This cannot be emphasized too strongly in any branch of science. The American physicist Freeman Dyson has pointed out that for Albert Einstein and J.Robert Oppenheimer in their later years, 'to discover the right equations was all that mattered'. One might say the same about some physicists working today to develop a 'theory of everything'. But if you take this view, then fluid dynamics was all sewn up once we could write down the Navier–Stokes equation. Yet if we had stopped there, we'd never have guessed at the rich variety of solutions that it held in store even for relatively simple experimental set-ups (take a look again at Fig. 7.38, for instance). Sometimes even knowing the

solutions is not enough—only through experiments can one interpret what they are telling us.

Finally, I should point out that not all changes of pattern or form take place via abrupt instabilities. I showed in Chapter 5 how a cluster of particles growing by diffusion-limited aggregation can be 'tuned' continuously between two characteristic branching forms by gradually changing the balance between noise and anisotropy (p. 121). But notice how, even in this case, the growth *forms* involve abrupt instabilities in the sense that a small protrusion can blossom into a side-branch. This extreme sensitivity to small fluctuations is another characteristic of continuous phase transitions, as we shall see.

Pattern selection

A linear stability analysis can reveal the point at which a non-equilibrium system is driven across the threshold of pattern formation. But can it tell us anything about the pattern that results? Exactly at the threshold, we've seen that (at least for the cases considered by Rayleigh and Taylor) there is a single 'marginally unstable' wave vector—a single characteristic length scale in the system. But how does this length scale manifest itself—as stripes, spots, travelling waves? And once the threshold is surpassed, an increasing number of perturbations with different wavelengths become able to grow. Which wavelength is selected?

This question is not peculiar to these examples in fluid dynamics; just about any system with pattern-forming potential faces choices. Like ourselves in a carpet warehouse, they have a gallery of designs from which to select. Soap molecules on the water surface (Fig. 2.22), metal deposits grown at electrodes (Figs 5.6 and 5.24), bacterial colonies (Figs 3.27, 5.31 and 5.32), jumping grains (Fig. 8.3)—they all find a catalogue of riches. Which to choose?

There is no universal way to answer this question; and in this respect we can see a major distinction between equilibrium and non-equilibrium systems. The former too often have several choices of pattern and form—but there is a simple rule (in principle) for deciding the best choice. Take, for example, a mineral such as calcium carbonate precipitating out of a supersaturated solution. There are three different forms of crystalline calcium carbonate, called calcite, aragonite and vaterite. Each has its own distinct pattern, defined in terms of the way that the calcium and carbonate ions are stacked together in the crystal. But left to its own devices, precipitating calcium carbonate will always choose to crystallize as calcite at room temperature and atmospheric pressure. The reason for this has been long known: calcite is the most thermodynamically stable of the options under these conditions. In other words, the energy (technically speaking, the *free* energy—p. 19) of calcite is lower than that of the other two crystal forms. At equilibrium, a system will always seek to adopt the configuration that has the lowest free energy. This means that balls will roll down hills, iron will rust in air, water below freezing point will turn to ice.

So when complex patterns form under equilibrium conditions, as is the case for example in lipid vesicles (Figs 2.19 and 2.20) or block co-polymers (Fig. 2.46), we know at least what the criterion for pattern selection is; and if we know the various contributing factors to the free energy, we can predict the most favourable pattern by finding the shape that minimizes this free energy. All of the structures in these figures can be understood on this basis.

During the 1960s and 1970s, Ilya Prigogine's group at Brussels held out the hope of finding a similar 'minimization principle' that would determine the choice of pattern adopted by systems away from equilibrium. That is to say, they hoped that there was some quantity analogous to free energy that non-equilibrium systems would seek to minimize. This would provide a universal criterion for pattern selection. For systems close to equilibrium Prigogine proposed the principle of minimum entropy production; but as I've said, this was subsequently ruled out as a general selection principle,

even in this relatively simple 'linear' regime, by Rolf Landauer. In doing so, Landauer seemed to show that there cannot even in principle be any all-encompassing minimization criterion that can be applied to pattern selection out of equilibrium. He considered the case of non-equilibrium systems with more than one locally stable state—more than one potential pattern, you could say. Such states can be compared to mountain reservoirs—water reaching any one reservoir will sit in it quite stably, even though it would be stabler still (that is, have less potential energy) at the foot of the mountain. For a system in equilibrium, it is a straightforward matter to calculate the relative probability that a particular locally-stable state will be occupied—this depends on its free energy. But away from equilibrium? Landauer's analysis of the problem built on the pioneering but often neglected work of the Soviet scientist R.L. Stratonovich, who studied these 'multistable' systems in the 1950s and 1960s. Landauer's results implied that, to identify the most favoured state, it is never enough simply to compare the characteristics of the system in the vicinity of the locally stable states (the shapes and altitudes of the reservoirs, you might say—or in Prigogine's case, the rates of entropy production in these states)—one has to consider what the terrain looks like along the pathways connecting the states. In other words, to address the problem of pattern selection, we are forced to consider the specific details of each system, including the nature of the randomizing 'noise' it experiences.

Bit by bit

Nonetheless, we can make a few generalizations, provided we bear in mind that these are not at all rigorous. Spontaneous pattern formation involves symmetry breaking; and as I've said, symmetry tends to break a bit at a time as the system is driven harder and harder. This alone helps us to understand why two types of pattern—stripes and hexagons—are particularly common. The minimal way to break the symmetry of a uniform two-dimensional system—that is, the way to break as little symmetry as possible—is to impose a periodic variation in just one dimension. What that means is that we impose a wavelike disturbance in one direction (Fig. 10.7), which creates parallel bands, stripes or rolls. Parallel to the stripes, symmetry is not broken: as we travel through the medium in this direction, we see no change in its character. It is only in the perpendicular direction that we can identify the broken symmetry—travelling in this direction, we see a periodic change from one state to another and back again. Thus, stripe-

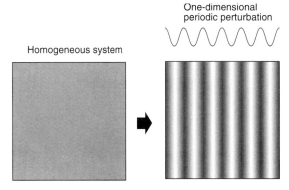

One-dimensional periodic perturbation

Homogeneous system

Fig. 10.7 The simplest way to break two-dimensional symmetry is to impose periodic variations in one dimension, creating stripe patterns.

like patterns are often the first to appear from a uniform system in two dimensions: we saw this in the formation of sand ripples and in the appearance of convection and Taylor–Couette rolls.

After breaking symmetry periodically in one dimension, the next 'minimal' pattern in a two-dimensional system involves breaking it in the other. This imposition of a second periodic variation breaks the system into discrete cells. If the state is to remain ordered and as symmetric as possible, there are only two options: to impose the periodic variation perpendicular to the rolls, creating square cells, or to impose two such variations at 60° angles, creating triangles or hexagons (Fig. 10.8). We saw right at the beginning of Chapter 2 that only these types of perfect polygonal tiles can be packed to fill space without gaps (Fig. 2.2). So the square, triangular and hexagonal patterns that we've seen in Turing patterns (Fig. 4.3), in convection (Figs 7.4*b* and 7.9) and

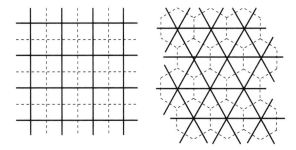

Fig. 10.8 When symmetry is broken in both directions in two dimensions, square or triangular cells result. In the latter case, the resulting pattern can be a triangular lattice or a hexagonal pattern of cells. In Rayleigh–Bénard convection, for example, the centres of the triangular lattice correspond to regions of upwelling, and the downwelling regions delineate hexagons (*dashed lines*).

in shaken sand (Fig. 8.3) are no mystery—they are a consequence of the *geometric* properties of space, which constrain the ways in which symmetry can be (minimally) broken.

These heuristic arguments aren't, however, a reliable guide for determining exactly what kind of broken symmetry will arise in any particular case. For example, the question of whether stripes/rolls or hexagons will be preferred has no universal answer. A linear stability analysis of convection won't help you either: you have to go to a more sophisticated level of theory to find that rolls are generally favoured. Qualitatively we can understand this on the grounds that rolls do not distinguish between upflow and downflow, whereas hexagonal cells do: upflow occurs in their centre and downflow at the edges. So rolls are geometrically symmetrical about the plane midway between the two plates of the convection chamber, but hexagonal cells are not. The initial patterned state tends to favour the retention of this symmetry, and so rolls are formed. But if the midplane symmetry is already broken in the unpatterned base state—for example, if the warmer fluid near the bottom is significantly less viscous than the cooler fluid towards the top (which is quite possible)—then hexagonal cells might appear instead. In the case of the Turing instability of reaction–diffusion systems, meanwhile, hexagonal spots are usually the default option; stripes have a tendency to break up into spots (see p. 87).

The only other choice to be made in pattern selection for these relatively simple cases is the *size* of the features—that is, the wavelength of the stripes or the separation of the hexagonal spots. Again, it's not possible to identify a single criterion that determines this. In some cases, such as Rayleigh–Bénard convection, I indicated above that linear stability analysis can be used to derive the wavelength of the instability at the onset of patterning (although if the simplifications in Rayleigh's analysis of convection, concerning the properties of the heated fluid, do not constitute good approximations, the critical wavenumber is different). But above the critical Rayleigh number there is a range of allowed wavenumbers, and then the wavelength of the roll pattern becomes dependent on the *history* of the system— how it reached the convecting state. Moreover, it is possible then for both the amplitude and the wavelength of the roll pattern to vary in time and in space throughout the system.

In an activator–inhibitor system, it is the relative ranges of the activator and inhibitor that is critical to the selection of pattern scale: recall that the inhibitor diffuses more rapidly, and so acts over long ranges. Just *how* long (and so how far away the next stripe has to be before activation can again dominate over inhibition) depends on the relative diffusion constants.

I should mention that there is one other possible mode of behaviour at the onset of pattern formation in systems like these that I have not yet mentioned. I said that below the threshold, wavy perturbations decay in time, while above it they grow (for a certain range of wavelengths). But they can also *oscillate* above the threshold. Then the pattern that appears is not stationary but is instead a travelling wave, commonly a spiral. The spiral patterns seen in the BZ reaction (Fig. 3.3) and in convection (Fig. 7.13) are examples of this type of instability.

So we can perhaps hope to understand a little about pattern selection close to the onset of spontaneous pattern formation; but further from the onset, when the system is being driven harder away from equilibrium, there really isn't any good theory that allows us to predict or rationalize the patterns that form. In most cases the only option is to resort to experiment, to characterize and categorize the taxonomy of dissipative structures that appear. If one knows enough about the basic physical interactions in the system, however, one might be able to devise a model that can be simulated on the computer, which can provide some insight into the essential ingredients of the pattern-forming process—some of the granular patterns described in Chapter 8 have been tackled this way.

Defects

Very often the patterns that appear further from the instability threshold are less than perfectly symmetrical; they are laced through with *defects*, sometimes to such an extent that all appearance of symmetry is lost. For example, we saw how the roll cells in convection or Turing structures can run into one another in dislocations (Figs 4.3*b*, 7.10 and 7.11), and how the hexagonal cells of Rayleigh-Bénard convection can exhibit a high degree of disorder (Fig. 7.2). Through an accumulation of such distortions, parallel stripes can become bent into more or less disordered wavy patterns, and a triangular lattice of spots can disintegrate into a more random scattering (Fig. 4.3*a*). This gives us the stripes of zebras, the spots of the leopard. Notice, however, that even in cases where disorder has overwhelmed all semblance of symmetry, we can still identify order of a kind: the average distance between spots or stripes remains more or less constant.

Rather than despairing at the disorder that defects engender, we can make some headway by shifting our attention from regularity to the defects themselves, developing taxonomies like that in Fig. 7.11 and asking how generic members of these schemes arise from characteristic deformations of the underlying pattern. There already exists a rich theory of defects to draw on here from studies of crystals and the structures of liquid-crystal materials (Fig. 7.12).

Close to the edge

All of this applies to infinite systems, by which I mean ones for which we ignore the boundaries. But of course no pattern-forming system is infinite—they always have edges. If the size of the system is vastly greater than that of the pattern's characteristic length scale, the effects of edges on the pattern as a whole may be negligible. Commonly, though, the system is not this large, and then the pattern may be influenced globally by the size or shape of the 'container'. We saw in Chapter 4, for example, how either hoops or spots could be selected from the same pattern-forming mechanism on animal tails, depending on the size and shape of the embryonic tail when the pattern is laid down during development. And more generally, the patterns of different animal pelts—a two-tone division of the whole body, a few large blotches or a multitude of small spots—can be determined by the relative size of the embryo at the patterning stage.

The shape of the boundary can occasionally change a pattern to something qualitatively different. In long, rectangular trays, convection rolls tend to form stripes (Fig. 7.7), whereas in circular dishes the rolls curl up into concentric circles (Fig. 7.8*c*). Moreover, the need for a whole number of pattern features to fit within the container may determine the wavelength, just as the wavelength and thus the frequency of an organ note is determined by the length of the pipe. In some systems, the pattern may change locally to adapt to the presence of a boundary—in Fig. 10.9, for example, concentric roll cells give way to short parallel rolls at the edges, so that the rolls can meet the boundary at the preferred right angles.

The race for dominance

In all of these cases, static factors—the geometry of regular divisions of space, the size and shape of boundaries—act to select a pattern. In cases where the pattern is *growing*, dynamic considerations may determine the choice made from the gallery of possible alter-

Fig. 10.9 Concentric convection rolls in carbon dioxide gas. The concentric pattern is surrounded by short, parallel rolls, which enable the pattern to adapt to the boundaries by meeting them at right angles. (From: Cross and Hohenberg 1993.)

natives. This is particularly the case for branching growth patterns. We saw in Chapter 5 that the characteristic scale of the tip and branching side-arms of a dendritic crystal is set by the condition that the growth speed just about balances the tendency for the tip to split—there is a unique, 'marginally stable' solution to the mathematics of the growth process. The question of what determines the pattern of a non-equilibrium electrodeposit or an expanding bacterial colony has been much debated. Why does the growing cluster adopt, say, an irregular fractal form, like a DLA cluster, rather than the dense-branched morphology (see Figs 5.31 and 5.32)? One proposal is that the pattern selected is simply that which grows fastest, and which therefore 'outruns' the others. There is some evidence to support this idea in certain cases, but it's not clear that it provides a robust criterion for branching growth in general. Another proposal is that the selected mode is that which *maximizes* the generation of entropy. At the moment, this issue remains unresolved.

Finally, we should include noise as a pattern-selecting influence. By noise I just mean the inevitable randomness in the environment, deriving from thermal fluctuations for instance. We saw in Chapter 5 that increasing noise can induce a change-over between branching growth modes from tip-splitting of broad, fat fingers to the ragged disorder of diffusion-limited aggregation (see Fig. 5.18). The message here is that noise does not necessarily affect all patterns equally—it may favour some over others. Rolf Landauer has argued that noise is central to

the transitions between different states of a non-equilibrium multistable system, and so is a critical (but commonly neglected) feature of pattern formation and selection. 'If we cannot characterize the noise', he says, 'we are going to be limited in our analysis of the system.' For Per Bak's self-organized critical states, meanwhile, much of the essential form of the system lies in the noise itself.

Increasing complexity with increasing driving

Although many pattern-forming systems have several faces, usually only a single pattern will be stable for a given set of conditions, and transitions between patterns take place as thresholds in control parameters are crossed. I have shown (in Chapter 7 in particular) that there is a common tendency for the patterns to become more ornate—we might say more complex—as the system is driven harder and harder. Shear flows progress from simple sine-like waves to embellished vortex streets to intermittently and then fully turbulent wakes as the Reynolds number (the flow velocity) increases (Fig. 7.29). So too do convection patterns progress from simple rolls to turbulent, dynamic structures. This sequence of increasing complexity is perhaps most clearly illustrated in the case of the oscillating Belousov–Zhabotinsky reaction in a continuous-flow stirred-tank reactor (Chapter 3). As the flow rate of chemicals through the vessel increases, the oscillations undergo a series of period-doubling bifurcations, so that the cycle repeats with every oscillation, then with every second oscillation, then with every fourth and so on. The same is true of the population model represented by the logistic equation (p. 227). This behaviour can be depicted as a *cascade* of pitchfork bifurcations (Fig. 10.10). One might liken this (albeit very loosely) to the excitation of additional harmonics as a trumpeter blows harder. Eventually the oscillations become chaotic (non-periodic), as if the system becomes overwhelmed with choices. Then the cascade loses its branched structure and breaks up into a dense forest of spots—and we lose sight of any order at all.

Correlations

One of the most striking things about many of the patterns (particularly the more ordered ones) that I've discussed in this book is that they seem to acquire a characteristic length scale out of thin air. Sand ripples and dunes have a particular wavelength—in the case of

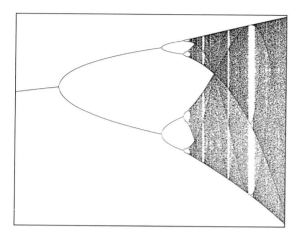

Fig. 10.10 Bifurcations in many non-equilibrium systems come in a sequential cascade as the system is driven further from equilibrium (here from left to right). At each bifurcation the number of fixed points doubles, leading to period doubling. The oscillations repeat every two, four, eight cycles and so on. The cascade structure gets increasingly finer and eventually gives way to non-periodic (chaotic) behaviour, seen here as a dense 'dust' of dots.

dunes especially, this is so far removed from the size of the grains themselves, or of their typical trajectory lengths, that one can't imagine these fundamental length scales having any influence on that which appears in the pattern. Just the same is true for Turing patterns. The size of the molecules and atoms in the chemical mixture, and the range of the interactions between them, is minuscule—about a tenth of a millionth of a millimetre. Yet the patterns have length scales big enough for us to see with our unaided eyes, perhaps several millimetres or so. How on earth can interactions on these unimaginably tiny scales give rise to patterns millions of times larger?

To put it another way: the appearance of these patterns implies that the components of the system must be able to communicate over distances much longer than those to which they are accustomed at equilibrium. Think of the rolls that appear in Rayleigh–Bénard convection. Before the onset of convection, the molecules are moving about throughout the quiescent fluid in a random, disorderly way; each molecule barely takes heed of what its immediate neighbours are doing, let alone what is happening a millimetre or so away, many millions of molecules distant. Yet above the critical Rayleigh number this independence has been lost, and the molecular motions have (on average) become *correlated* over these vast distances. That is to say, if we

were to observe the molecular motions on the descending edge of one of the roll cells, we would know that statistically identical motions were being executed by molecules one wavelength away—and two, three and so forth throughout the container. This kind of long-ranged correlation, according to which molecules behave coherently over distances that far outstrip the sphere of their own influence, is characteristic of many pattern-forming systems.

How is it made possible? Perhaps the molecules are able to relay their individual, tiny influences from neighbour to neighbour over such scales? Forget it—in the frenzied environment of a hot liquid, that is like trying to play Chinese whispers at a rock concert. Yet long-ranged correlations do imply an extraordinary degree of cooperativity amongst the molecular constituents.

The appearance of long-ranged correlations in systems undergoing abrupt changes in behaviour is not unique to non-equilibrium systems—it has been long recognized in equilibrium phase transitions too. The key to such behaviour, both at equilibrium and away from it, is that the system loses all sense of scale. Long-ranged correlations may develop when a system becomes *scale-invariant*, and so is able to support fluctuations over scales ranging from those of intermolecular forces to those millions of times longer.

This is what happens to iron at the Curie point, which is an example of a *critical point* (p. 213). All continuous phase transitions happen at these special locations in phase space. I mentioned earlier that at a critical point, fluctuations occur on all length scales. Take a look at Fig. 8.19: here the black and white regions represent regions of gas and liquid as a fluid is taken through its liquid–vapour critical point. But they could equally well represent regions in a piece of iron where the magnetization points in one direction or the other. Above the Curie temperature the directions are random (the image would look grey from a distance); below it they are predominantly in a single direction. But exactly at the Curie temperature there are black and white regions over all scales up to the size of the entire system. At this point, there is no telling whether black or white will prevail upon cooling. The critical system is infinitely sensitive to fluctuations—the slightest imbalance suffices to tip it one way or the other.

As a system like this approaches its critical point, each element feels the influence of more and more of the rest of the system. Far from the critical point, only the behaviour of a magnetic atom's nearest neighbours matters—if these all point in one direction, the atom in question will be inclined to follow suit. But as the critical point is approached, each atom's sphere of influence (called its correlation length) extends wider and wider. And exactly at the critical point, the correlation length becomes as big as the entire system. I should point out that this does not imply that the magnetic field set up by each atomic magnet becomes stronger, or reaches out further; rather, the behaviour of the atoms becomes more *collective*, so that progressively larger groups will behave cooperatively.

To this extent, non-equilibrium pattern formation can resemble a critical phase transition. But there are important differences too, particularly in terms of the structures that result. For in our piece of iron, the ordering that results as we pass through the phase transition has a characteristic length scale that reflects the scale of interatomic forces: the magnetized iron adopts a regular structure in which the periodicity of the magnetic alignment occurs on the same scale as the periodicity of the atoms. We could have guessed this length scale at the outset, for it is of the same order as the range of each atom's magnetic influence. In non-equilibrium patterns this is not so: the scale of ordering vastly exceeds the range of interaction of the constituents, and there is no obvious hint of a length scale of this magnitude in the microscopic physics of the unpatterned state. This is why we can regard such patterns as global *emergent* properties of the system, which are likely to remain hidden to a highly reductionistic analysis.

Power laws and scaling

In most of the discussion above I've been concerned with patterns that form in systems that are essentially *deterministic*, which is to say that at least in principle we can write down equations (such as the Navier–Stokes equation) that describe the behaviour exactly. As we saw, that doesn't by any means imply that we can solve the equations, but it follows that, once the initial and boundary conditions are specified, we have all the ingredients of the process.

Some of the patterns that I've talked about in the book do not share this deterministic character; the equations that describe them contain a random (stochastic) element that is unpredictable and impossible to formulate in anything other than statistical, average terms. Diffusion-limited aggregation is like this—the particles become attached to the growing DLA aggregate only after executing a random walk, for example of the kind generated by the jostling of air molecules, so

that their precise trajectories are indeterminate in advance. Sand piles formed by the sequential addition of grains are like this too. Noise is a major ingredient of the form-determining process in these cases.

The universal patterns and forms of noise-dominated systems are commonly 'hidden'—they become apparent only in 'mathematical space'. We saw in Chapter 8, for example, how fluctuations or variations in space and time that look at first glance to be random can contain the hidden form of self-organized criticality, distinguished by power laws that follow a $1/f$ law or something close to it. And the most robust feature of disordered fractals like DLA clusters or city shapes is their power-law scaling behaviour, which reveals a fractional exponent that tells us much about the similarity or otherwise of structures that bear no *particular* visible features in common. Some researchers believe that power laws hold the key to much of the complex behaviour exhibited by non-equilibrium systems, and that long-ranged correlations and critical-like behaviour are a natural and inevitable consequence of this. It is likely, however, that self-organized criticality and power-law behaviour are the exception rather than the rule. Nonetheless, it is clear that noise, power-law behaviour, scale invariance, avalanche behaviour and fractal forms are intimately connected in some deep way that remains to be fully explored and unravelled.

That universal power-law scaling arises in self-organized critical systems is at least in one sense no surprise—yet again, it is found in equilibrium continuous phase transitions too. As a system approaches its critical point, the variables that describe its behaviour—the correlation length, the density differences between liquid and gas, or more technical quantities such as the magnetic susceptibility or the compressibility—start to obey power laws. That is to say, their value is proportional to the distance from the critical point (say, the temperature difference $T-T_c$, where T_c is the critical temperature) raised to a power corresponding to some 'critical exponent'. For the correlation length ξ the relationship might be $\xi \propto (T-T_c)^{\nu}$, with ν being the critical exponent. What this relationship implies is that the behaviour of these variables no longer depends on the details of the system—whether the fluid approaching its critical point is water or carbon dioxide, whether the magnet is iron or cobalt. Instead, everything is determined by the *universality class* of the phase transition: those in the same class have the same critical exponents. This scaling behaviour therefore reveals an underlying universality in critical phenomena: apparently very different kinds of system can undergo critical transitions in the same universality class, and so their scaling laws, as the critical point is approached, are identical. Conversely, this means that we can find out all about the critical behaviour of one system by studying another in the same universality class—even if one system is a fluid separating into liquid and gas phases and the other is a magnetic material! We can see an echo here of the kind of universality in pattern formation that I have implied throughout the book: apparently very different systems can exhibit the same pattern-forming sequences.

Look and see

Well, I suspect that all this may seem like heavy going in comparison with the previous chapters, and I'm afraid that overviews can tend to be a bit like that. In order to pull out unifying themes, we have to forgo the par-

ticulars of tangible examples and become somewhat abstract (and lose the attractive pictures in the process). I hope you have been able to tolerate that, because I believe it is one of the principal messages of this book that we can map many of nature's tapestries onto some universal blueprints, in which specifics cease to matter. At the same time, I want to stress that it *is* a mapping that is being performed here, and that 'the map is not the territory'. Maps have a fascination of their own, but that's nothing compared to the real thing. This is why I hope you can try to create some of these patterns for yourself, with the recipes given in the appendices, and why too I hope you will discover that the most exciting, the most profound experience of them is to be found through direct encounter. These self-made patterns are everywhere—in the vegetable patch, in the coffee cup, on mountain tops and in the city streets. I hope you enjoy them.

APPENDIX 1
SOAP-FILM STRUCTURES

Plateau's rules for intersecting soap films give rise to some striking symmetrical patterns when the films are suspended on wire frames of regular geometrical shapes. These are fairly easy to make even for someone with a soldering technique as laughable as mine. I show in Figs 2.10 and 2.11 some of the simplest possibilities, but the options are limited only by your skill at fabrication. I have used copper wire of 1-mm diameter, and made figures with sides 5-cm long. The task is made easier by bending a continuous piece of wire to construct as much of the frame as you can, rather than trying to solder together lots of 5-cm lengths. And remember to include an extra appendage to hold on to.

These frames can be dipped into a solution of water and washing-up liquid. Note that the solution loses some of its film-making ability over time. For more complex frames, like the octahedron, the pattern of films is not unique, and you can induce rearrangements by blowing gently. If small extraneous bubbles get trapped along some vertices, these can be removed by careful pricking.

To make a catenoid (see p. 36), you can construct a tweezer-like frame with two circular loops at the ends.

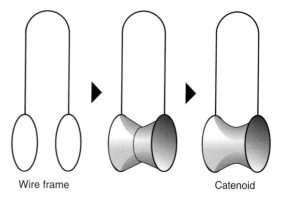

Wire frame   Catenoid

Dipping the ends will often generate two half-catenoids with a circular film in between them; this can be converted to a catenoid by pricking the central film. You'll find that the films are surprisingly resilient when freshly made.

For a wider range of experiments with bubbles and foams, I highly recommend C.V. Boys' *Soap Bubbles* (see Bibliography: Bubbles and Foams).

APPENDIX 2
OSCILLATING
CHEMICAL REACTIONS

There is a variety of reliable oscillatory chemical reactions described in the chemistry literature, including many accessible recipes in books intended for teaching or for a general readership. One of the most striking in terms of the colour change is the *iodate/iodine/peroxide oscillator*. The recipe that I have tested for myself is as follows:

Solution A: 200 ml of potassium iodate (KIO_3) solution—made by adding 42.8 g KIO_3 and 80 ml of 2M sulphuric acid to distilled water to make a total volume of 1 litre.

Solution B: 200 ml of malonic acid/manganese sulphate ($MnSO_4$) solution—made by adding 15.6 g malonic acid and 4.45 g $MnSO_4$ to distilled water to a total of 1 litre.

Solution C: 40 ml of 1% starch solution—made by adding a slurry of 'soluble' starch to boiling water.

Solution D: 200 ml of 100 vol. (about 30%) hydrogen peroxide (H_2O_2) solution.

Mix solutions A, B and C together in a conical flask and then initiate the reaction by adding solution D. Mix well using a magnetic stirrer. After a minute or two the solution, which is initially blue (owing to the formation of iodine, which reacts with starch to form a blue compound), turns a pale yellow (as the iodine intermediate disappears), and then abruptly blue again to begin another cycle. The colour changes persist for about 15–20 min, but finally run out of steam because some of the initial reagents are consumed in each cycle and not replenished.

After a few minutes the mixture begins to bubble, as carbon dioxide gas is generated from the oxidation of malonic acid.

If the mixture is not stirred, the colour changes still take place but grow from filamentary patches throughout the solution.

It is important that the malonic acid solution is not prepared too far in advance—it begins to decompose over the course of several weeks.

The most famous oscillatory reaction is the *Belousov–Zhabotinsky reaction*, for which various recipes are available in the literature. Here's one that I have seen work:

Solution A: 400 ml of 0.5M malonic acid (52.1 g malonic acid in a litre of water).

Solution B: 200 ml of 0.01M cerium(IV) sulphate ($Ce(SO_4)_2$) in 6M sulphuric acid.

Solution C: 0.25M potassium bromate ($KBrO_3$) (41.8 g $KBrO_3$ in 1 litre of water).

Mix solutions A and B in a magnetically stirred conical flask, and then add solution C to initiate the reaction. After about 3 min, the solution starts to alternate between colourless and yellow. The oscillations last for 10–15 min.

This is the colour change that Belousov first saw; but it can be made more dramatic by adding 1 ml of an indicator called ferroin (iron tris(phenanthroline)), which makes the solution change between blue and a purplish red. The chemistry behind these oscillations is described in Chapter 3.

I have taken these recipes from the chemical demonstrations leaflet of the chemistry department of University College, London, and am extremely grateful to Graeme Hogarth and Andrea Sella for help in performing these experiments and those in the following two appendices.

There are many other oscillating reactions, and variants of these two recipes, to be found in:

B.Z. Shakhashiri (1985). *Chemical Demonstrations: A Handbook for Teachers of Chemistry*. University of Wisconsin Press, Madison.

H.W. Roesky & K. Möckel (1996). *Chemical Curiosities*. VCH, Weinheim.

L.A. Ford (1993). *Chemical Magic*. Dover, New York.

APPENDIX 3
CHEMICAL WAVES IN THE BZ REACTION

The target patterns of the inhomogeneous Belousov–Zhabotinsky (BZ) reaction always looked to me so extraordinary that I found it hard to believe they would be easy to reproduce. I was thrilled to find, when I tried the reaction first-hand, that this was not the case. This is a recipe that seems very reliable:

Solution A: 2 ml sulphuric acid + 5 g sodium bromate ($NaBrO_3$) in 67 ml water.

Solution B: 1 g sodium bromide (NaBr) in 10 ml water.

Solution C: 1 g malonic acid in 10 ml water.

Solution D: 1 ml of ferroin (25 mM phenanthroline ferrous sulphate).

Solution E: 1 g Triton X-100 (a kind of soap) in 1 litre of water.

Put 6 ml of solution A into a Petri disk about 3 inches in diameter, add 1–2 ml of solution B and 1 ml of solution C. The solution turns a brownish colour as bromine is produced. Make sure you do not inhale deeply over the dish—bromine is noxious! After a minute or so the brown colour will disappear. Once the solution has become clear, add 1 ml of solution D (which will turn the liquid red) and a drop of solution E. Swirl the Petri dish gently to mix the solutions (it will turn blue as you do so, but then quickly back to red), then leave to stand. Gradually, blue spots will appear in the quiescent red liquid, and these will slowly expand as circular wave fronts. New wave fronts will be initiated behind the expanding waves. Typically there will be one to a dozen or so separate target-wave centres, and the blue fronts annihilate one another as they collide.

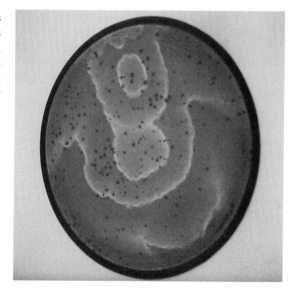

This reaction is most impressively seen when the dish is placed on an overhead projector (see above). The heat of the projector will warm the solution and accelerate the wave fronts somewhat. After some time, bubbles (of carbon dioxide) will start to appear. These can begin to obscure or disrupt the pattern, but you can get rid of them and restart the process by swirling the solution around a little.

This recipe is taken from the chemical demonstrations leaflet of the chemistry department of University College, London.

APPENDIX 4
LIESEGANG BANDS

This is a wonderful experiment, but takes several days. The bands are zones of precipitation of an insoluble compound, which occur at intervals down a column filled with a gel, through which one of the reagents of the precipitation reaction diffuses from above.

You can use a burette as the column (about 1-cm diameter), although ideally a glass tube without gradation markings is best. The recipe I have used involves the reaction between cobalt chloride and ammonium hydroxide, which precipitates bluish bands of cobalt hydroxide. The cobalt chloride is dispersed in a gelatin gel: mix 1.5 g of fine-grained gelatin and 1 g of hydrous cobalt chloride ($CoCl_2.6H_2O$) with 25 ml of distilled water and heat to boiling point for five minutes. Then transfer this mixture immediately to the glass column, cover the top of the column with plastic film, and allow to stand for 24 h to set at room temperature (22°C).

Then add 1.5 ml of concentrated ammonia solution to the top of the solidified gel using a pipette. Cover the tube again and leave it to stand.

After several days, the bands begin to appear down the column. They are closely spaced—about a millimetre apart, although the spacing is not constant (see p. 62). You have to get on eye level with the bands to see them clearly, but they should be sharp and well defined (see figure).

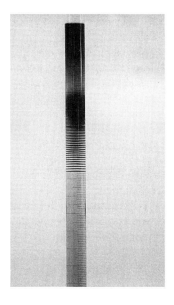

This recipe is taken from:

R. Sultan and S. Sadek (1996). Patterning trends and chaotic behaviour in Co^{2+}/NH_4OH Liesegang systems. *Journal of Physical Chemistry* **100**, 16912.

References to other systems are given in Henisch (1988) (see Bibliography: Waves).

APPENDIX 5
THE HELE-SHAW CELL

The cell is basically two clear, rigid plates separated by a small gap. The plates are in fact trays, having raised edges to contain the liquid. Glass is recommended, but clear plastic (Perspex) works fine and is easier to work with. I have taken my design from:

Tamás Vicsek (1988). Construction of a radial Hele–Shaw cell. In *Random Fluctuations and Pattern Growth*, (ed. H.E. Stanley and N. Ostrowsky), p. 82. Kluwer Academic Publishers, Dordrecht.

The top tray measures 27 × 27 cm, and the bottom one 34 × 34 cm; the Perspex is 4 mm thick. The pieces are glued with epoxy resin.

The top plate is separated from the lower one by flat spacers at each corner—British pennies give about the right separation. The viscous liquid is glycerine, purchased from a pharmacist; the viscous fingering patterns are clearer if the glycerine is coloured with food colouring. (Using glycerine rather than oil makes the assembly easier to clean in water.) Air is injected through a small hole in the top plate. A 3-mm hole is recommended, but I simply used the empty ink tube from a ball-point pen, which is closer to 2 mm in

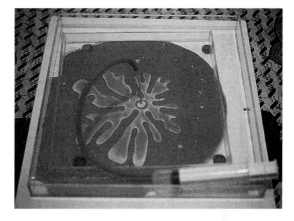

internal diameter. This was glued in place in the central hole. The air can be injected through a large plastic syringe if you can get one; but it is just as good to blow. Remember that the viscous fingering pattern is a non-equilibrium shape, so that you should blow quite sharply rather than slowly to ensure that the bubble grows out of equilibrium.

APPENDIX 6
BÉNARD CONVECTION

Polygonal convection cells will appear in a thin layer of a viscous liquid heated gently from below. This is a classic 'kitchen' experiment, since it really does not involve much more than heating oil in a saucepan on a cooker. The base of the pan must be flat and smooth, however, and preferably also thick to distribute the heat evenly. A skillet works well. The oil layer need be only about 1 or 2 mm deep. The flow pattern can be revealed by sprinkling a powdered spice such as cinnamon onto the surface of the oil.

For a more controlled experiment, silicone oil can be used: this is available commercially in a range of viscosities, and a viscosity of 0.5 cm^2/s is generally about right. The convection cells can be seen more clearly if metal powder is suspended in the fluid (see Plate 1). Bronze powder can be obtained from hardware shops or arts suppliers. Aluminium flakes can be extracted from the pigment of 'silver' model paints, by decanting the liquid and then washing the residual flakes in acetone (nail-varnish remover). These powders will settle in silicone oil if left to stand.

These procedures are based on:

S.J. VanHook and Michael Schatz (1997). Simple demonstrations of pattern formation. In *Physics Teacher*, October 1997.

This paper provides the names and addresses of some US suppliers of the substances involved.

APPENDIX 7
GRAIN STRATIFICATION IN THE MAKSE CELL

My Makse cell is not a masterpiece of engineering, because I was impatient to put it together and see if the experiment worked. No doubt far more elegant varieties could be devised. The main feature I wanted to include was that the Perspex plates be detachable, so that they might be cleaned. Ideally they should also be treated with an anti-static agent, like those used on vinyl records, to prevent grains from sticking to the surface, but I haven't found this essential.

The plates are 20 × 30 cm, with a gap of 5 mm between them (see figure). (I am told that the Boston team have made a cell 2-ft high for lecture demonstrations, but I haven't seen this in action.) The cells described in the original paper by Makse *et al.* are left open at one end, but I have preferred to secure the plates to an endpiece at both ends. Partly this helps to ensure that they remain parallel (which is otherwise trickier to ensure if the plates are not glued to the base), but it also means that the striped layers can be deposited to fill the cell completely, which I think makes for a more attractive effect.

The prettiest results are achieved with coloured grains, but granulated sugar and sand (purchased from a pet shop) work well. The important factor is that the grains be both of different sizes and of different shapes—the sugar grains are larger and more square. (Table salt, which is more similar to the sand in both size and shape, didn't work at all.) And the best results are obtained by pouring the 50 : 50 mixture of grains at a slow and steady rate into one corner of the cell. To ensure this, I used an A5 envelope as a funnel, with the tip of one corner cut off.

This is one of the most satisfying experiments—a dramatic result for rather little effort.

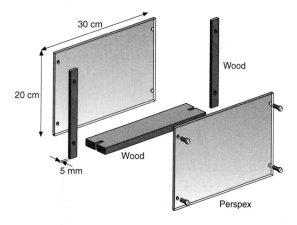

a

b

BIBLIOGRAPHY

Pattern formation, form and complexity

S. Camazine (1998). *Self-Organized Biological Super-structures*. Princeton University Press.

J. Cohen and I. Stewart (1994). *The Collapse of Chaos*. Penguin, London.

P. Coveney and R. Highfield (1995). *Frontiers of Complexity*. Faber and Faber, London.

R. Dawkins (1990). *The Selfish Gene*. Oxford University Press.

R. Dawkins (1996). *The Blind Watchmaker*. W.W. Norton, New York.

M. Ghyka (1977). *The Geometry of Art and Life*. Dover, New York.

S.J. Gould (1991). *Wonderful Life*. Penguin, London

J.P. Grotzinger and D.H. Rothman (1996). An abiotic model for stromatolite morphogenesis. *Nature* **383**, 423.

D. McKay *et al.* (1996). Search for past life on Mars: Possible relict biogenic activity in Martian meteorite ALH84001. *Science* **273**.

W. Schopf (ed.) (1991). *Earth's Earliest Biosphere*. Princeton University Press.

P.S. Stevens (1974). *Patterns in Nature*. Penguin, London.

I. Stewart (1998). *Life's Other Secret. The New Mathematics of the Living World*. Wiley, New York.

I. Stewart and M. Golubitsky (1993). *Fearful Symmetry*. Penguin, London.

D'A. Thompson (1961). *On Growth and Form*. Cambridge University Press. See also the Complete Revised Edition: Dover, New York, 1992.

M.M. Waldrop (1992). *Complexity*. Penguin, London.

H. Weyl (1969). *Symmetry*. Princeton University Press.

Bubbles and foams

J.H. Aubert, A.M. Kraynik and P.B. Rand (1986). Aqueous foams. *Scientific American* **254**(5), 58.

C.V. Boys (1959). *Soap Bubbles*. Dover, New York.

B. Fourcade, M. Mutz and D. Bensimon (1992). Experimental and theoretical study of toroidal vesicles. *Physical Review Letters* **68**, 2551.

P. Harting (1872). On the artificial production of some of the principal organic calcareous formations. *Quarterly Journal of the Microscopy Society* **12**, 118.

S. Hildebrandt and A. Tromba (1996). *The Parsimonious Universe*. Springer-Verlag, New York.

S. Hyde, S. Andersson, K. Larsson, Z. Blum, T. Landh, S. Lidin and B. Ninham (1997). *The Language of Shape*. Elsevier, Amsterdam.

R. Lipowsky (1991). The conformation of membranes. *Nature* **349**, 475.

S. Mann and G.A. Ozin (1996). Synthesis of inorganic materials with complex form. *Nature* **382**, 313.

X. Michalet and D. Bensimon (1995). Vesicles of toroidal topology: observed morphology and shape transformations. *J. Phys. II France* **5**, 263.

G.A. Ozin and S. Oliver (1995). Skeletons in the beaker: synthetic hierarchical inorganic materials. *Advanced Materials* **7**, 943.

J. Prost and F. Rondelez (1991). Structures in colloidal physical chemistry. *Nature* **350** (Supplement), 11.

E.L. Thomas, D.M. Anderson, C.S. Henkee and D. Hoffman (1988). Periodic area-minimizing surfaces in block copolymers. *Nature* **334**, 598.

K. von Frisch (1975). *Animal Architecture*. Hutchinson and Co., London.

D. Weaire (ed.) (1996). *The Kelvin Problem*. Taylor and Francis, London.

D. Weaire and R. Phelan (1994). Optimal design of honey-combs. *Nature* **367**, 123.

D. Weaire and R. Phelan (1994). The structure of monodisperse foam. *Philosophical Magazine Letters* **70**, 345.

D. Weaire and R. Phelan (1994). A counter-example to Kelvin's conjecture on minimal surfaces. *Philosophical Magazine Letters* **69**, 107.

A. Winter and W.G. Siesser (eds) (1994). *Coccolithophores.* Cambridge University Press.

W. Wintz, H.G. Dobereiner and U. Seifert (1996). Starfish vesicles. *Europhysics Letters* **33**, 403.

Waves

K. Agladze, J.P. Keener, S.C. Müller and A. Panfilov (1994). Rotating spiral waves created by geometry. *Science* **264**, 1746.

E. Ben-Jacob, I. Cohen, O. Shochet, I. Aranson, H. Levine and L. Tsimring (1995). Complex bacterial patterns. *Nature* **373**, 566.

M.P. Brenner, L.S. Levitov and E.O. Budrene (1997). Physical mechanisms for chemotactic pattern formation in bacteria. *Biophysical Journal* (submitted).

E.O. Budrene and H. Berg (1991). Complex patterns formed by motile cells of *Escherichia coli. Nature* **349**, 630.

E.O. Budrene and H. Berg (1995). Dynamics of formation of symmetrical patterns by chemotactic bacteria. *Nature* **376**, 49.

B. Chopard, P. Luthi and M. Droz (1994). Reaction–diffusion cellular automata model for the formation of Liesegang patterns. *Physical Review Letters* **72**, 1384.

I.R. Epstein and K. Showalter (1996). Nonlinear chemical dynamics: oscillations, patterns, and chaos. *Journal of Physical Chemistry* **100**, 13132.

G. Ertl (1991). Oscillatory kinetics and spatio-temporal self-organization in reactions at solid surfaces. *Science* **254**, 1750.

L. Glass (1996). Dynamics of cardiac arrhythmias. *Physics Today* August 1996, p. 40.

M. Gorman, M. El-Hamdi and K.A. Robbins (1994). Experimental observation of ordered states of cellular flames. *Combustion Science and Technology* **98**, 37.

R.A. Gray and J. Jalife (1996). Spiral waves and the heart. *International Journal of Bifurcation and Chaos* **6**, 415.

P. Heaney and A. Davis (1995). Observation and origin of self-organized textures in agates. *Science* **269**, 1562.

H.K. Henisch (1988). *Crystals in Gels and Liesegang Rings.* Cambridge University Press.

S. Jakubith, H.H. Rotermund, W. Engel, A. von Oertzen and G. Ertl (1990). Spatiotemporal concentration patterns in a surface reaction: propagating and standing waves, rotating spirals, and turbulence. *Physical Review Letters* **65**, 3013.

R. Kapral and K. Showalter (eds) (1995). *Chemical Waves and Patterns.* Kluwer Academic Publishers, Dordrecht.

J. Lechleiter, S. Girard, E. Peralta and D. Clapham (1991). Spiral calcium wave propagation and annihiliation in *Xenopus laevis* oocytes. *Science* **252**, 123.

F. Mertens and R. Imbihl (1994). Square chemical waves in the catalytic reaction $NO + H_2$ on a rhodium(110) surface. *Nature* **370**, 124.

M. Markus and B. Hess (1990). Isotropic cellular automaton for modelling excitable media. *Nature* **347**, 56.

P.J. Ortoleva (1994). *Geochemical Self-Organization.* Oxford University Press.

H.G. Pearlman and P.D. Ronney (1994). Self-organized spiral and circular waves in premixed gas flames. *Journal of Chemical Physics* **101**, 2632.

T. Pynchon. (1987). *Gravity's Rainbow.* Penguin, London.

S. Scott (1991). Clocks and chaos in chemistry. In *Exploring Chaos. A Guide to the New Science of Disorder* (ed. N. Hall). W.W. Norton, New York.

S.K. Scott (1992). Chemical reactions as nonlinear systems. *Nonlinear Science Today* **2**(3), 1.

S.K. Scott (1994). *Oscillations, Waves, and Chaos in Chemical Kinetics.* Oxford University Press.

L. Smolin (1996). *Galactic disks as reaction–diffusion systems.* (Preprint.)

R. Tyson, S.R. Lubkin and J.D. Murray (1997). A minimal mechanism of bacterial pattern formation. (Preprint).

A.T. Winfree (1987). *When Time Breaks Down.* Princeton University Press.

Bodies

J. Boissonade, E. Dulos and P. De Kepper (1995). Turing patterns: from myth to reality. In *Chemical Waves and Patterns* (ed. R. Kapral and K. Showalter). Kluwer Academic Publishers, Dordrecht.

P.M. Brakefield *et al.* (1996). Development, plasticity and evolution of butterfly eyespot patterns. *Nature* **384**, 236.

V. Castets, E. Dulos, J. Boissonade and P. De Kepper (1990). Experimental evidence of a sustained standing Turing-type nonequilibrium chemical pattern. *Physical Review Letters* **64**, 2953.

A.H. Church (1904). *On the Relation of Phyllotaxis to Mechanical Laws.* Williams & Norgate, London.

S. Douady and Y. Couder (1992). Phyllotaxis as a physical self-organized growth process. *Physical Review Letters* **68**, 2098.

A. Gierer and H. Meinhardt (1972). A theory of biological pattern formation. *Kybernetik* **12**, 30.

B. Goodwin (1994). *How the Leopard Changed Its Spots*. Weidenfeld and Nicolson, London.

A.J. Koch and H. Meinhardt (1994). Biological pattern formation: from basic mechanisms to complex structures. *Reviews of Modern Physics* **66**, 1481–1507.

P.A. Lawrence (1992). *The Making of a Fly*. Blackwell Scientific Publications.

K.-J. Lee, W.D. McCormick, J.E. Pearson and H.L. Swinney (1994). Experimental observation of self-replicating spots in a reaction-diffusion system. *Nature* **369**, 215.

H. Meinhardt (1982). *Models of Biological Pattern Formation*. Academic Press, London.

H. Meinhardt (1995). Dynamics of stripe formation. *Nature* **376**, 722–723.

H. Meinhardt (1995). *The Algorithmic Beauty of Sea Shells*. Springer, New York.

J.D. Murray (1988). How the leopard gets its spots. *Scientific American* **258**(3), 62.

J.D. Murray (1990). *Mathematical Biology*. Springer, Berlin.

H.F. Nijhout (1991). *The Development and Evolution of Butterfly Wing Patterns*. Smithsonian Inst. Press, Washington.

C. Nüsslein-Volhard (1996). Gradients that organize embryo development. *Scientific American* August 1996, p. 54.

Q. Ouyang and H.L. Swinney (1991). Transition from a uniform state to hexagonal and striped Turing patterns. *Nature* **352**, 610.

Q. Ouyang and H.L. Swinney (1995). Onset and beyond Turing pattern formation. In *Chemical Waves and Patterns* (ed. R. Kapral and K. Showalter). Kluwer Academic Publishers, Dordrecht.

S. Kondo and R. Asai (1995). A reaction–diffusion wave on the skin of the marine angelfish *Pomacanthus*. *Nature* **376**, 765.

I. Stewart (1995). *Nature's Numbers*. Weidenfeld and Nicolson, London.

A. Turing (1952). The chemical basis of morphogenesis. *Philosophical Transactions of the Royal Society* B **237**, 37.

Branches

E. Ben-Jacob (1993). From snowflake formation to growth of bacterial colonies. Part I: diffusive patterning in azoic systems. *Contemporary Physics* **34**, 247–273.

E. Ben-Jacob (1997). From snowflake formation to growth of bacterial colonies. Part II: cooperative formation of complex colonial patterns. *Contemporary Physics* **38**, 205.

E. Ben-Jacob and P. Garik (1990). The formation of patterns in non-equilibrium growth. *Nature* **343**, 523.

E. Ben-Jacob, O. Shochet, I. Cohen, A. Tenenbaum, A. Czirok and T. Vicsek (1995). Cooperative strategies in formation of complex bacterial patterns. *Fractals* **3**, 849.

E. Ben-Jacob, O. Shochet, A. Tenenbaum, I. Cohen, A. Czirok and T. Vicsek (1994). Generic modelling of co-operative growth patterns in bacterial colonies. *Nature* **368**, 46.

W.A. Bentley and W.J. Humphreys (1962). *Snow Crystals*. Dover, New York.

R.M. Brady and R.C. Ball (1984). Fractal growth of copper electrodeposits. *Nature* **309**, 225.

B. Chopard, H.J. Hermann and T. Vicsek (1991). Structure and growth mechanism of mineral dendrites. *Nature* **353**, 409.

F. Family, B.R. Masters and D.E. Platt (1989). Fractal pattern formation in human retinal vessels. *Physica D* **38**, 98.

J.M. Garcia-Ruiz, E. Louis, P. Meakin and L.M. Sander (eds) (1993). *Growth Patterns in Physical Sciences and Biology*. Plenum Press, New York.

M. Gottlieb (1993). Angiogenesis and vascular networks: complex anatomies from deterministic non-linear physiologies. In *Growth Patterns in Physical Sciences and Biology*, (ed. J.M. Garcia-Ruiz, E. Louis, P. Meakin and L.M. Sander). Plenum Press, New York.

A.J. Hurd (ed.) (1989). *Fractals. Selected Reprints*. American Association of Physics Teachers, College Park.

D. Kessler, J. Koplik and H. Levine (1988). Pattern selection in fingered growth phenomena. *Advances in Physics* **37**, 255.

H. Lauwerier (1991). *Fractals*. Princeton University Press.

B. Mandelbrot (1984). *The Fractal Geometry of Nature*. W.H. Freeman, New York.

M. Matsushita and H. Fukiwara. Fractal growth and morphological change in bacterial colony formation. In *Growth Patterns in Physical Sciences and Biology*, (ed. J.M. Garcia-Ruiz, E. Louis, P. Meakin and L.M. Sander). Plenum Press, New York.

W.W. Mullins and R.F. Sekerka (1964). Stability of a planar interface during solidification of a dilute binary alloy. *Journal of Applied Physics* **35**, 444.

J. Nittmann and H.E. Stanley (1986). Tip splitting without interfacial tension and dendritic growth patterns arising from molecular anisotropy. *Nature* **321**, 663.

J. Nittmann and H.E. Stanley (1987). Non-deterministic approach to anisotropic growth patterns with continuoulsy tunable morphology: the fractal properties of some real snowflakes. *Journal of Physics A* **20**, L1185.

B. Perrin and P. Tabeling (1991). Les dendrites. *La Recherche* May 1991, 656.

P. Prusinkiewicz and A. Lindenmayer (1990). *The Algorithmic Beauty of Plants.* Springer, New York.

L.M. Sander (1987). Fractal growth. *Scientific American* **256**(1), 94–100.

H.E. Stanley and N. Ostrowsky (eds) (1986). *On Growth and Form.* Martinus Nijhoff, Dordrecht.

H.E. Stanley and N. Ostrowsky (eds) (1988). *Random Fluctuations and Pattern Growth.* Kluwer, Dordrecht.

G.B. West, J.H. Brown and B.J. Enquist (1997). A general model for the origin of allometeric scaling laws in biology. *Science* **276**, 122.

Breakdowns

M.G. Anderson (ed.) (1988). *Modelling Geomorphological Systems.* John Wiley, New York.

P. Cowie (1997). Cracks in the Earth's surface. *Physics World* February 1997, p. 31.

A. Czirok, E. Somfai and T. Vicsek (1993). Experimental evidence for sef-affine roughening in a micromodel of geomorphological evolution. *Physical Review Letters* **71**, 2154.

A. Czirok, E. Somfai and T. Vicsek (1994). Self-affine roughening in a model experiment in geomorphology. *Physica* A **205**, 355.

R. Dawkins (1996). *River Out of Eden.* Weidenfeld and Nicolson, London.

J.E. Gordon (1991). *The New Science of Strong Materials.* Penguin, London.

E. Ijjasz-Vasquez, R.L. Bras and I. Rodriguez-Iturbe (1993). Hack's relation and optimal channel networks: the elongation of river basins as a consequence of energy minimization. *Geophysical Research Letters* **20**, 1583.

J.W. Kirchner (1993). Statistical inevitability of Horton's laws and the apparent randomness of stream channel networks. *Geology* **21**, 591.

B.B. Mandelbrot, D.E. Passoja and A.J. Paullay (1984). Fractal character of fracture surfaces of metals. *Nature,* **308**, 721.

M. Marder (1993). Cracks take a new turn. *Nature* **362**, 295.

M. Marder and J. Fineberg (1996). How things break. *Physics Today* September 1996, p. 24.

P. Meakin (1988). Simple models for colloidal aggregation, dielectric breakdown and mechanical breakdown patterns. In *Random Fluctuations and Pattern Growth* (ed. H.E. Stanley and N. Ostrowsky). Kluwer, Dordrecht.

P. Meakin *et al.* (1990). Simple stochastic models for material failure. In *Disorder and Fracture* (eds J.C. Charmet, S. Roux and E. Guyon). Plenum Press, New York.

A.B. Murray and C. Paola (1994). A cellular model of braided rivers. *Nature* **371**, 54.

L. Niemeyer, L. Pietronero and H.J. Wiesmann (1984). Fractal dimension of dielectric breakdown. *Physical Review Letters* **52**, 1033.

I. Rodriguez-Iturbe and A. Rinaldo (1997). *Fractal River Basins. Chance and Self-Organization.* Cambridge University Press.

I. Rodriguez-Iturbe, A. Rinaldo, R. Rigon, R.L. Bras, E. Ijjasz-Vasquez and A. Marani (1992). Fractal structures as least energy patterns: the case of river networks. *Geophysical Research Letters* **19**, 889.

A. Skjeltorp (1988). Fracture experiments on monolayers of microspheres. In *Random Fluctuations and Pattern Growth* (ed. H.E. Stanley and N. Ostrowsky). Kluwer, Dordrecht.

K. Sinclair and R.C. Ball (1996). Mechansim for global optimization of river networks from local erosion rules. *Physical Review Letters* **76**, 3360.

C. Stark (1991). An invasion percolation model of drainage network evolution. *Nature* **352**, 423.

H. van Damme and E. Lemaire (1990). From flow to fracture and fragmentation in colloidal media. In *Disorder and Fracture* (eds J.C. Charmet, S. Roux and E. Guyon). Plenum Press, New York.

A. Yuse and M. Sano (1993). Transitions between crack patterns in quenched glass plates. *Nature* **362**, 329.

Fluids

C. Allègre (1988). *The Behaviour of the Earth.* Harvard University Press.

M. Assenheimer and V. Steinberg (1994). Transition between spiral and target states in Rayleigh-Bénard convection. *Nature* **367**, 345.

V.V. Beloshapkin *et al.* (1989). Chaotic streamlines in pre-turbulent states. *Nature* **337**, 133.

D.S. Cannell and C.W. Meyer (1988). Introduction to convection. In *Random Fluctuations and Pattern Growth,* (ed. H.E. Stanley and N. Ostrowsky). Kluwer, Dordrecht.

J.P. Gollub (1994). Spirals and chaos. *Nature* **367**, 318.

G. Houseman (1988). The dependence of convection planform on mode of heating. *Nature* **332**, 346.

W.B. Krantz, K.J. Gleason, and N. Caine (1988). Patterned ground. *Scientific American* **259**(6), 44.

L.D. Landau and E.M. Lifshitz (1959). *Fluid Mechanics*. Pergamon Press, Oxford.

V. L'Vov and I. Procaccia (1996). Turbulence: a universal problem. *Physics World* August 1996, 35.

P. Machetel and P. Weber (1991). Intermittent layered convection in a model mantle with an endothermic phase change at 670 km. *Nature* **350**, 55.

J.-B. Manneville and P. Olson (1996). Convection in a rotating fluid sphere and banded structure of the Jovian atmosphere. *Icarus* **122**, 242.

P.S. Marcus (1988). Numerical simulation of Jupiter's Great Red Spot. *Nature* **331**, 693.

S.W. Morris, E. Bodenschatz, D.S. Cannell and G. Ahlers (1993). Spiral defect chaos in large aspect ratio Rayleigh–Bénard convection. *Physical Review Letters* **71**, 2026.

T. Mullin (1991). Turbulent times for fluids. In *Exploring Chaos. A Guide to the New Science of Disorder*, (ed. N. Hall). W.W. Norton, New York.

D. Ruelle (1993). *Chance and Chaos*. Penguin, London.

R. Scorer and A. Verkaik (1989). *Spacious Skies*. David and Charles, Newton Abbott.

J. Sommeria, S.D. Meyers and H.L. Swinney (1988). Laboratory simulation of Jupiter's Great Red Spot. *Nature* **331**, 689.

P.J. Strykowski and K.R. Sreenivasan (1990). On the formation and suppression of vortex 'shedding' at low Reynolds numbers. *Journal of Fluid Mechanics* **218**, 71.

P.J. Tackley, D.J. Stevenson, G.A. Glatzmaier and G. Schubert (1993). Effects of an endothermic phase transition at 670 km depth in a spherical model of convection in the Earth's mantle. *Nature* **361**, 699.

D.J. Tritton (1988). *Physical Fluid Dynamics*. Oxford University Press.

G.J.F. van Heijst and J.B. Flór (1989). Dipole formation and collisions in a stratified fluid. *Nature* **340**, 212.

M.G. Verlarde and C. Normand (1980). Convection. *Scientific American* **243**(1), 92.

G.M. Zaslavsky, R.Z. Sagdeev, D.A. Usikov and A.A. Chernikov (1991). *Weak Chaos and Quasi-Regular Patterns*. Cambridge University Press.

Grains

R.S. Anderson (1996). The attraction of sand dunes. *Nature* **379**, 24.

R.S. Anderson and K.L. Bunas (1993). Grain size segregation and stratigraphy in aeolian ripples modelled with a cellular automaton. *Nature* **365**, 740.

R.A. Bagnold (1941). *The Physics of Blown Sand and Desert Dunes*. Methuen, London.

P. Bak (1997). *How Nature Works*. Oxford University Press.

P. Bak, C. Tang and K. Wiesenfeld (1987). Self-organized criticality. An explanation of $1/f$ noise. *Physical Review Letters* **59**, 381.

P. Bak and M. Paczuski (1993). Why Nature is complex. *Physics World* December 1993, p. 39.

J.D. Barrow (1995). *The Artful Universe*. Penguin, London.

S.B. Forrest and P.K. Haff (1992). Mechanics of wind ripple stratigraphy. *Science* **255**, 1240.

V. Frette, K. Christensen, A. Malthe-Sørenssen, J. Feder, T. Jøssang and P. Meakin (1996). Avalanche dynamics in a pile of rice. *Nature* **379**, 49.

H.M. Jaeger and S.R. Nagel (1992). Physics of the granular state. *Science* **255**, 1523.

H.M. Jaeger, S.R. Nagel and R.P. Behringer (1996). The physics of granular materials. *Physics Today* April 1996, p. 32.

R. Jullien and P. Meakin (1992). Three-dimensional model for particle-size segregation by shaking. *Physical Review Letters* **69**, 640.

J.B. Knight, H.M. Jaeger and S.R. Nagel (1993). Vibration-induced size separation in granular media: the convection connection. *Physical Review Letters* **70**, 3728.

N. Lancaster (1995). *Geomorphology of Desert Dunes*. Routledge, London.

H.A. Makse, S. Havlin, P.R. King and H.E. Stanley (1997). Spontaneous stratification in granular mixtures. *Nature* **386**, 379.

F. Melo, P.B. Umbanhowar and H.L. Swinney (1995). Hexagons, kinks, and disorder in oscillated granular layers. *Physical Review Letters* **75**, 3838.

G. Metcalfe, T. Shinbrot, J.J. McCarthy and J.M. Ottino (1995). Avalanche mixing of granular solids. *Nature* **374**, 39.

W.G. Nickling (1994). Aeolian sediment transport and deposition. In *Sediment Transport and Depositional Processes*, (ed. K. Pye). Blackwell Scientific Publications.

M. Schroeder (1991). *Fractals, Chaos, Power Laws*. W.H. Freeman, New York.

T. Shinbrot (1997). Competition between randomizing impacts and inelastic collisions in granular pattern formation. *Nature* **389**, 574.

P.B. Umbanhowar, F. Melo and H.L. Swinney (1996). Localized excitations in a vertically vibrated granular layer. *Nature* **382**, 793.

P.B. Umbanhowar, F. Melo and H.L. Swinney (1997). Periodic, aperiodic, and transient patterns in vibrated granular layers. *Physica* A (submitted).

R.F. Voss and J. Clarke (1975). '1/f noise in music and speech. *Nature* **258**, 317.

B.T. Werner (1995). Eolian dunes: computer simulations and attractor interpretation. *Geology* **23**, 1057.

J.C. Williams and G. Shields (1967). Segregation of granules in vibrated beds. *Powder Technology* **1**, 134.

O. Zik, D. Levine, S.G. Lipson, S. Shtrikman and J. Stavans (1994). Rotationally induced segregation of granular materials. *Physical Review Letters* **73**, 644.

Communities

R. Axelrod (1984). *The Evolution of Cooperation.* Basic Books, New York.

P. Bak (1997). *How Nature Works.* Oxford University Press.

P. Bak, K. Chen and M. Creutz (1989). Self-organized criticality in the 'Games of Life'. *Nature* **342**, 780.

M. Batty and P.A. Longley (1994). *Fractal Cities: a Geometry of Form and Function.* Academic Press.

M.P. Hassell, H.N. Comins and R.M. May (1991). Spatial structure and chaos in insect population dynamics. *Nature* **353**, 255.

K. Higgins, A. Hastings, J.N. Sarvela and L.W. Botsford (1997). Stochastic dynamics and deterministic skeletons: population behaviour of Dungeness crab. *Science* **276**, 1431.

P.M. Kareiva (1990). Stability from variability. *Nature* **344**, 111.

P. Kareiva (1987). Habitat fragmentation and the stability of predator-prey interactions. *Nature* **326**, 388.

P. Kareiva and U. Wennergren (1995). Connecting landscape patterns to ecosystem and population processes. *Nature* **373**, 299.

S. Levy (1992). *Artificial Life.* Jonathon Cape, London.

M.P. Lombardo (1985). Mutual restraint in tree swallows: a test of the TIT FOR TAT model of reciprocity. *Science* **227**, 1363.

H.A. Makse, S. Havlin and H.E. Stanley (1995). Modelling urban growth patterns. *Nature* **377**, 608.

R. May (1991). The chaotic rhythms of life. In *Exploring Chaos. A Guide to the New Science of Disorder*, (ed. N. Hall). W.W. Norton, New York.

R.M. May (1976). Simple mathematical models with very complicated dynamics. *Nature* **261**, 459.

M. Milinski (1987). TIT FOR TAT in sticklebacks and the evolution of cooperation. *Nature* **325**, 433.

M.A. Nowak and R.M. May (1992). Evolutionary games and spatial chaos. *Nature* **359**, 826.

M.A. Nowak and R.M. May (1993). The spatial dilemmas of evolution. *International Journal of Bifurcation and Chaos* **3**, 35.

M. Nowak and K. Sigmund (1993). A strategy of win-stay, lose-shift that outperforms tit-for-tat in the Prisoner's Dilemma game. *Nature* **364**, 56.

M.A. Nowak, R.M. May and K. Sigmund (1995). The arithmetics of mutual help. *Scientific American* June 1995, p. 50.

W. Poundstone (1992). *Prisoner's Dilemma.* Oxford University Press.

R.V. Solé and J.M.G. Vilar (1997). Turing, noise and pattern in ecology. (Preprint.)

J. von Neumann and O. Morgenstern (1944). *Theory of Games and Economic Behavior.* Princeton University Press.

R.H. Whittaker (1975). *Communities and Ecosystems,* 2nd edn. Macmillan, New York.

G.S. Wilkinson (1984). Reciprocal food sharing in the vampire bat. *Nature* **308**, 181.

Principles

C. Bowman and A.C. Newell (1997). Natural patterns and wavelets. *Reviews of Modern Physics Colloquium Series.* (In press.)

P. Coveney and R. Highfield (1990). *The Arrow of Time.* W.H. Allen, London.

M.C. Cross and P. Hohenberg (1993). Pattern formation outside of equilibrium. *Reviews of Modern Physics* **65**, 851.

R. Landauer (1978). Stability in the dissipative steady state. *Physics Today* 1978.

R. Landauer (1975). Inadequacy of entropy and entropy derivatives in characterizing the steady state. *Physical Review A* **12**, 636.

G. Nicolis (1989). Physics of far-from-equilibrium systems and self-organisation. In *The New Physics*, (ed. P. Davies). Cambridge University Press.

I. Prigogine (1980). *From Being to Becoming.* W.H. Freeman, San Francisco.

H. Swinney (1997). Emergence and evolution of patterns. In *Critical Problems in Physics*, (ed. V.L. Fitch, D.R. Marlow and M.A.E. Dementi). Princeton University Press.

INDEX